REAL-TIME DIGITAL SIGNAL PROCESSING

from MATLAB® to C with the TMS320C6x DSK

REAL-TIME DIGITAL SIGNAL PROCESSING

from MATLAB® to C with the TMS320C6x DSK

Thad B. Welch
United States Naval Academy, Annapolis, Maryland

Cameron H.G. Wright
University of Wyoming, Laramie, Wyoming

Michael G. Morrow
University of Wisconsin, Madison, Wisconsin

Taylor & Francis
Taylor & Francis Group
Boca Raton   London   New York

A CRC title, part of the Taylor & Francis imprint, a member of the
Taylor & Francis Group, the academic division of T&F Informa plc.

MATLAB and Simulink are registered trademarks of The MathWorks, Inc.

Code Composer Studio and DSP/BIOS are trademarks of Texas Instruments, Inc.

Windows is a trademark of Microsoft Corporation.

Published in 2006 by
CRC Press
Taylor & Francis Group
6000 Broken Sound Parkway NW, Suite 300
Boca Raton, FL 33487-2742

© 2006 by Taylor & Francis Group, LLC
CRC Press is an imprint of Taylor & Francis Group

No claim to original U.S. Government works
Printed in the United States of America on acid-free paper
10 9 8 7 6 5 4 3 2 1

International Standard Book Number-10: 0-8493-7382-4 (Hardcover)
International Standard Book Number-13: 978-0-8493-7382-4 (Hardcover)

Library of Congress Cataloging-in-Publication Data

Catalog record is available from the Library of Congress

Visit the Taylor & Francis Web site at
http://www.taylorandfrancis.com

and the CRC Press Web site at
http://www.crcpress.com

To Donna...

To Robin...

To Jan...

Foreword

Digital signal processing is at the "heart" of most technologies that we use today. Our cell phones use digital signal processing to generate the DTMF (dual tone multi-frequency) tones that are sent to wireless networks. Our noise-canceling headphones use adaptive digital signal processing to cancel the noise in the environment around us. Digital cameras use digital signal processing to compress images into JPEG formats for efficient storage so that we can store hundreds of images in a single memory card. It is digital signal processing that allows us to play compressed music in our iPODs. Digital signal processing controls even the anti-lock brakes in our cars today. And these are just a few of the examples of real-time signal processing in the world around us.

There are many good textbooks today to teach digital signal processing—but most of them are content to teach the theory, and perhaps some MATLAB simulations. This book has taken a bold step forward. It not only presents the theory, it reinforces it with simulations, and then it shows us how to actually use the results in real-time applications. This last step is not a trivial step, and that is why so many books and courses present only theory and simulations. With the combined expertise of the three authors of this text—Thad Welch, Cam Wright, and Mike Morrow—the reader can step into the real-time world of applications with a text that presents an accessible path.

I have been fortunate to co-author several papers with the authors of this text, and can speak from first-hand experiences of their dedication to engineering education. They go the extra mile to continue to expand their understanding and their abilities to present complex material in a logical, straightforward manner. They attend conferences on engineering education; they chair sessions on engineering education; they write papers on engineering education; they live engineering education! So, I am delighted to be able to have an opportunity to tell the readers of this text that they are in for, in the authors' own words, "a ride...".

Delores M. Etter
Professor, Electrical Engineering
Distinguished Chair in Science and Technology
United States Naval Academy
Annapolis, Maryland

(Dr. Etter is member of the National Academy of Engineering and a Fellow of the IEEE and the American Society of Engineering Education. She served as the Deputy Under Secretary of Defense for Science and Technology from 1998–2001. She is also the author of a number of engineering textbooks, including several on MATLAB.)

About the Authors

Thad B. Welch, Ph.D., P.E., is an Associate Professor and Permanent Military Professor in the Department of Electrical and Computer Engineering at the U.S. Naval Academy (USNA), Annapolis, Maryland, and previously taught at the U.S. Air Force Academy (USAFA) in the Department of Electrical Engineering. He is a Commander in the U.S. Navy, and won the 1997 Clements Outstanding Educator Award at USAFA, the 2001 ECE Outstanding Educator Award, 2002 Raouf Award for Excellence in the Teaching of Engineering, and the 2003 ECE Outstanding Researcher Award at USNA. Dr. Welch is the former chairman and a founding member of the Technical Committee on Signal Processing Education for the Institute of Electrical and Electronic Engineers (IEEE) Signal Processing Society. He is a senior member of the IEEE and a member of the American Society for Engineering Education (ASEE).

Cameron H. G. Wright, Ph.D., P.E., is an Assistant Professor in the Department of Electrical and Computer Engineering at the University of Wyoming, and previously taught for nearly ten years at the U.S. Air Force Academy (USAFA) in the Department of Electrical Engineering where he was Professor and Deputy Department Head. A retired Lieutenant Colonel in the U.S. Air Force, he won the Brigadier General R. E. Thomas Award for Outstanding Contributions to Cadet Education in 1992 and 1993. In 2005, he won the IEEE Student Choice Award for Outstanding Professor of the Year and the Mortar Board "Top Prof" Award at the University of Wyoming. Dr. Wright is a founding member of the Technical Committee on Signal Processing Education for the IEEE Signal Processing Society. He is a senior member of the IEEE, and a member of ASEE, the National Society of Professional Engineers, and the International Society of Optical Engineers.

Michael G. Morrow, M.Eng.E.E., P.E., is a Faculty Associate in the Department of Electrical and Computer Engineering at the University of Wisconsin–Madison. A retired Lieutenant Commander in the U.S. Navy, he previously taught in the Electrical Engineering Department at the U.S. Naval Academy. Mr. Morrow won both the 2002 Department of Electrical and Computer Engineering Outstanding Educator Award and the 2003 Gerald Holdridge Teaching Excellence Award at the University of Wisconsin–Madison, and was designated a Microsoft Most Valuable Professional (MVP). He is the founder and president of Educational DSP (eDSP), LLC, developing affordable DSP education solutions. He is a member of the Technical Committee on Signal Processing Education for the Institute of Electrical and Electronic Engineers (IEEE) Signal Processing Society, a senior member of the IEEE, and a member of the American Society for Engineering Education (ASEE).

Contents

List of Figures

List of Tables

Program Listings

Preface

THIS book is intended to be used by both students and working engineers who need a straightforward, practical background in real-time digital signal processing (DSP). In the past, there has been a formidable "gap" between theory and practice with regard to real-time DSP. This book bridges that gap using methods proven by the authors.

We anticipate that the reader will use this book in conjunction with a more traditional text if this is their first exposure to DSP. This book provides a practical, step-by-step framework that reinforces basic DSP theory (what the authors refer to as *enduring fundamentals*). This framework utilizes a series of demonstrations, exercises, and projects that each begin with the theory, progress to MATLAB, and ultimately run in real-time on some of the latest high-performance DSP hardware. Be sure to check out the appendices—some people think they are worth the price of this book all by themselves!

Ideally, the reader should either be enrolled in, or have already taken, an introductory DSP (or discrete-time signals and systems) class. However, we have had success using various parts of this book with students who have not yet had a DSP class. The text coverage is broad enough to accommodate both undergraduate and graduate level topics. A *basic* familiarity with MATLAB and the C programming language is expected—but you don't have to be an expert in either. To take full advantage of this book, the reader should have access to a modest collection of inexpensive tools. In particular, recommended items include a standard PC running Windows, a copy of MATLAB and its Signal Processing Toolbox, a Texas Instruments TMS320C6713 DSK (or TMS320C6711 DSK) and the software development tools (Code Composer Studio™ version 2.1 or higher) that are shipped with the DSK.[1] Some other miscellaneous items, such as a signal source (a CD player works well), amplified speakers (the type typically attached to a PC work well), and 3.5 mm stereo patch cables (sometimes called 1/8th inch stereo phono plug cables) will be useful. For processing the input and output signals, several different codecs for the DSK are supported (see Chapter 1). Access to some common test equipment such as an oscilloscope, a spectrum analyzer, and a signal generator allows even more flexibility, but we show how a second inexpensive DSK or a PC's soundcard can be used to substitute for these items if desired.

This book was written in response to the many requests by both students and faculty at a variety of universities. When the authors presented the concepts that appear in this book at various conferences, we were besieged by an audience trying to "bridge the gap" from theory to practice (using real-time hardware) on their own. This book collects in a single source our unified step-by-step transition to get across that "gap." We hope you enjoy the ride...

T.B.W., C.H.G.W., M.G.M.

---

[1]The acronym "DSK" stands for "**DSP Starter Kit**." DSKs can be purchased from authorized Texas Instruments (TI) distributors or directly from TI (see `dspvillage.ti.com`).

Acknowledgments

THIS book would not have been possible without the support and assistance of Texas Instruments (TI), Inc. In particular, the authors would like to extend a sincere thank-you to Cathy Wicks, whose tireless efforts for TI's University Program has helped make DSP affordable for countless students and professors. Cathy's predecessors Christina Peterson, Maria Ho, and Torrence Robinson also contributed to our efforts that eventually resulted in this book. TI's support of DSP education is unsurpassed in the industry and the authors greatly appreciate such a forward-looking corporate vision.

We would also like to thank Nora Konopka, Helena Redshaw, and Katy Smith of CRC Press (part of the Taylor & Francis Group), whose enthusiasm for this book was inspiring, and who helped guide the project to completion. Their ready help, quick responses, and never-failing sense of humor should be the model to which other publishers aspire.

Our appreciation also extends to Debbie Rawlings as well as to Susan and Alan Czarnecki for the generous use of their waterfront vacation homes for some of our more productive "geek-fests" during which key parts of this book were written.

We would be remiss if we omitted an acknowledgment related to the mechanics of writing the text: this book was typeset using LATEX, a wonderfully capable document preparation system developed by Leslie Lamport as a special version of Donald Knuth's TEX program. LATEX is ideally suited to technical writing, and is well supported by the worldwide members of the TEX Users Group (TUG); investigate `http://www.tug.org/` for details. Both LATEX and TEX are *freely available* in the public domain (the name TEX is a trademark of the American Mathematical Society). We used the excellent WinEdt shareware editor (see `http://www.winedt.com/`) as a front-end to the free MikTeX distribution of LATEX (see `http://www.miktex.org/`).

Enduring Fundamentals

Chapter 1

Introduction and Organization

1.1 Why Do You Need This Book?

IF you want to learn about real-time[1] digital signal processing (DSP), this book can save you many hours of frustration and help you avoid countless dead ends. In the past, "bridging the gap" from theory to practice in this area has been challenging. We wrote this book to eliminate the impediments that were preventing our own and our colleagues' students from learning about this fascinating subject. When these barriers are removed, as this book will do for you, we believe that you will find real-time DSP to be an exciting field that is relatively straightforward to understand. The expected background of the reader and the tools needed to get the most out of this book are listed in the Preface.

Real-time DSP can be one of the "trickiest" topics to master in the field of signal processing. Even if your algorithm is perfectly valid, the actual implementation in real-time may suffer from problems that have more to do with computer engineering and software engineering principles than anything related to signal processing. While becoming an *expert* in real-time DSP typically requires many years of experience and learning, such skills are in very high demand. This book was written to start you on the path toward becoming such an expert.

1.1.1 Other DSP Books

There are dozens of books that eloquently discuss and explain the various theoretical aspects of digital signal processing. Texts such as [1–4], written primarily for electrical engineering students, are all excellent. For a less mathematical treatment, [5,6] are good choices. It has been shown that computer-based demonstrations help students grasp various DSP concepts much more easily [7–14]. To take advantage of this fact, a number of books also include software programs that help the student more clearly understand the underlying concepts or mathematical principles that the author is trying to relate. In recent years, as MATLAB has become an integral part of engineering education at most institutions, this software has increasingly been provided as MATLAB programs (often called *m-files*) delivered via floppy disks, CD-ROMs, or the worldwide web. Textbooks such as [1, 4, 15] are popular examples of comprehensive theoretical DSP texts that include MATLAB software. Some books are less theoretical but provide many MATLAB demonstrations [6, 16–18]. These are often used along with one of the more in-depth texts listed previously. Finally, there are books that

---

[1]The phrase *real-time* means that the system is responding "fast enough" to some external event or signal to allow proper functioning. CD players, digital cellular telephones, GPS receivers, and aircraft digital flight controls are common examples of real-time DSP.

are aimed at helping the reader learn to make the best use of MATLAB for DSP and other technical pursuits [19, 20].

1.1.2   Demos and DSP Hardware

Demonstrations using MATLAB are extremely valuable, and we use them extensively in our own courses. However, they typically use previously stored signal files and cannot be considered "real-time" demonstrations. Some MATLAB programs, using a PC sound card or data acquisition card, have a fairly limited ability to bring signals in from the "real world" and perform some processing using the general purpose CPU of the PC, but we have found this to be inadequate for teaching real-time DSP. Students need to be introduced to some of the more common aspects of specialized hardware used for real-time DSP, but in a way that minimizes the many frustrations that past students and faculty have encountered.

While there are other books available that include discussions of using real-time DSP hardware (e.g., [21–23]), we found that these other books don't really meet our students' needs. These other books just don't provide a smooth transition for a reader unfamiliar with real-time DSP or specialized programming concepts, and many require fairly expensive DSP hardware to run the included programs. In response to that need, we created a set of tools that could be used to learn real-time DSP in a series of reasonable steps, beginning with the well-known winDSK6 program, progressing to the familiar MATLAB environment, and finally making the transition to actual real-time hardware using inexpensive DSP Starter Kits (DSKs). When these tools became known [24–45] to our colleagues at various universities, we were inundated with requests to consolidate these tools into a book. You are now holding the result in your hands.

1.1.3   Philosophy of This Book

This book is designed to be used with any and all of the previously mentioned DSP texts. What sets this book apart is that it allows the reader to take the next step in mastering DSP: we take a concept and show the reader how to easily progress from a demonstration in MATLAB to running a similar demonstration in real-time code on actual DSP hardware. Until now, the learning curve in moving to real-time DSP hardware has been too steep for most students and too time-intensive for most faculty. This book overcomes these problems in both a methodical and practical way.

The reader is guided through examples and exercises which demonstrate various DSP principles. In each case we begin using familiar interfaces such as MATLAB, then lead the reader in a step-by-step fashion to the point where the algorithms are running in real-time on industry-standard DSP hardware. It's important to note that unlike most other DSP hardware-related books, the software for this book allows the transition to DSP hardware without requiring the reader to first learn assembly language or obscure C code libraries to implement the examples and exercises in real-time. Some examples don't even require a knowledge of MATLAB or C.

1.2   Real-Time DSP

An underlying assumption of most digital signal processing operations is that we have a sampled signal, our *digital signal*, that we wish to process. In an educational environment, these signals are often stored for subsequent retrieval or synthesized when needed. While this storage or synthesis method is very convenient for classroom demonstrations, computer-based assignments, or homework exercises, it does not allow for real-time processing of a signal. Our students get much more excited about DSP when we incorporate real-time signal processing into our classroom presentations and the associated laboratory exercises. This increased excitement leads to greatly increased learning opportunities for our students.

We use the term *real-time processing* to mean that the processing of a particular sample must occur within a given time period or the system will not operate properly. In a *hard real-time* system, the system will fail if the processing is not done in a timely manner. For example, in a gasoline engine control system, the calculations of fuel injection and spark timing must be completed in time for the next cycle or the engine will not operate. In a *soft real-time* system, the system will tolerate some failures to meet real-time targets and still continue to operate, but with some degradation in performance. For example, in a hand-held audio player, if the decoding for the next output sample is not completed in time, the system could simply repeat the previous sample instead. As long as this happened infrequently, it would be imperceptible to the user. Although general purpose microprocessors can be employed in many situations, the performance demands and power constraints of real-time systems often mandate specialized hardware. This may include specialized microprocessors optimized for signal processing (digital signal processors or DSPs), programmable logic devices, application specific integrated circuits (ASICs), or a combination of any or all of them as required to meet system constraints. *Please note that we have now used the acronym "DSP" in two different ways*—a very common occurrence in the digital signal processing field! In the first case, "DSP" meant "digital signal *processing*." In the second case, "DSP" referred to a "digital signal *processor*." The intended use of the acronym DSP should be clear from the context in which it is used.

1.3   How to Use This Book

This book is designed to allow someone with a basic understanding of DSP theory to rapidly transition from the familiar MATLAB environment to performing DSP operations on a realistic hardware target. Our target of choice is a member of the high-performance Texas Instruments (TI) C6000 family. Specifically, we chose the TMS320C6713 digital signal processing starter kit (called a C6713 DSK). This book also supports the older but similar C6711 DSK. See Figure 1.1 for photos of the C6713 and C6711 DSKs. The fixed-point C6211 DSK, which was replaced by the floating-point C6711 DSK, can also be used with many of the programs accompanying this book, but Code Composer Studio™ version 2.1 or higher is still needed. The C67xx DSK target was selected because of its relatively inexpensive purchase price, widespread utilization and compatibility with designs used in industry, and a feature rich software development tool set (Code Composer Studio) that is included with the DSK.

Typically, real-time DSP hardware has to communicate with the "outside" world. This is usually accomplished on the input side with an analog-to-digital converter (ADC) and on the output side with a digital-to-analog converter (DAC). Integrated circuit (IC) chips that combine the ADC and DAC functions in one device are called *codec* chips, which is an acronym for "*coder* and *decoder*." This book supports several different codecs for the DSK. We support the C6713 DSK's built-in stereo codec (TLV320AIC23) that is capable of up to 24-bits per sample and a sample frequency of 96 kHz. The C6711 DSK's built-in codec (TLC320AD535) is fully supported, although the fixed 8 kHz sampling frequency and monaural design greatly limit the effects we can demonstrate. CD-quality stereo codecs for the C6711 DSK such as the PCM3003-based codec available from TI and the PCM3006-based codec available from eDSP [46] (see Figure 1.2) are supported to allow more interesting DSP effects.

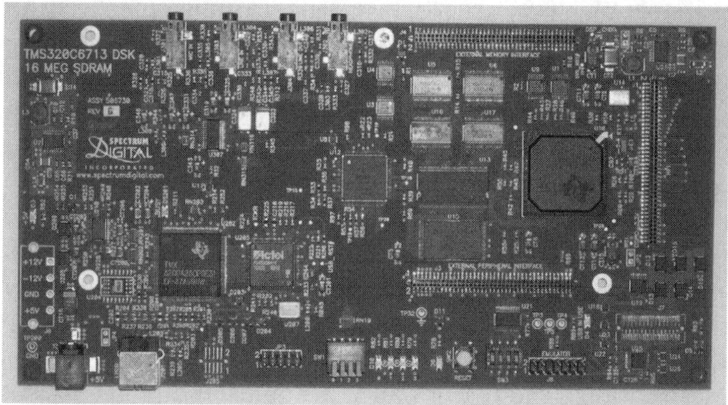

(a) The C6713 DSK.

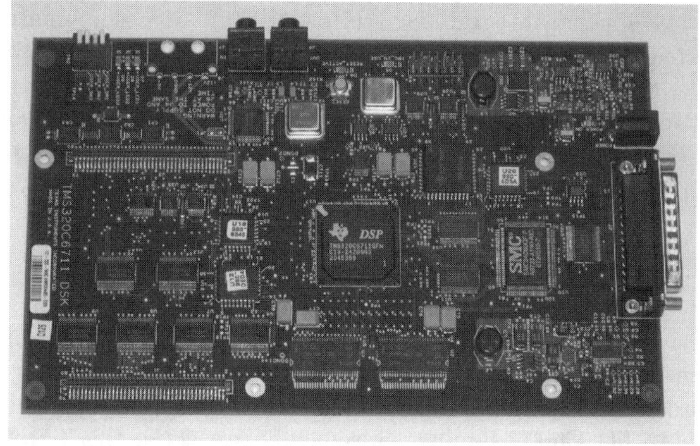

(b) The C6711 DSK.

Figure 1.1: The C671x DSK circuit boards.

Figure 1.2: The PCM3006 stereo codec from eDSP.

(a) Top view of the HPI interface board on the C6713 DSK.

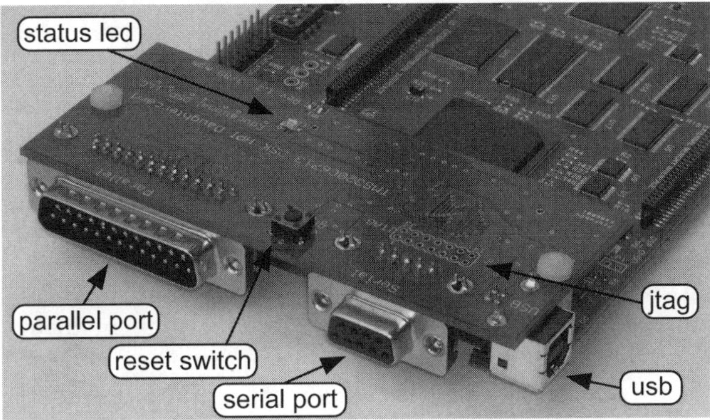

(b) Side view of the HPI interface board on the C6713 DSK.

Figure 1.3: The HPI interface board on the C6713 DSK.

When Texas Instruments replaced the C6711 DSK with the C6713 DSK, it provided a much more capable, stable, and robust DSP development environment for DSP education. However, while this new 6713 DSK had many improvements over the 6711 DSK, it did not include any way to transfer data to and from the host computer except through the JTAG debugger interface, which is extremely limited in bandwidth and requires that the TI Code Composer Studio (CCS) software tools be available. This means that the existing suite of winDSK6 [29, 47, 48] demonstration software and other software tools cannot run on the 6713 DSK, denying educators a valuable teaching and classroom demonstration resource. Also, there is no way to interface an application on the host PC directly to the new DSK, limiting the ability of students to create stand-alone, interactive projects using the DSK. To solve this problem, the authors created a small inexpensive add-on interface board for the TMS320C6713 DSK that uses the Host Port Interface (HPI) to provide both a means for a PC host application to boot software onto the DSK, and to permit the transfer of data between the DSK and the host PC application [45]. Figure 1.3 shows this interface board installed on the C6713 DSK. We also created a software package that makes it possible for students to create stand-alone Windows applications that communicate directly with this new DSK via the interface board. In addition to parallel port communication, the HPI interface board provides USB, RS-232, and digital input/output ports as user selectable resources available to the DSK software (see the eDSP web site [46] for more information). Using the HPI interface board on the C6713 DSK permits full use of all the winDSK6 features that appear throughout this book.

1.3.1 Transition to Real-Time

For each DSP concept in this book, we will take a four-step approach. Specifically, we will follow the approach listed below.

- Briefly review the relevant DSP theory.

- Demonstrate the concept with an easy-to-use tool called winDSK6. With winDSK6, you can program and manipulate the real-time hardware *without* having to write a program.

- Explain and demonstrate how MATLAB techniques can be used to implement the concept (not necessarily in real-time, but in a way most students find easy to understand).

- Provide and explain the C code necessary for you to implement your own real-time program using a DSK and its software development tools.

For most readers of this book, the first step should serve only as a refresher and to set in context the overall discussion. The next step permits the reader to further explore the concept and facilitates "what if" experimentation unencumbered by the need to program any code. The next step, using MATLAB examples, helps to reinforce your understanding of the underlying DSP theories. These examples use standard MATLAB commands that occasionally require the Signal Processing Toolbox.[2] Our well-commented MATLAB code is written such that the algorithm is clearly evident; optimizations that may obscure the underlying concept are avoided. Once the reader has worked through this non-real-time DSP experience, the final step is the key to "bridging the gap" to real-time operation. From the discussion in the book, the reader will be able to confidently implement the same algorithms in C using state-of-the-art real-time DSP hardware. Each chapter ends with a list of follow-on challenges which the reader should now be prepared to implement as desired.

A cautionary note: some of our students have tried to "save time" by skipping the MATLAB step and jumping right into the C code. Don't do it! It's been demonstrated time and time again that the students who first work with the algorithms in MATLAB consistently get the C version to work correctly. Those who skip the MATLAB step have a much harder time, and often can't get their code to work properly at all. Don't say we didn't try to warn you!

1.3.2 Chapter Coverage

The first nine chapters of this book are what we believe are the *enduring fundamentals* of DSP, presented in the context of real-time operation. The experience you gain while studying these chapters is crucial to being ready for the DSP projects that are presented in the remaining chapters. A special mention needs to be made here regarding the appendices. While most other DSP books force you to track down and look up various other pieces of information that you need for real-time DSP from a plethora of sources, the appendices at the end of this book collect in a single location a distilled and simplified version of all the important topics that are needed to work effectively with DSP hardware such as the C67xx DSK. The appendices by themselves are probably worth the price of this book!

---

[2] The Signal Processing Toolbox is an optional product for MATLAB, also available from The MathWorks. You may also want to explore graphical programming environments suitable for DSP (i.e., SIMULINK from The MathWorks and LabVIEW from National Instruments).

Table 1.1: Organization of the book's CD-ROM. CCS stands for Code Composer Studio, the software development environment that accompanies the Texas Instruments DSK.

File or Directory	Comment
ReadMe.txt	Read this file before using any of the software
winDSK	Directory containing winDSK6 software
test_signals	Directory containing test signals for exercises and Projects
chapter_xx\matlab	Directory containing MATLAB files for Chapter xx
chapter_xx\ccs	Directory containing CCS files for Chapter xx
appendix_x	Directory containing files related to Appendix x
common_code	Directory containing files related to all CCS projects

1.3.3 Hardware and Software Installation

The CD-ROM that accompanies this book contains a great deal of useful software (see Table 1.1). This software is an integral part of the book, and the remaining chapters assume that you have already installed it all properly, along with the DSK itself and the software that accompanies the DSK. To install the DSK hardware and software, complete the "Hardware Getting Started" guide and the "Software Getting Started" guide that come with your DSK. To install the software that accompanies this book, follow the procedure outlined below.

1. Put the bookware CD-ROM into your computer's CD drive.

2. View the ReadMe file for the latest information.

3. Run the setup program contained on the bookware CD-ROM.

A setup program is provided to make the overall software installation much easier. Just copying files yourself from the CD-ROM to your hard disk may not result in the files being copied to the appropriate directories.

After you have completed the hardware and software installation discussed above, launch winDSK6 by double-clicking on its Windows icon. Ensure the correct selections have been made for "DSK and Host Configuration" from the appropriate pulldown menus[3] *prior* to initiating the DSK Confidence Test. A successfully completed DSK confidence test will be one of your best indicators that you have properly installed the DSK. You can also use the winDSK6 confidence test in the future to easily verify the proper operation of your DSK.

1.3.4 Reading Program Listings

Important: Some of the code listings in this book include lines that, despite our best efforts, are too long to fit within the book's margins and yet still use meaningful variable and function names. So be watchful when reading program listings for those instances where a line wrap occurred in the listing due only to page margins. In that case, the wrapped part of the line is indented, and the characters "[+]" show up to identify the beginning of the wrapped part of the line. The "[+]" characters are *not* part of the program, which you'll confirm if you compare the listing as printed in the book to the actual program file on the CD-ROM. Note that the line numbers shown at the left edge of the listings do not increment for the wrapped part of a line.

---

[3]If using a parallel port, right click the winDSK6 parallel port mode window to see an explanation of the available modes. In general, try to use the fastest mode that works with your computer.

1.4 Get Started

Once the software is installed and you begin reading the remainder of this book, we encourage you to stop frequently and try out various programs and examples as they are mentioned. As Sophocles once said, "One must learn by doing the thing; for though you think you know it, you have no certainty until you try it." Real-time DSP can be tremendous fun; we hope this book helps you find it as much fun as we do ...

Chapter 2

Sampling and Reconstruction

2.1 Theory

WHENEVER we wish to obtain a real world signal in order to process it digitally, we must first convert it from its natural analog form to the more easily manipulated digital form. This involves grabbing, or "sampling," the signal at certain instants in time. We assume the sampling instants are equally spaced in time (T_s), so that the *sampling frequency* (F_s) is equal to $1/T_s$. Each individual sample represents the amplitude of the signal at that instant in time, and the *number of bits* per sample that we use to store this amplitude determines how accurately we can represent it. More bits means better fidelity, but it also means greater storage and processing requirements. The effect of changing the number of bits will be discussed later in the chapter.

2.1.1 Choosing a Sampling Frequency

One potential problem that can occur during sampling is called *aliasing*, which results in samples that do not properly represent the original signal. Once aliasing has crept into your data, no processing in the world can "fix" your samples so that the original signal can be recovered. To prevent aliasing, the sample frequency, F_s, of the ADC (analog to digital converter) must be greater than twice the maximum frequency f_h contained in the analog input signal.[1] Often the sample frequency is considerably higher than $2f_h$. Typically, some form of input signal conditioning (such as an analog lowpass filter) ensures that the maximum frequency contained in the analog input signal is less than $F_s/2$. The effect of changing the sample frequency will be discussed later in the chapter.

2.1.2 Input/Output Issues: Samples or Frames?

While it is easier to understand DSP theory and operations on a sample-by-sample basis, in reality this is often an inefficient way to configure the actual input and output of data. Just as a computer hard disk transfers data in blocks of many bytes rather than one byte or word at a time, many DSP systems transfer data in "blocks" known as *frames*. We will discuss both methods in this book, beginning with sample-by-sample processing. Chapter 6 discusses frame-based processing in more detail.

---

[1]This assumes a lowpass, or baseband, signal. For a bandpass signal having bandwidth BW, we find $F_s \geq 2BW$ but F_s must also satisfy other conditions. See [49] for more detail.

Figure 2.1: A generic DSP system.

2.1.3   The Talk-Through Concept

A block diagram of a generic DSP system is shown in Figure 2.1. This figure shows a very basic block diagram of a DSP system. However, for this chapter we can simplify this diagram even more. Our goal at this point in the text is simply to pass the signal entering the ADC directly to the DAC (digital to analog converter), with no actual processing being performed. This process is routinely called *talk-through*, and is often used as the first test of a DSP system to determine if it is working, if the input and output connections are correct, and to familiarize the user with the system and its software tools. Talk-through is also invaluable for showing the user what the underlying limitations of the ADC and DAC process will impose on more complicated applications. The DSP algorithm for talk-through is simply to pass the sample from the ADC directly to the DAC. A block diagram of this very basic operation is just a simplified version of Figure 2.1, and is shown in Figure 2.2. Note that the DSP talk-through algorithm is so simple that the entire DSP algorithm block can be omitted.

Ideally, the reconstructed analog signal coming out of the DAC should be identical to the analog input signal that went into the ADC. Once we have implemented a talk-through system similar to Figure 2.2, we can explore some of the capabilities and limitations of the hardware we are using. This may seem like a trivial step, but having a more detailed understanding of this process is very helpful before we move on to incorporating more complicated DSP algorithms into the system.

2.2   winDSK6 Demonstration

2.2.1   Starting winDSK6

If you double click on the winDSK6 icon (as shown in Figure 2.3), the winDSK6 application will launch, and a window similar to Figure 2.4 will appear.

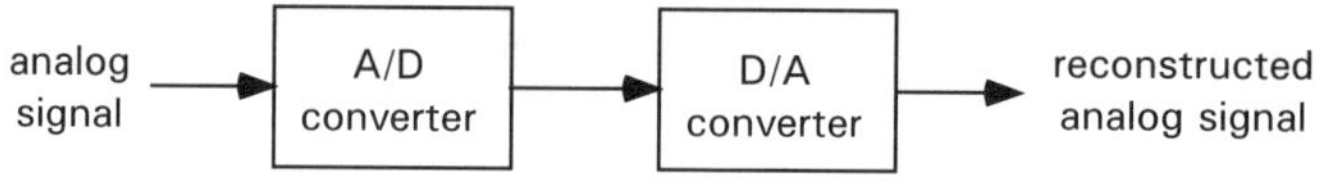

Figure 2.2: A talk-through system.

Figure 2.3: winDSK6 desktop icon.

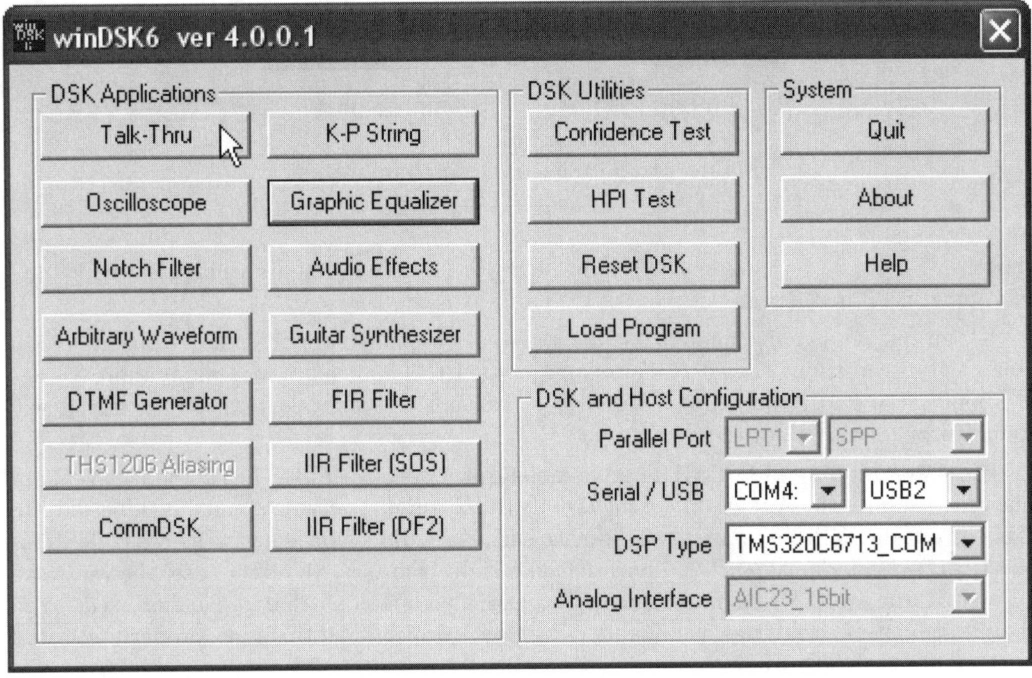

Figure 2.4: winDSK6 ready to load the Talk-Thru application.

2.2.2 Talk-Thru Application

Clicking on the winDSK6 Talk-Thru button will load the Talk-Thru (talk-through) program into the attached DSK, and a window similar to Figure 2.5 will appear. If you have an audio source (e.g., a CD player) connected to the DSK's audio input, and a pair of speakers (common powered PC speakers work fine) connected to the audio output, whatever music you are playing on your CD player should now be audible. If you are using the headphone jack of the CD player (instead of the "line out" jack) to send the signal to the DSK, you may need to adjust your CD player volume level for a proper result.

If you are experiencing difficulty (no audio heard), verify that your CD player and speakers are functioning properly by connecting the speakers directly to the CD player. At this point, adjust the system volume to your desired level. When everything is functioning properly, reconnect the CD player and the speakers to the DSK. One other common cause of no signal is using a monaural (mono) audio cable instead of a stereo audio cable. As

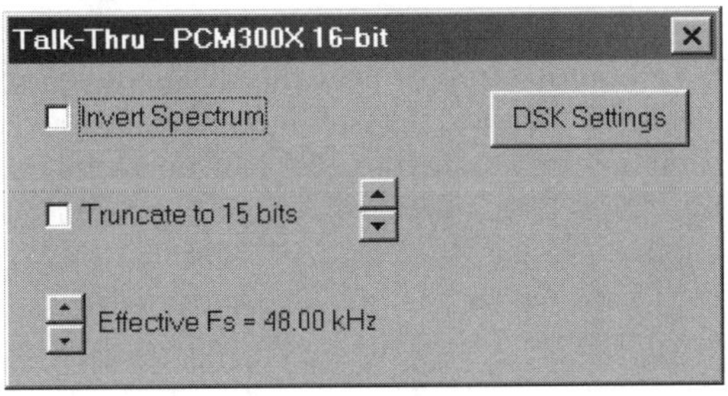

Figure 2.5: winDSK6 running the Talk-Thru application.

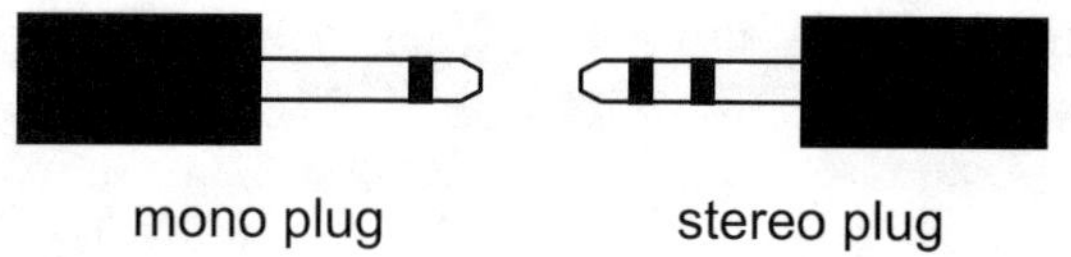

Figure 2.6: A mono and stereo mini-plug.

shown in Figure 2.6, a mono cable's mini-plug has 2 metal segments while a stereo cable's mini-plug has 3 metal segments.

If your output signal sounds fuzzy, distorted, or clipped (basically, not as you expected), you may be overdriving the ADC. For example, the PCM 3006 has a maximum input voltage of about ±1 volt. If you input a signal that exceeds this allowed range, your signal will sound fuzzy, distorted, or clipped.[2]

If you are using a C6711 DSK's native monaural codec, the Talk-Thru demonstration application samples the Analog Interface Circuit (AIC) input and writes the incoming data to the AIC output. With a multichannel device such the C6713 DSK's native stereo codec, this effect is created independently on both left and right channels. In either case, the winDSK6 "Codec Hardware" setting must match the audio connections that are being used. You should also ensure that the "DSK and Host Configuration" settings are properly selected in the main winDSK6 window.

The application permits demonstration of three basic effects:

1. Quantization — The effect of different bit-length conversions can be shown by the variable truncation of the audio data, reducing the effective resolution of the codec converters to a minimum of one bit. From DSP theory, we know that the signal to quantization noise ratio (SQNR) is proportional to the number of bits used (approximately 6 dB increase in SQNR for each added bit of resolution).

2. Spectral Inversion — By selecting the Invert Spectrum checkbox, the sign bit of every other sample is changed (every other sample is multiplied by -1). This is equivalent to modulating the input signal with a frequency of one half the sample frequency. The effect is that the frequency spectrum is flipped around a frequency equal to the sample frequency divided by four, effectively "scrambling" the signal as it would be heard by a listener. If the resulting signal is then passed through a second DSK performing the same operation, the original signal will be recovered.

3. Aliasing — The *effective* sample rate (or sample frequency F_s) can be varied by using a variable decimation factor.[3] A single input sample is repeatedly transmitted for an integral number of output samples, which reduces the effective sample rate of the converter by that integral number. In this way, aliasing can be easily demonstrated even when using a sigma-delta converter. When operating with a reduced effective sample rate, the Invert Spectrum effect is modified to give inversion about the effective F_s, not the actual codec F_s.

---

[2] How your ADC deals with this overdriven condition depends on how it was designed. The most common results are saturation or wrap around. Some ADCs can be programmed either to saturate or wrap around.

[3] All of the supported codecs use sigma-delta conversion, which internally oversamples the signal. This means that simply changing the basic sample frequency won't necessarily result in the expected aliasing.

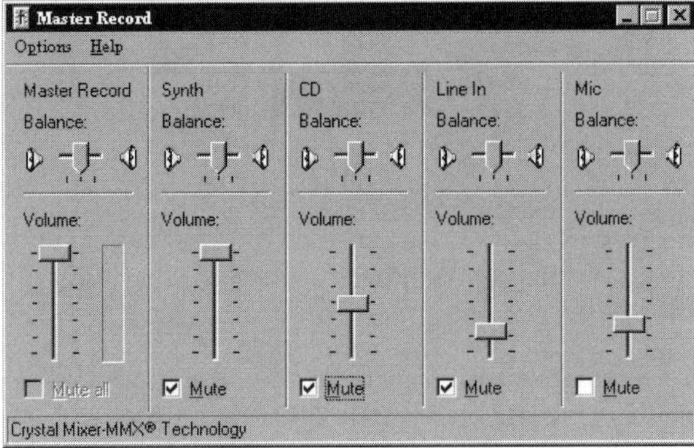

Figure 2.7: A typical sound recording mixer (from Crystal).

2.3 Talk-Through Using Windows

If you have a personal computer (PC), you almost certainly have the ability to record, store, and play back sound files. In Microsoft Windows,[4] the recorded files end with a `.wav` extension. If you haven't worked with wav-files before, click on Start, then Find (or Search), then Files or Folders... (or All files and folders) and conduct a search using `*.wav` in the "Named:" (or All or part of the file name) area of the dialog box. Be sure to specify the location you want searched. Searching `C:\WINDOWS\MEDIA` is a good place to start if you're using Windows 98, ME, or XP. Searching the entire `C` drive is very effective, but will take considerably longer.

Once the search is complete, the name and file details associated with all of the files ending with the `.wav` extension will be displayed. Double-clicking on a file name will play back the wav-file through the PC's sound system. Note that this is really only the DAC half of the talk-through operation; the ADC has already been performed, and the resulting samples were stored in the wav-file.

In order to create your own wav-files, you'll need a microphone that can be connected to your PC's sound card input. Most sound cards have at least three connectors: "microphone in," "line out," and "speaker out." The "microphone in" jack is where you connect your microphone. Different sound cards are shipped with different software to support them, but you should ensure that the Mic Balance setting is not muted (Figure 2.7 shows a typical sound recording mixer program) or no signal will be sent to the ADC. Note that some computers, notably laptop computers, have the sound card circuitry integrated on the main board of the computer rather than as a separate card; however, the connectors and operation should be the same.

Be sure to adjust the Master Record Balance so that while you are recording, the system does not saturate (overdrive the input). The status of the input level is usually indicated by a vertically oriented colored bar next to the Master Record Balance Volume adjust slider. As the input level increases a green bar grows in length. As the input increases further, the bar continues to grow in length, but the color changes to yellow (a warning), then to red (saturation or clipping is occurring). You will need to adjust your system using trial and error. While you do not want saturation (red bar) to occur (resulting in a distorted signal), you also don't want the system gain set so low that little or no green is displayed while you are recording (resulting in a low signal-to-noise ratio).

---

[4]Microsoft Windows is a trademark of Microsoft Corporation.

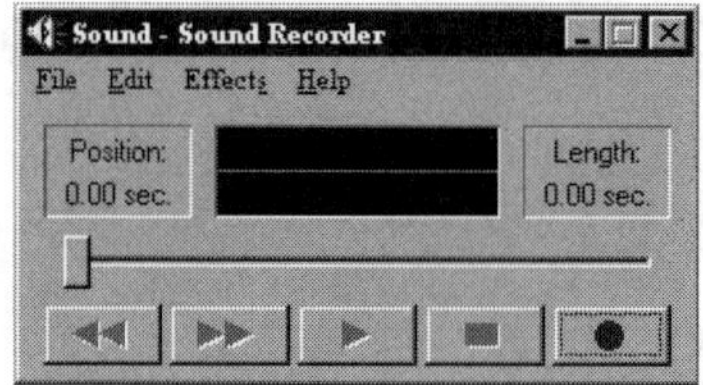

Figure 2.8: The Windows sound recorder applet.

To actually record a wav-file, click on Start, Programs, Accessories, Multimedia, and Sound Recorder. This sequence may be slightly different on your PC. When you have found and launched the Sound Recorder program, a window similar to Figure 2.8 will appear. The usage of the Sound Recorder program is straightforward and pull-down help is available. Once you have recorded a wav-file, you can play back the file. You can also save your wav-file using the "File, Save" pull-down menu.

You now have the ability to record, save, and playback a wav-file using only your PC. Playback using the Sound Recorder program can also use special effects such as changing the playback speed, adding an echo, or playing the file backwards. Options within the Sound Recorder program also allow you to vary the sample rate and the number of bits used to represent each sample. Three common combinations of these settings are shown below.

Radio quality	8 bits/sample	mono	11,025 samples/sec
Telephone quality	8 bits/sample	mono	22,050 samples/sec
CD quality	16 bits/sample	stereo	44,100 samples/sec

Most systems also allow for other recording specifications and for encoding formats other than the default Pulse Code Modulation (PCM).

While the Sound Recorder greatly simplifies the basic tasks of recording, storing, and playing back wav-files, you are limited to the features built into the program and by the ADC and DAC specifications of your PC's sound card. Obviously, this is not a real-time operation.

2.4   Talk-Through Using MATLAB and Windows

MATLAB has a number of data manipulation functions [50]. Executing the command

```
help audiovideo
```

will display the functions associated with data manipulation.[5] For the PC platform, the following edited results of the help command are of interest for the talk-through operation:

```
Audio input/output objects.
   audioplayer    - Windows audio player object.
   audiorecorder  - Windows audio recorder object.

Audio hardware drivers.
   sound          - Play vector as sound.
   soundsc        - Autoscale and play vector as sound.
   wavplay        - Play sound using Windows audio output device.
```

---

[5]Use `help audio` for earlier versions of MATLAB.

```
wavrecord      - Record sound using Windows audio input device.


Audio file import and export.
  wavread        - Read Microsoft WAVE (".wav") sound file.
  wavwrite       - Write Microsoft WAVE (".wav") sound file.
```

Help on these or any of the individual MATLAB functions can be accessed, as before, by typing the command

```
help functionname
```

where `functionname` is the name of the MATLAB function for which you desire help. Additionally, you can access a number of help options using the pull-down help menu directly from the MATLAB menu bar (see Figure 2.9). Given these audio file related MATLAB functions, we will now discuss a solution to the talk-through problem. We can import an existing wav-file using the `wavread` command, then play the file back using the sound command. Example code to do this would be

Listing 2.1: Reading and playing back a wav-file using MATLAB.

```
1  [Y, Fs, Nbits, Opts] = wavread('c:\windows\media\tada.wav');
   sound(Y, Fs)
```

The first command reads a wav-file named `tada.wav`, which is located in the `c:\windows\media` directory. The outputs Y, Fs, $Nbits$, and $Opts$ of the wavread command represent the file data, sample frequency, number of bits per sample, and the optional file information, respectively. Typing `whos` at the MATLAB command line returns a screen output similar to

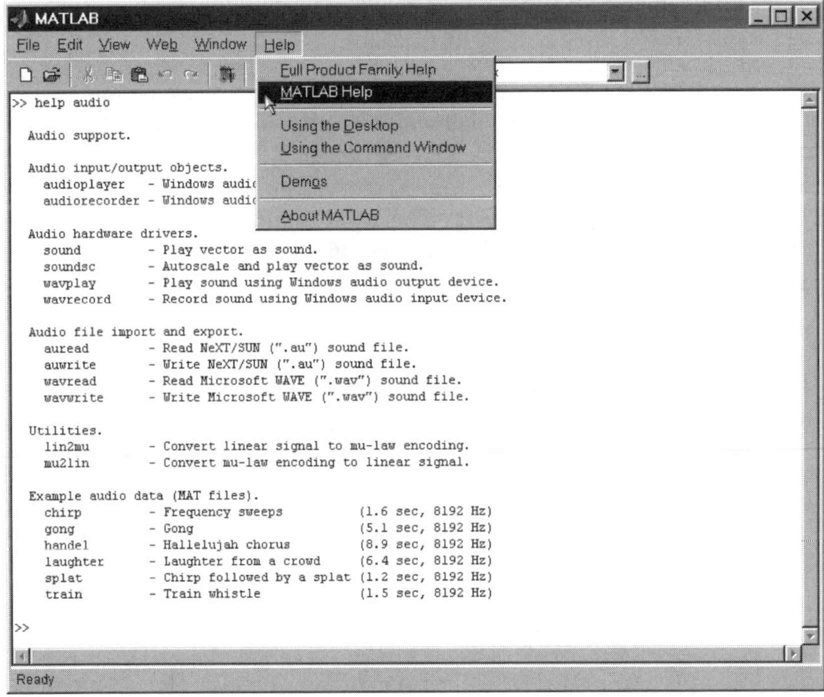

Figure 2.9: MATLAB command window with the help pull-down menu open.

```
  Name          Size          Bytes            Class

  Fs            1x1               8        double array
  Nbits         1x1               8        double array
  Opts          1x1            1048        struct array
  Y          27459x1          219672       double array


  Grand total is 27479 elements using 220736 bytes
```

Typing `Fs`, `Nbits`, or `Opts` will allow you to determine that the sample frequency is 22,050 Hz, the number of bits per sample is 8 bits/sample, and `Opts` is a structured array that contains significant amount of information about the `tada.wav` file. Finally, the sound command plays back the vector Y at a sample frequency of Fs through the PC's sound card.

A more compact form could be

```
[Y, Fs] = wavread('c:\windows\media\tada.wav');
sound(Y, Fs)
```

If you execute these two commands the wav-file will sound as it did before. An even more compact form would be

```
[Y] = wavread('c:\windows\media\tada.wav');
sound(Y)
```

If you execute these two commands the wav-file will not sound as it did before. In fact, the sound command default sample frequency is 8192 Hz. The reduction of the playback sample frequency from 22050 Hz to 8192 Hz results in a significantly increased playback time and the subsequent distortion of the intended information within the wav-file. This playback speed problem can be corrected by using the command

```
sound(Y, 22050)
```

While this solution seems straightforward, it is much easier to recover the sample frequency, F_s, using the `wavread` command, and include this value in the sound playback command. This Windows and MATLAB technique has the added advantage that a wav-file can be created using the Sound Recorder program and then either played back unmodified, or processed off-line and played back later. The results, after MATLAB processing, can be stored in the `*.mat` format by using the `save` command or in the `*.wav` format by using the `wavwrite` command. As with any computer file, it can also be stored on CD-R media for playback using a CD player.

2.4.1 Talk-Through Using MATLAB Only

There are several ways that will allow a simple talk-through operation to be performed using only MATLAB and the PC sound card. Since our goal is to easily transition to DSP hardware implementation of this and other DSP algorithms, we will limit our discussion to MATLAB's built-in `audiorecorder` function and the SIMULINK program (also from The MathWorks).

MATLAB's `audiorecorder.m` function

Recent versions of MATLAB provide the `audiorecorder.m` feature for 32-bit Windows PCs. This feature allows for sound recording and playback using the PC's soundcard *without* the need for MATLAB's DAQ toolbox. Figure 2.10 depicts MATLAB interfacing to the sound

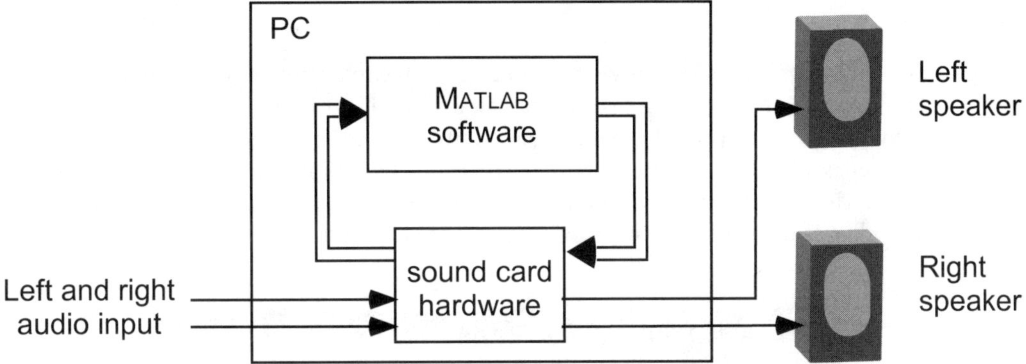

Figure 2.10: MATLAB interface to the PC sound card.

card. The MATLAB help associated with the `audiorecorder` is provided below. Included in this help is a complete example related to sound recording and playback.

```
>> help audiorecorder

AUDIORECORDER Audio recorder object.
    AUDIORECORDER creates an 8000 Hz, 8-bit, 1 channel AUDIORECORDER
    object. A handle to the object is returned.

    AUDIORECORDER(Fs, NBITS, NCHANS) creates an AUDIORECORDER object with
    sample rate Fs in Hertz, number of bits NBITS, and number of channels
    NCHANS. Common sample rates are 8000, 11025, 22050, and 44100 Hz. The
    number of bits must be 8, 16, or 24 on Windows, 8 or 16 on UNIX.  The
    number of channels must be 1 or 2 (mono or stereo).

    AUDIORECORDER(Fs, NBITS, NCHANS, ID) creates an AUDIORECORDER object
    using audio device identifier ID for input. If ID equals -1 the default
    input device will be used.  This option is only available on Windows.

    Example:  Record your voice on-the-fly.  Use a sample rate of 22050 Hz,
              16 bits, and one channel.  Speak into the microphone, then
              pause the recording.  Play back what you've recorded so far.
              Record some more, then stop the recording. Finally, return
              the recorded data to MATLAB as an int16 array.

        r = audiorecorder(22050, 16, 1);
        record(r);      % speak into microphone...
        pause(r);
        p = play(r);    % listen
        resume(r);      % speak again
        stop(r);
        p = play(r);    % listen to complete recording
        mySpeech = getaudiodata(r, 'int16'); % get data as int16 array

    See also audioplayer, audiodevinfo, methods, audiorecorder/get,
             audiorecorder/set.
```

SIMULINK

SIMULINK contains a huge variety of modeling and simulation environment tools and toolboxes for use with MATLAB. As shown in the example in Figure 2.11, the diagrams can look quite simple but still provide remarkable flexibility and versatility.

By double clicking on the *From Wave Device* SIMULINK block, all of the user adjustable parameters for that block are available for viewing and modification. This can be seen in Figure 2.12. To start the simulation, click on the start simulation button as shown in

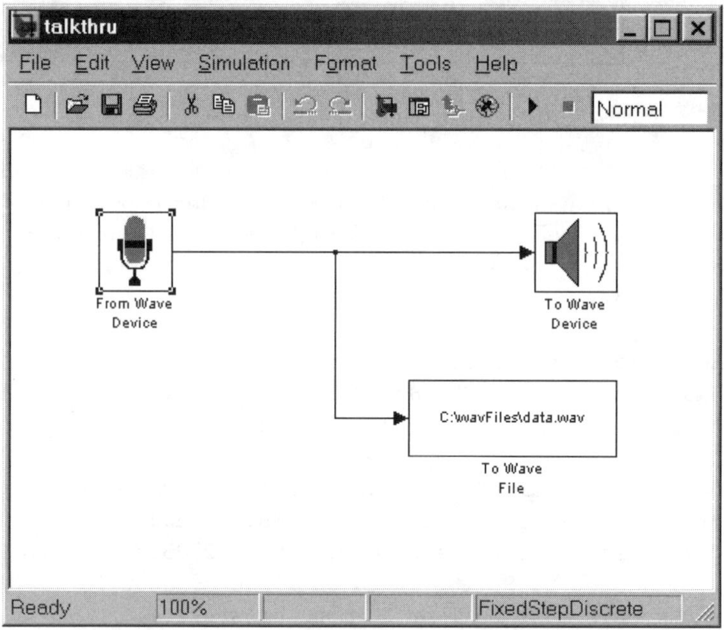

Figure 2.11: SIMULINK model of a PC-based talk-through system.

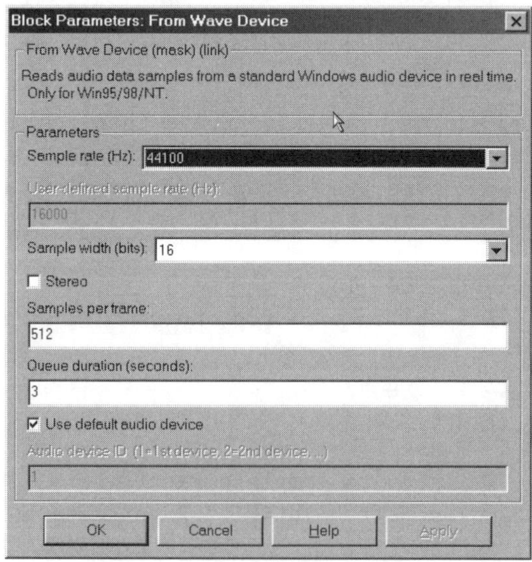

Figure 2.12: Block parameters from the wave device in Figure 2.11.

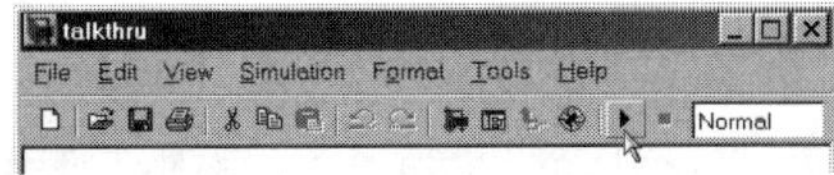

Figure 2.13: Click on the **start simulation** icon on the SIMULINK toolbar.

Figure 2.13. This example is talk-through with the added feature of performing a recording to a wav-file. The wav file, `data.wav` can be read back into MATLAB using the `wavread` function at anytime.

2.4.2 Talk-Through Using MATLAB

Controlling a DSK from MATLAB is not a trivial task, so we've provided a support library for you. The files necessary to run this application are on the CD-ROM that accompanies this book in the `matlab` directory of Appendix E. Using this MATLAB-to-DSK interface software (see details in Appendix E), a single MATLAB m-file allows a simple talk-through program to be constructed which uses the DSK rather than the PC sound card. In this instance, we are reading frames (groups of samples) from the DSK into a MATLAB variable, then writing that same data back out to the DSK. The frame size in this example is 500 samples each. As shown below, the m-file consists of an initial set-up phase (lines 1–8), then reads a single frame of data (line 10) before entering a forever loop (lines 12–15) where the "SwapFrame" function is used to do the actual talk-through operation. This is your first actual *real-time* operation! The code shown below is written for a C6211 or C6711 DSK connected to the PC via a parallel port; to use a different configuration see Appendix E.

Listing 2.2: A simple MATLAB m-file for DSK talk-through.

```
1   c6x_daq('Init', '320ad535.out', 'dsk6211_spp_lpt1');
    c6x_daq('FrameSize', 500);
3   c6x_daq('QueueSize', 2000);
    Fs = c6x_daq('SampleRate', 8000);
5   numChannels = c6x_daq('NumChannels', 1);
    c6x_daq('TriggerMode', 'Immediate'); % disables triggering
7   c6x_daq('LoopbackOff'); % turn off the direct DSK loopback
    c6x_daq('FlushQueues'); % flush the DSK's queues
9
    data = c6x_daq('GetFrame'); % read frame to prime for SwapFrame
11
    while 1 % begin forever loop
13    c6x_daq('SwapFrame', data); % send/receive data
      % data = data * 10; % add gain
15  end
```

Due to the bandwidth limitation of the PC parallel port and the overhead associated with C6211/C6711 DSK's interface, this demonstration is intended to be run using the DSK's onboard codec (TLC320AD535), not one of the higher speed stereo codecs.

To verify that the data is actually being transferred through MATLAB, you can activate line 14 in the forever loop by removing the comment symbol (%). This adds gain to the signal.

2.5   DSK Implementation in C

The files necessary to run this application are in the ccs\MyTalkThrough directory of Chapter 2 on the CD-ROM that accompanies this book. The primary file of interest is ISRs.c, which contains the interrupt service routines. This file includes the necessary variable declarations and performs a swap of the left and right channel data. This swap of left and right channel data is used so that you actually have a few lines of code to view.

The code listings are given below.

Listing 2.3: Talk-through declarations.

```
1  #define LEFT   0
   #define RIGHT  1

3
   float temp;
```

An explanation of Listing 2.3 follows.

1. (Lines 1–2): Define LEFT and RIGHT for user convenience.

2. (Line 4): Declares a temporary variable that is used to allow for channel swapping.

Listing 2.4: Talk through code to swap left and right channels.

```
   /* I added my routine here   */
2
   temp=CodecData.channel[RIGHT];                        // R to temp
4  CodecData.channel[RIGHT]=CodecData.channel[LEFT];     // L to R
   CodecData.channel[LEFT]=temp;                         // temp to L
6
   /* end of my routine   */
```

An explanation of Listing 2.4 follows.

1. (Line 3): Assigns the right channel data to the variable temp.

2. (Line 4): Assigns the left channel data to the right channel.

3. (Line 5): Assigns temp to the left channel.

Unlike the C6711 DSK (with or without an enhanced codec), the C6713 DSK's input circuitry contains a voltage divider that reduces the input voltage level by a factor of 2 (i.e., a −6 dB change). To counteract this signal level decrease for our programs, the DSK_Support.c file (in the common_code directory on the book's CD-ROM) automatically inserts +6 dB of input gain whenever the C6713 DSK is selected.

Now that you understand the code...

Go ahead and copy all of the files into a separate directory. Open the project in Code Composer Studio (CCS) and select "Rebuild All." Once the build is complete, select "Load Program" to load the binary code into the DSK and then click on "Run." Your talk-through system is now running on the DSK. Remember, the DSK or the enhanced codecs that are available for the DSK do not contain audio power amplifiers to drive the connected loads. For the best results, use amplified speakers with your DSK (e.g., powered speakers used with PCs).

2.6 Follow-On Challenges

Consider extending what you have learned, using the C compiler and the DSK.

1. Experiment with scaling the output values (multiplying the input value, e.g., by 0.3, 1.6, and so on...) that you pass to the DAC. Are there limits associated with this scaling?

2. Implement your own program for spectral inversion on the DSK.

3. Change the sign bit of every sample that you pass to the DACs. Describe the effect of this sign bit change on how the output signal sounds.

4. Modify your talk-through code to output *only* the left or right channel value.

5. Modify your talk-through code to combine the left and right channel inputs and send the result to *both* the left and right channel outputs. Are there any limitations associated with combining the left and right channels?

6. Reduce the number of bits used by the DAC by using only the 8 most significant bits (MSBs). Can you hear the difference when only the 8 MSBs are used?

7. Reduce the number of bits used by the DAC by using only the 4 most significant bits (MSBs). Can you hear the difference when only the 4 MSBs are used?

8. Reduce the number of bits used by the DAC by using only the 2 most significant bits (MSBs). Can you hear the difference when only the 2 MSBs are used?

9. Reduce the number of bits used by the DAC by using only the most significant bit (MSB). Can you hear the difference when only the MSB is used?

Chapter 3

FIR Digital Filters

3.1 Theory

$\mathbf{F}$ILTERING is one of the most common DSP operations. Filtering can be used for noise suppression, signal enhancement, removal or attenuation of a specific frequency, or to perform a special operation such as differentiation, integration, or the Hilbert transform [1]. While this is not a complete list of all of the possible applications of filters, it may serve to remind us of the importance of filtering.

Filters can be thought of, designed, and implemented in either the sample domain or the frequency domain. This chapter, however, will only deal with sample domain filter implementation on a sample-by-sample basis. Frame-based processing and frequency domain filter implementation are discussed in Chapters 6 and 7, respectively.

3.1.1 Traditional Notation

The notation used in many continuous-time signals and systems texts[1] is to label the input signal as $x(t)$, the output signal as $y(t)$, and the impulse response of the system as $h(t)$. These time domain descriptions have frequency domain equivalents; they are obtained using the Fourier transform, which is shown as $\mathcal{F}\{\ \}$. The Fourier transform of $x(t)$ is $\mathcal{F}\{x(t)\} = X(j\omega)$; similarly the Fourier transform of $y(t)$ is $Y(j\omega)$ and that of $h(t)$ is $H(j\omega)$. $H(j\omega)$ is also called the frequency response of the system. These Fourier transform pairs are summarized below.

$$x(t) \overset{\mathcal{F}}{\longleftrightarrow} X(j\omega)$$

$$y(t) \overset{\mathcal{F}}{\longleftrightarrow} Y(j\omega)$$

$$h(t) \overset{\mathcal{F}}{\longleftrightarrow} H(j\omega)$$

The most common notation used in discrete-time signals and systems texts is to label the input signal samples as $x[n]$, the output signal samples as $y[n]$, and the impulse response as $h[n]$. Note that the discrete-time impulse response $h[n]$ is called the *unit sample response* in some texts. In this book, parentheses "( )" will be used to denote continuous-time, while square brackets "[ ]" will be used to denote discrete-time. Discrete-time descriptions (such as $x[n]$, $y[n]$, and $h[n]$) have frequency domain equivalents that are obtained using the

---

[1] As in essentially all subject areas covered by engineering and science texts, there is *no* universally agreed upon standard notation. We use the notation of the majority here, but your favorite book may be different.

discrete-time Fourier transform (DTFT), which we will also abbreviate as $\mathcal{F}\{\}$. The DTFT of $x[n]$ is $\mathcal{F}\{x[n]\} = X\left(e^{j\omega}\right)$, of $y[n]$ is $Y\left(e^{j\omega}\right)$, and of $h[n]$ is $H\left(e^{j\omega}\right)$. $H\left(e^{j\omega}\right)$ is also called the frequency response of the system. These discrete-time Fourier transform pairs are summarized below.

$$x\left[n\right] \overset{\mathcal{F}}{\longleftrightarrow} X\left(e^{j\omega}\right)$$

$$y\left[n\right] \overset{\mathcal{F}}{\longleftrightarrow} Y\left(e^{j\omega}\right)$$

$$h\left[n\right] \overset{\mathcal{F}}{\longleftrightarrow} H\left(e^{j\omega}\right)$$

Notice that the abbreviation used for the continuous-time Fourier transform and the DTFT are the same because the context should make it clear which transform is used. For example, if the signal or system being transformed is a discrete-time signal or system, it will be implied by the square bracket notation and thus the DTFT should be inferred. Also of interest is the fact that the DTFT of a discrete-time signal, e.g., $x[n]$, results in a continuous-frequency function, $X\left(e^{j\omega}\right)$, since ω is a continuous variable. A summary of the continuous-time and discrete-time notation is provided in Figure 3.1.

3.1.2 FIR Filters Compared to IIR Filters

The title of this chapter contains the acronym FIR, which stands for **F**inite **I**mpulse **R**esponse. All FIR filters are, by definition, discrete-time filters (there is no such thing as a continuous-time FIR filter). If we excite an FIR filter with a unit sample (a sample of value one) followed by an infinite number of zero-valued samples, we will have excited the system with the discrete-time version of an impulse function (sometimes called a unit sample function). Exciting an Nth order FIR filter with an impulse will result in $N + 1$ output terms before all the remaining terms will have a value of exactly zero (since the filter has $N + 1$ coefficients). Thus the impulse response is finite.

Another type of filter, called an IIR filter, has an **I**nfinite **I**mpulse **R**esponse. If you have ever designed analog filters, you have designed IIR filters. Think about the output of an analog filter that has an impulse for its input. Mathematically, the output of the system never fully decays to (and remains) at the exact value of zero. IIR filters are discussed in more detail in Chapter 4.

3.1.3 Calculating the Output of a Filter

To calculate the output of a continuous-time system that has been given a continuous-time input signal we need to convolve the input signal with the system's impulse response. Since this involves continuous signals, we use integration (discrete signals use summation instead of integration). Thus to calculate the output we need to evaluate the convolution integral. This is an operation that many beginning students find to be mysterious and intimidating

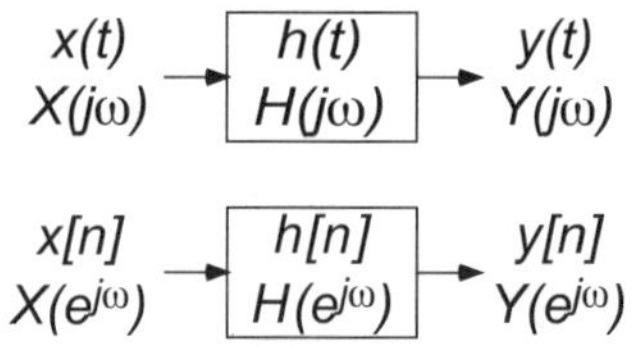

$x(t)$ → | $h(t)$ | → $y(t)$
$X(j\omega)$ | $H(j\omega)$ | $Y(j\omega)$

$x[n]$ → | $h[n]$ | → $y[n]$
$X(e^{j\omega})$ | $H(e^{j\omega})$ | $Y(e^{j\omega})$

Figure 3.1: A summary of the continuous-time and discrete-time notation.

(but get used to convolution—it comes up over and over again). The general form of the convolution integral is

$$y(t) = \int_{-\infty}^{\infty} h(\tau)x(t - \tau)\,d\tau.$$

If we restrict our discussion to realizable signals and systems, then due to causality (i.e., we can't calculate the output based on an input that hasn't arrived yet) the convolution integral becomes

$$y(t) = \int_{0}^{\infty} h(\tau)x(t - \tau)\,d\tau.$$

Similarly, to calculate the output of a discrete-time system that has a discrete-time input signal we use the convolution sum. The general form of the convolution sum is

$$y[n] = \sum_{k=-\infty}^{\infty} h[k]x[n - k].$$

If we again restrict our discussion to realizable signals and systems, then due to causality the convolution sum becomes

$$y[n] = \sum_{k=0}^{\infty} h[k]x[n - k].$$

For an FIR system, the filter coefficients are the individual terms that make up the impulse response of the system. These FIR filter coefficients are commonly called the b coefficients. In MATLAB, when all of the b coefficients are formed into a vector, it is called the B vector. Making this substitution (b for h) and remembering that an FIR filter of order N has $N+1$ coefficients, the convolution sum takes on the general form of the FIR difference equation, namely,

$$y[n] = \sum_{k=0}^{N} b[k]x[n - k].$$

This equation tells us that in order to calculate a value for the current output, $y[0]$, we must perform the dot product of $B \cdot X$, where $B = \{b[0], b[1], \ldots, b[N]\}$ and X represents the current and past values of the input, $X = \{x[0], x[-1], \ldots, x[-N]\}$. That is,

$$y[0] = \sum_{k=0}^{N} b[k]x[-k] = b[0]x[0] + b[1]x[-1] + \cdots + b[N]x[-N].$$

The block diagram associated with implementing the FIR difference equation, which is another way of saying an FIR filter, is shown in Figure 3.2. The blocks containing z^{-1} are delay blocks that store the value in the block for one sample period. The delay blocks may be thought of as synchronous shift registers that have their clocks tied to the ADC and DAC's sample clock, but are typically just memory locations accessed by the DSP CPU.

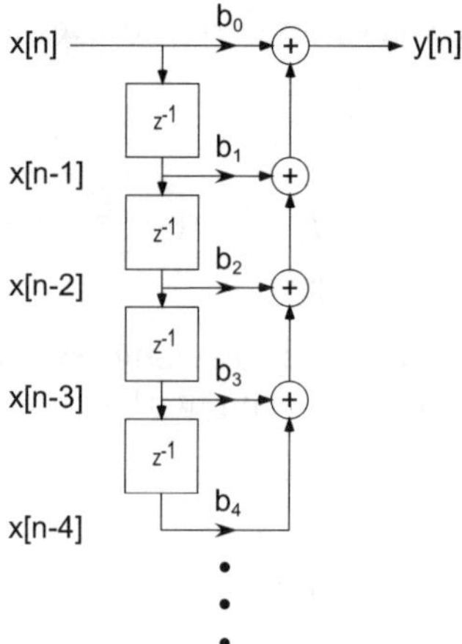

Figure 3.2: Block diagram associated with the implementation of an FIR filter.

3.2   winDSK6 Demonstration

If you double click on the winDSK6 icon, the winDSK6 application will launch, and a window similar to Figure 11.3 will appear. Before proceeding, be sure to select the appropriate values for "DSK and Host Configuration." These selections are "sticky," such that the next time you run winDSK6, the choices you made will still be selected.

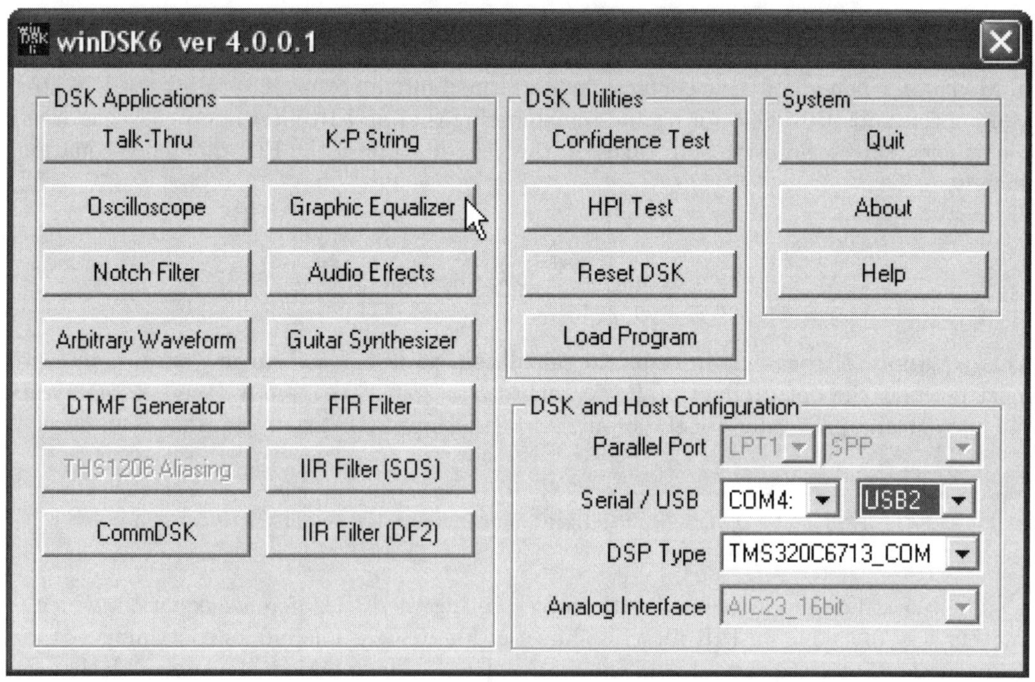

Figure 3.3: winDSK6 ready to load the Graphic Equalizer application.

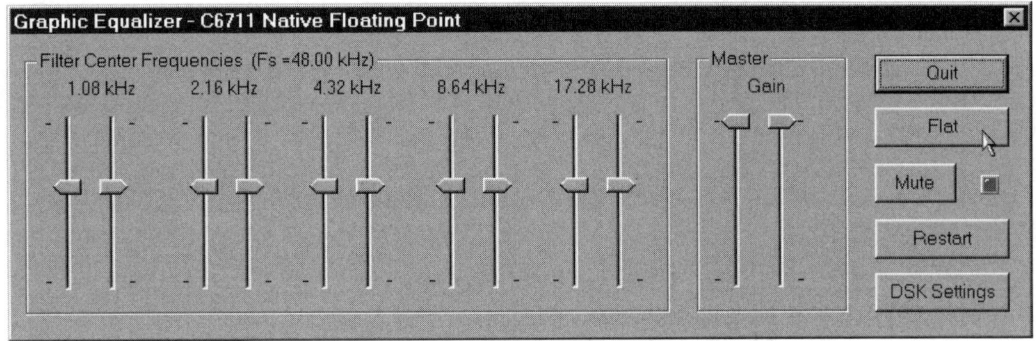

Figure 3.4: winDSK6 running the Graphic Equalizer application.

3.2.1 Graphic Equalizer Application

Clicking on the winDSK6 Graphic Equalizer button will load that program into the attached DSK, and a window similar to Figure 11.4 will appear. The Graphic Equalizer application implements a five-band audio equalizer as shown in the signal flow shown in Figure 3.5. If your DSK has a stereo codec attached (and you have selected that codec from the winDSK6 main window), an independently adjustable equalizer is active on both left and right channels.

The equalizer uses five FIR filters (a lowpass (LP) filter, 3 bandpass (BP) filters, and a highpass (HP) filter) operating in parallel. The gain sliders (A_1 to A_5) in the dialog box operate on memory locations used to control the gains of each filter and the overall system gain. The 5 FIR filters are designed as high order ($N = 128$) filters; the resulting steep roll-off of these filters can be seen in Figure 3.6.

There are a number of ways you can experience the effect of the graphic equalizer filtering. For example, you could connect the output of a CD player to the signal input of the DSK, and connect the DSK signal output to a set of powered speakers. Play some familiar music while you adjust the graphic equalizer slider controls and listen to the result. A more objective experiment would be to play the track of additive white Gaussian noise

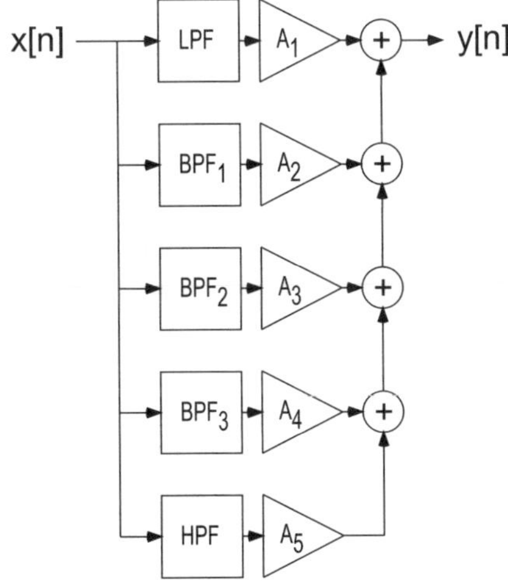

Figure 3.5: Block diagram associated with winDSK6's 5-band Graphic Equalizer application.

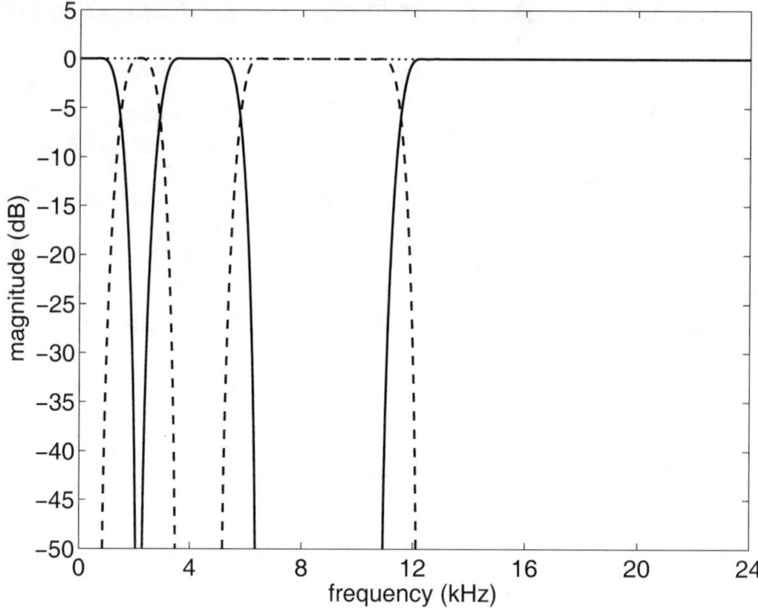

Figure 3.6: Frequency response of winDSK6's 5-band Graphic Equalizer application. The dotted straight line at 0 dB represents the sum of all five bands.

(AWGN) on the CD-ROM that accompanies this book (in directory test_signals play file awgn.wav), which theoretically contains all frequencies. If the DSK signal output is then connected to a spectrum analyzer, you could observe which band of frequencies is affected, and how much it is affected, as you adjust the slider controls. If you don't have a spectrum analyzer available, a second DSK running winDSK6 can be used in its place (select the "Oscilloscope" button from the main screen, select "Spectrum Analyzer" from the next screen, and select "Log10" to display the result in decibels). Alternatively, you can use your computer's sound card to gather a portion of the DSK's output. This can be accomplished using the Windows sound recorder, MATLAB's data acquisition (DAQ) toolbox, or the audio recorder that is part of MATLAB (version 6.1 or later). This recorded data can be analyzed and displayed using MATLAB.

3.2.2 Notch Filter Application

The winDSK6 Notch Filter application actually implements a second order IIR filter, but we can make it seem to be an FIR filter. Clicking on the Notch Filter button will load the program into the attached DSK, and a window similar to Figure 3.7 will appear. If you decrease the Q adjustment (r) until it reaches zero, you have put the poles of the filter at the origin of the z-plane and the system will behave just like an FIR filter (this concept will be discussed further in Chapter 4). This adjustment is shown on the bottom slider bar in Figure 3.7. Also notice that the "Filter Type" was changed from "Bandpass" to "Notch". The frequency responses associated with four different settings of the notch filter are overlaid and shown in Figure 3.8. Ideally, an infinite amount of attenuation is present at the notch frequency. This explains why a properly adjusted (tuned) notch filter can completely remove an interfering tone.

The effect of the Notch Filter application can be heard by adding a sinusoidal signal (tone) to a music signal. Most computer sound cards will perform this summing for you.

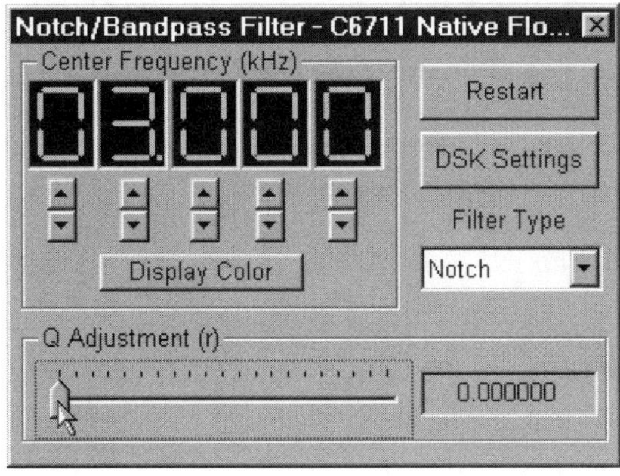

Figure 3.7: winDSK6 running the Notch Filter application with $r = 0$.

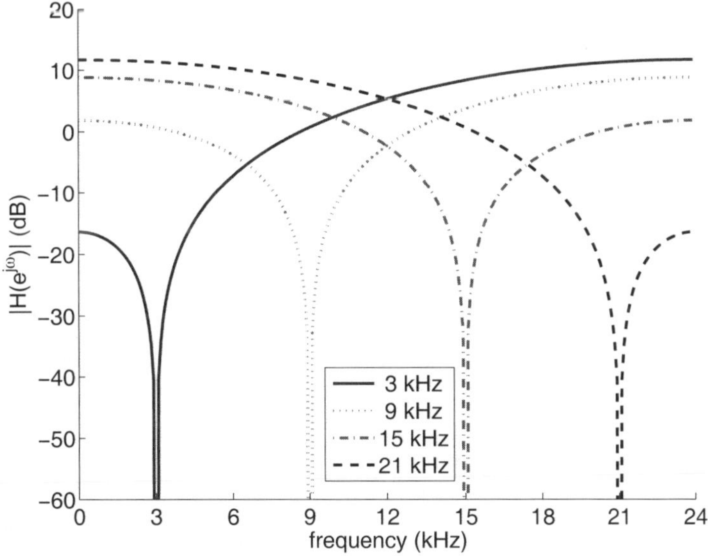

Figure 3.8: The frequency response of four different notch filters having notch frequencies of 3, 9, 15, and 21 kHz respectively.

You will need to experiment with your sound card's mixer controls to determine exactly how your particular system responds. Most systems are capable of summing an audio CD signal and another signal by playing the CD on the drive installed in the computer and injecting the second signal from a function generator through the sound card "line input" or "microphone input" connector. The sound card "line output" or "headphone output" is then connected to the signal input of the DSK. As before, the DSK signal output is connected to a set of powered speakers. If a function generator is not available, you may use one of the audio test tones in directory `test_signals` on the CD-ROM that accompanies this book and play them with a second, external CD player. You can also easily create your own audio test tones in MATLAB, then record them on to CD-R or CD-RW media and play them in the same way on an external CD player (this concept will be discussed further in Chapter 5).

When the center frequency of the notch filter equals the frequency of the injected tone, you will hear the sound of the tone disappear from the speakers.

3.2.3 Audio Effects Application

The winDSK6 Audio Effects application contains a mixture of both FIR and IIR applications. Clicking on the Audio Effects button will load the program into the attached DSK, and a window similar to Figure 3.9 will appear. The flanging and chorus effects are both implemented with FIR filters, so they provide good examples for this chapter.

The block diagram of the flanging effect is shown in Figure 3.10, where α is a scale factor and $\beta[n]$ is a periodically varying delay described by

$$\beta[n] = \frac{R}{2}\left(1 - \cos\left(\omega_0 n\right)\right).$$

R is the maximum number of sample delays and ω_0 is some low frequency. In winDSK6, you can adjust α with the **Alpha** slider, ω_0 with the **Frequency** slider, and R with the **Delay** slider enclosed in the **Flanger** portion of the winDSK6 Audio Effects window (see Figure 3.9). Thus the delay time $\beta[n]$ varies sinusoidally from a minimum of 0 to a maximum of R. Flanging is a special sound effect often used by musicians (particularly guitarists); it sounds as if the musical instrument has taken on a sort of "whooshing" sound up and down the frequency scale. More about this and other special effects can be found in Chapter 10.

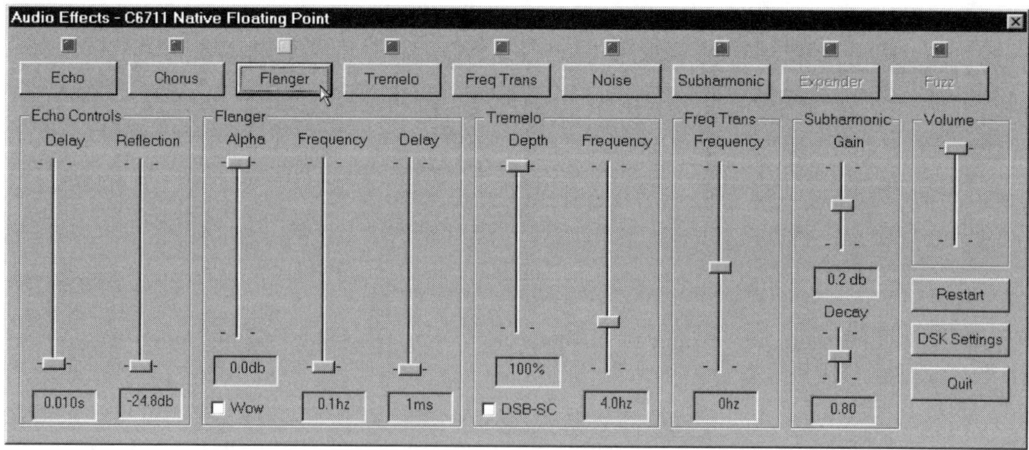

Figure 3.9: winDSK6 running the Audio Effects application.

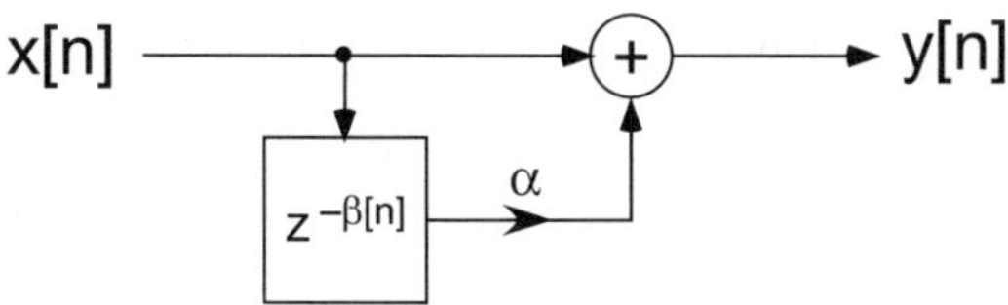

Figure 3.10: The block diagram of the flanging effect.

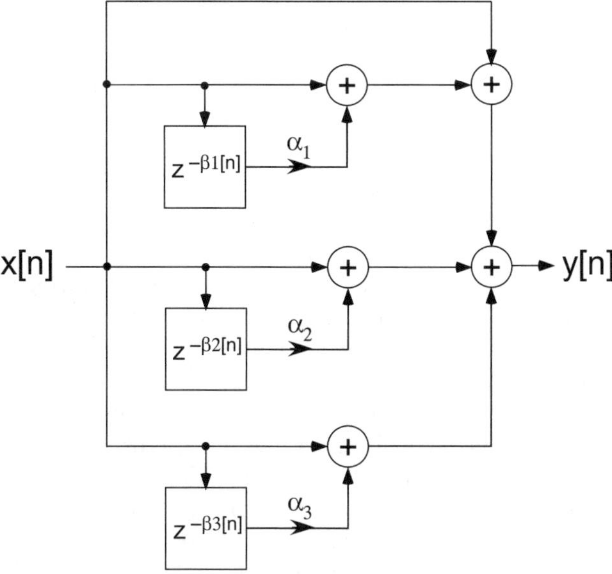

Figure 3.11: The block diagram of the chorus effect.

The block diagram of another musical special effect, the chorus effect, is shown in Figure 3.11. To generate the chorus effect, three separately flanged signals are summed with the original signal. For a proper chorus effect, each of the β's and α's should be independent.

3.3 MATLAB Implementation

MATLAB has a number of ways of performing the filtering operation. In this chapter we will only discuss two of them. The first is the built-in `filter` function and the second is to build your own routine to perform the FIR filtering operation. The built-in function allows us to filter signals almost immediately, but does very little to prepare us for real-time filtering using DSP hardware.

3.3.1 Built-In Approach

As mentioned previously, MATLAB has a built-in function called `filter.m`. This function can be used to implement both an FIR filter (using only the numerator (B) coefficients) and an IIR filter (using both the denominator (A) and the numerator (B) coefficients). The first few lines of the on-line help associated with the `filter` command are provided below. This and any other MATLAB help is available from the command line by typing,

```
help MATLAB function or command name
```

In the case of the `filter` command,

```
>> help filter

 FILTER One-dimensional digital filter.
    Y = FILTER(B,A,X) filters the data in vector X with the
    filter described by vectors A and B to create the filtered
    data Y.  The filter is a "Direct Form II Transposed"
    implementation of the standard difference equation:

    a(1)*y(n) = b(1)*x(n) + b(2)*x(n-1) + ... + b(nb+1)*x(n-nb)
                          - a(2)*y(n-1) - ... - a(na+1)*y(n-na)
```

Notice that in the difference equation discussion of the MATLAB `filter` command, the A and B coefficient vector indices start at 1 instead of at 0. MATLAB does not allow for an index equal to zero. While this may only seem like a minor inconvenience, improper vector indices account for a significant number of the errors that occur during MATLAB algorithm development. In our classes, we typically create another vector, say n, which is composed of integers with the first element equal to zero (i.e., `n=0:15` creates $n = \{0, 1, 2, 3, \ldots, 15\}$), and use this n vector to "fool" MATLAB into counting from zero for things such as plot axes. See the code given below for an example of this technique.

The MATLAB code shown below will filter the input vector x using the FIR filter coefficients in vector B. Notice that the input vector x is zero padded (line 6) to flush the filter. This technique differs slightly from the direct implementation of the MATLAB `filter` command in which for M input values there will be M output values. Our technique assumes that the input vector is both preceded and followed by a large number of zeros. This implies that the filter is initially at rest (no initial conditions) and will relax or flush any remaining values at the end of the filtering operation.

Listing 3.1: Simple MATLAB FIR filter example.

```matlab
1  % Simulation inputs
   x = [1  2  3  0  1  -3  4  1];            % input vector x
3  B = [0.25  0.25  0.25  0.25];            % FIR filter coefficients B

5  % Calculated terms
   PaddedX=[x zeros(1,length(B)-1)];  % zero pad x to flush filter
7  n=0:(length(x) + length(B) - 2);   % plotting index for the output
   y=filter(B, 1, PaddedX);           % performs the convolution
9
   % Simulation outputs
11 stem(n, y)                         % output plot generation
   ylabel('output values')
13 xlabel('sample number')
```

The output for this example follows.

```
 y =
   Columns 1 through 8
     0.2500    0.7500    1.5000    1.5000    1.5000    0.2500    0.5000    0.7500
   Column 9
     0.5000    1.2500    0.2500
```

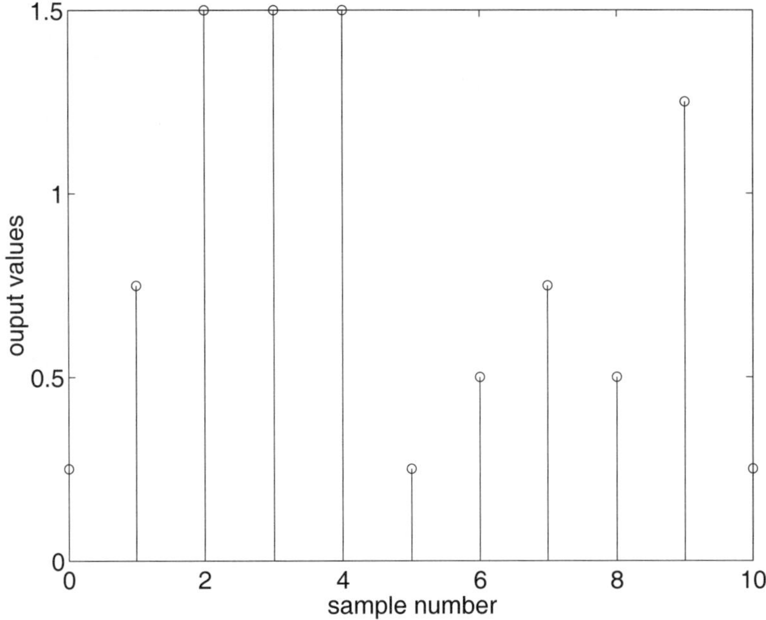

Figure 3.12: Stem plot of the filtering of x with B.

The stem plot from the example is shown in Figure 3.12.

In this example, eight input samples were filtered and the results were returned all at once. Notice that when an 8-element vector x is filtered by a 4-element vector B that 11 elements were returned ($8 + 4 - 1 = 11$). This is an example of the general result that states that the length L of the sequence resulting from the convolution (filtering) of x and B is $L = \ \text{length}(x) + \ \text{length}(B) - 1$.

The FIR filter coefficients associated with this filter were $B = [0.25\ 0.25\ 0.25\ 0.25]$. Since there are four coefficients for the filter, this is a third order filter (i.e., $N = 3$). The filtering effect in this case is an averaging of the most recent 4 input samples (i.e., the current sample and the previous three samples). This type of filter is called a moving average (MA) filter and is one type of lowpass FIR filter. Figure 3.13 shows the frequency response associated with MA filters of order $N = 3$, 7, 15, and 31, for a sample frequency of 48 kHz.

All of the MA filters shown in Figure 3.13 have a 0 Hz (D.C.) gain of 1 (which equals 0 dB). To insure that any filter has a D.C. gain of 0 dB, the impulse response $h[n]$ must sum to 1. The relationship between the D.C. response and the impulse response can be shown quickly by recalling the z-transform for a causal system described by $h[n]$,

$$H(z) = \sum_{n=0}^{\infty} h[n]z^{-n}.$$

To convert $H(z)$ into the frequency response $H(e^{j\omega})$, the variable substitution $z = e^{j\omega}$ is required.[2] Making this substitution,

$$H\left(e^{j\omega}\right) = \sum_{n=0}^{\infty} h[n]\left(e^{j\omega}\right)^{-n}.$$

---

[2]The expression $e^{j\omega}$ can be thought of as a vector with a magnitude of 1 and an angle of ω.

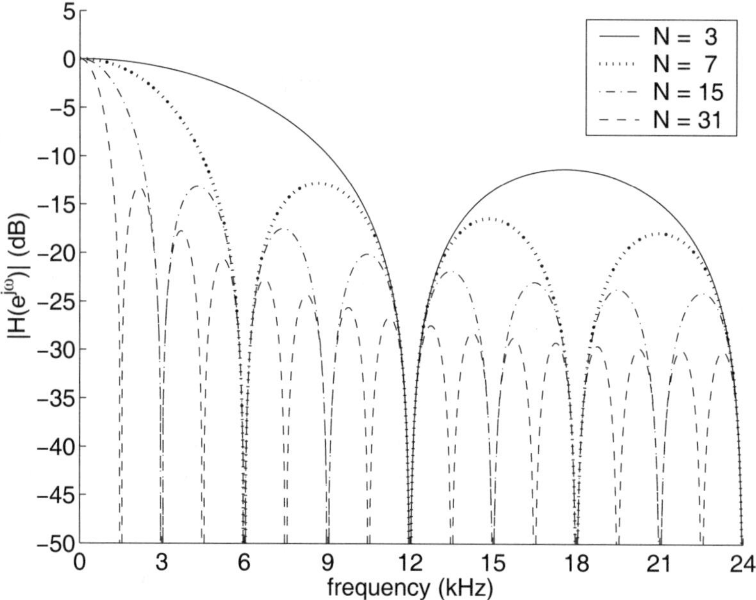

Figure 3.13: Magnitude of the frequency response for MA filters of order 3, 7, 15, and 31.

To evaluate the D.C. response of an Nth order FIR filter, set $\omega = 0$ and the upper limit of summation equal to N. This results in

$$H\left(e^{j\omega}\right)\big|_{\omega=0} = H(1) = \sum_{n=0}^{N} h[n](1)^{-n} = \sum_{n=0}^{N} h[n].$$

This relationship explains why, for $N = 3$, each of the four $h[n]$ terms associated with the MA filter were defined as $1/4 = 0.25$. Similarly, for $N = 31$, each of the $h[n]$ terms would be $1/32 = 0.03125$ to ensure a D.C. response equal to 1 (i.e., 0 dB).

A Real World Filtering Example

There are an unlimited number of data sets or processes that can be filtered. For example, what if we wanted to know the 4-day or 32-day average value of a stock market's closing value? If the closing values are filtered, much of the day-to-day market variations can be removed. The cutoff frequency of the filter used to filter the closing values would control the amount of the remaining variations. As shown in Figure 3.13 for an MA filter, the cutoff frequency is inversely related to the filter order. Figure 3.14 shows filtered and unfiltered closing values of the NASDAQ composite index for calendar year 2001.

The upper portion of the figure plots the raw data and the results of filtering the raw data with both a 4-term (3rd order) and a 32-term (31st order) MA filter. This subplot was created using the MATLAB function `filtfilt`. This function implements a zero-phase forward and reverse filter (that is, there is zero group delay). This forward/reverse technique to eliminate group delay *cannot* be used in real-time filtering. For additional information on the MATLAB function `filtfilt`, type `help filtfilt` from the MATLAB command prompt.

The lower portion of the figure plots the raw data and the results of filtering the raw data with both a 4-term (3rd order) and a 32-term (31st order) MA filter using the MATLAB function `filter`. Since realizable MA filters have nonzero group delay, the filtered data lags

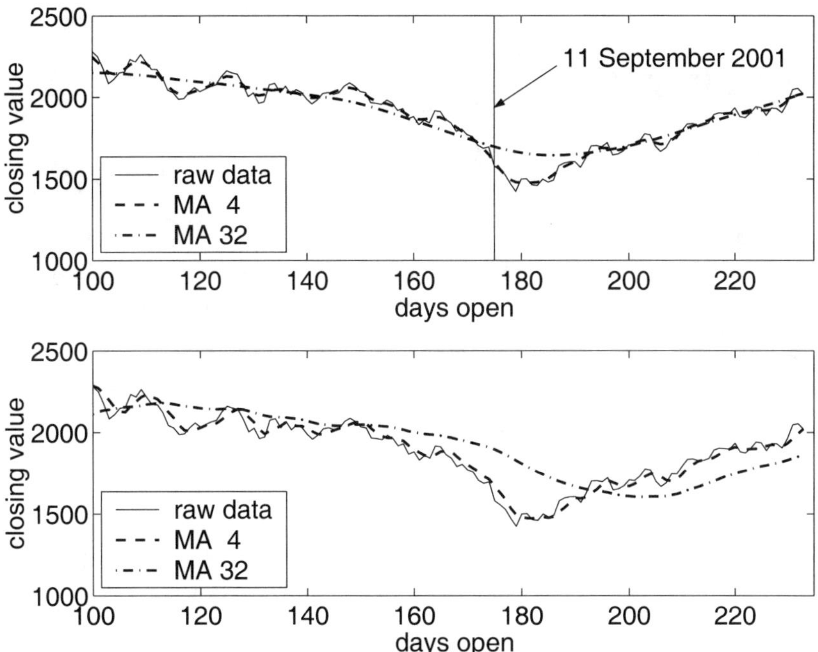

Figure 3.14: Filtered and unfiltered closing values of the NASDAQ composite index for calendar year 2001. The upper plot was generated with the MATLAB function `filtfilt`. The lower plot was generated with the MATLAB function `filter`.

the raw data by an amount of time equal to the group delay G_D multiplied by the sample period T_s.

3.3.2 Creating Your Own Filter Algorithm

The previous example helped us with MATLAB-based filtering, but the built-in function `filter.m` is of little use to us in performing real-time FIR filtering with DSP hardware. The next MATLAB example more closely implements the algorithm needed for a real-time process. This code will calculate a single output value based upon the current input value and the three previous input values.

Listing 3.2: MATLAB FIR filter adjusted for real-time processing.

```
1  %  This m-file is used to convolve x[n] and B[n] without
   %  using the MATLAB filter command.  This is one of the
3  %  first steps toward being able to implement a real-time
   %  FIR filter in DSP hardware.
5  %
   %  In sample-by-sample filtering, you are only trying to
7  %  accomplish two things,
   %
9  %  1.  Calculate the current output value, y(0), based on
   %      just having received a new input sample, x(0).
11 %  2.  Setup for the arrival of the next input sample.
   %
13 %  This is a BRUTE FORCE approach!
```

```matlab
%
15
   % Simulation inputs
17 x = [1  2  3  0];                   % input x = [x(0) x(-1) x(-2) x(-3)]
   N = 3;                             % order of the filter = length(B) - 1
19 B = [0.25  0.25  0.25  0.25];      % FIR filter coefficients B


21 % Calculated terms
   y = 0;                            % initializes the output value y(0)
23 for  i = 1:N+1                     % performs the dot product of B and x
        y = y + B(i)*x(i);
25 end


27 for  i = N:-1:1                    % shift stored x samples to the right so
        x(i+1) = x(i);                % the next x value, x(0), can be placed
29 end                               % in x(1)


31 % Simulation outputs
   x                                 % notice that x(1) = x(2)
33 y                                 % average of the last four input values
```

The input and output vectors from this FIR moving average filter program are shown below.
Note the four input samples result in a single output sample, as expected of a third order
filter. Also, the displayed version of vector x shows the effect of shifting the values to the
"right" to make room for the next sample, as discussed below.


```
x =

     1     1     2     3


y =
   1.5000
```


A few items need to be discussed concerning this example.

1. The filter order N was declared (line 21) despite the fact that MATLAB can determine
   the filter order based solely on the length of the vector B (that is, `N=length(B)-1`).
   Declaring both the filter order N and the FIR filter coefficients B (line 22), increases
   the portability of the C/C++ code we will derive from the MATLAB code. Increased
   code portability may also be thought of as decreased machine dependance, which is
   generally a sought-after code attribute.

2. Only four values of x were stored (line 20). FIR filtering involves the dot product of
   only $N + 1$ terms. Since in this example $N = 3$, only four x terms are required.

3. The example is called a "brute force" approach, which is based largely on the shifting of
   the stored x values within the x vector (lines 30–32), to make room for the next sample
   that would overwrite the value at $x(0)$. This unnecessary operation wastes resources
   which may be needed for other operations. Since our ultimate goal is efficient real-
   time implementation in DSP hardware, more elegant and efficient solutions to this
   problem will be discussed in the next section.

3.4 DSK Implementation in C

Several modifications of the MATLAB thought process are needed as we transition toward efficient real-time programming.

1. A semantic change is required since in MATLAB B is often called a vector, but in the C/C++ programming language, B is called an array.

2. The zero memory index value which does not exist in a MATLAB vector *does* exist in the C/C++ programming language and it is routinely used in array notation.

3. The DSP hardware must process the data from the analog-to-digital converter (ADC) in real-time. Therefore, we cannot wait for all of the message samples to be received prior to beginning the algorithmic process.

4. Real-time DSP is inherently an interrupt driven process and the input samples should only be processed using interrupt service routines (ISRs). Given this observation, it is incumbent upon the DSP programmer to ensure that the time requirements associated with periodic sampling are met. More bluntly, if you do not complete the algorithm's calculation before another input sample arrives, you have not met your real-time schedule, and your system will fail. This leads to the observation that, "the correct answer, if it arrives late, is wrong!"

5. Even though the DSP hardware has a phenomenal amount of processing power, this power should not be wasted.

6. The input and output ISRs are *not* magically linked! Nothing will come out of your DSP hardware unless you program the device to do so.

7. The digital portion of both an ADC and a digital-to-analog converter (DAC) are inherently *integer* in nature. No matter what the ADC's input range is, the analog input voltage is mapped to an integer value. For a 16-bit converter using two's complement representation, the possible integer values range from $+32,767$ to $-32,768$.

8. For clarity and understandability, declarations and assignments of variables, e.g., FIR filter coefficients, can be moved into `.c` and `.h` files.

3.4.1 Brute Force FIR Filtering in C: Part 1

The first version we will examine of FIR filter implementation code in C takes a brute force approach, similar to the last MATLAB example. The purpose of this first approach is understandability, which comes at the expense of efficiency.

The files necessary to run this application are in the Chapter 3 `ccs\FIRrevA` directory. The primary file of interest is `ISRs.c`, which contains the interrupt service routines. This file includes the necessary variable declarations and performs the actual FIR filtering operation. To allow for the use of a stereo codec (e.g., the native C6713 codec, the PCM3006-based daughtercard codec for the C6711, etc.), the program implements independent Left and Right channel filters. However, for clarity only the Left channel will be discussed below. In the code example, N is the filter order, the B array holds the FIR filter coefficients, the $xLeft$ array holds both the current input value ($x[0]$) and the past input values ($x[-1], x[-2]$, and $x[-3]$). The variable $yLeft$ is the current output value of the filter, $y[0]$. The integer i is used as an index counter in the `for` loops.

Listing 3.3: Brute force FIR filter declarations.

```
#define N 3

float B[N+1] = {0.25, 0.25, 0.25, 0.25};
float xLeft[N+1];
float yLeft;

int i;
```

The code shown below performs the actual filtering operation. The program instructions that move a sample value from the appropriate ADC register to CodecDataIn. Channel[LEFT] in the input ISR and from CodecDataOut.Channel[LEFT] to the appropriate DAC register in the output ISR are not shown here. The five main steps involved in the filtering operation will be discussed following the code listing.

Listing 3.4: Brute force FIR filtering for real-time.

```
/* I added my routine here */
xLeft[0] = CodecDataIn.Channel[LEFT];       // current input value
yLeft = 0;                        // initialize the output value

for (i = 0; i <= N; i++) {   // x is length N+1
    yLeft += xLeft[i]*B[i]; // perform the dot-product
}

for (i = N; i > 0; i--) {
    xLeft[i] = xLeft[i-1];  // shift for the next input
}

CodecDataOut.Channel[LEFT] = yLeft; // output the value
/* end of my routine */
```

The five real-time steps involved in brute force FIR filtering

An explanation of Listing 3.4 follows.

1. (Line 2): The most current sample from the ADC side of the codec is assigned to the current input array element, xLeft[0].

2. (Line 3): The current output of this filter is given the name yLeft. Since this same variable will be used in the calculation of each output value of the filter, it must be reinitialized to zero before each dot product is performed.

3. (Lines 5–7): These 3 lines of code perform the dot product of x and B. The equivalent operation is,

$$yLeft = xLeft[0]B[0] + xLeft[-1]B[1] + xLeft[-2]B[2] + xLeft[-3]B[3].$$

4. (Lines 9–11): These 3 lines of code shift all of the values in the x array one element to the right. The equivalent operation is,

$$xLeft[2] \rightarrow xLeft[3]$$
$$xLeft[1] \rightarrow xLeft[2]$$
$$xLeft[0] \rightarrow xLeft[1].$$

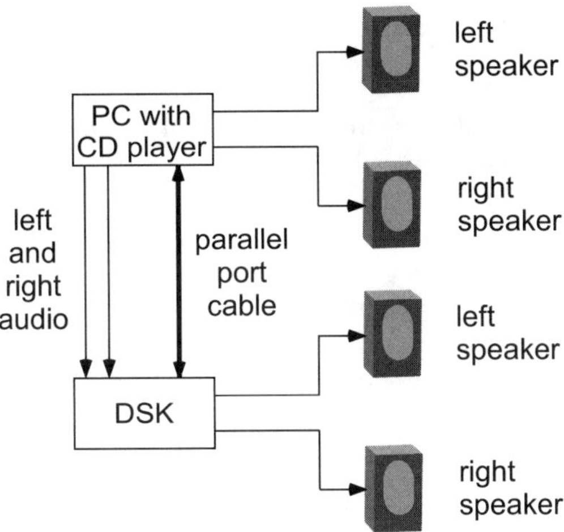

Figure 3.15: One method of listening to the unfiltered and filtered audio signals.

After the shift to the right is complete, the next incoming sample, $x[0]$ can be written into the $xLeft[0]$ memory location without a loss of information. Also notice that $xLeft[3]$ was overwritten by $xLeft[2]$. You might expect the operation $xLeft[3] \rightarrow xLeft[4]$ and so on should be performed, but there is no $xLeft[4]$ or higher because $xLeft$ only contains 4 elements. In summary, the "old" $xLeft[3]$ is no longer needed and is therefore overwritten.

5. (Line 13): This line of code completes the filtering operation by transferring the result of the dot product, $yLeft$, to the `CodecDataOut.Channel[LEFT]` variable for transfer to the DAC side of the codec via the transmit ISR.

Now that you understand the code...

Go ahead and copy all of the files into a separate directory. Open the project in CCS and "Rebuild All." Once the build is complete, "Load Program" into the DSK and click on "Run." Your FIR LP filter is now running on the DSK. Remember this program would typically be used for audio filtering, so a good way to experience the effects of your filter is to listen to unfiltered and filtered music.[3] Figure 3.15 shows a method to listen to the original (unfiltered) music and the filtered music simultaneously. This technique does require a second set of speakers, but it is much more convenient than the alternative of using only one set of speakers. Remember, the DSK or the enhanced codecs that are available for the DSK do not contain audio power amplifiers to drive the connected loads. For the best results, use amplified speakers with your DSK (e.g., powered speakers used with PCs).

3.4.2 Brute Force FIR Filtering in C: Part 2

The previous section introduced a *brute force* approach to FIR filtering. While this implementation was straightforward and relatively easy to understand, it suffers from two major problems.

1. Most FIR filters use a considerably higher order than the fourth order filter discussed in the previous example. Most filters also require more than a few digits of numerical

---

[3]You may need to increase the order of the filter to really hear a difference in the filtered output. Try a 31st order MA filter by setting $N = 31$ and making all the values of B equal to $1/32 = 0.03125$.

precision to accurately specify the B coefficients. These facts make manual entry of the B coefficients very inconvenient.

2. Step four in the five real-time steps involved in the brute force FIR filtering section that was discussed above shifted all of the values in the x array one element to the right after each dot product operation. This "manual" shifting is a very inefficient use of the DSK's computational resources.

These problems will be addressed in the sections that follow.

The declaration for array `B[N+1]` shown below represents just the first 12 lines required to initialize a 30th order FIR filter that was designed using MATLAB.

```
float B[N+1] = {
 2 {-0.031913481327},   /* h[0] */
   {0.000000000000},    /* h[1] */
 4 {-0.026040505746},   /* h[2] */
   {-0.000000000000},   /* h[3] */
 6 {-0.037325855883},   /* h[4] */
   {0.000000000000},    /* h[5] */
 8 {-0.053114839831},   /* h[6] */
   {-0.000000000000},   /* h[7] */
10 {-0.076709627018},   /* h[8] */
   {0.000000000000},    /* h[9] */
12 {-0.116853446730},   /* h[10]*/
```

Do you really want to enter all of these coefficient values manually? What about a 200th order filter or a higher order filter? Even if you had the time and inclination to do so, do you actually believe you could enter these coefficients without making a typographical error?

If you are designing your FIR filters using MATLAB, a solution to this problem could be copying and pasting coefficients from the MATLAB window to your C program editor screen.[4] An even better solution involves a single MATLAB script file that will create a `coeff.h` and a `coeff.c` file for use in your CCS project. The script file is named `FIR_DUMP2C.m` and it can be found in the Chapter 3 `matlab` directory (and also the Appendix E `\MatlabExports` directory). The MATLAB help associated with this file is shown below.

```
>> help FIR_dump2c

    function FIR_DUMP2C(filename, varname, coeffs, FIR_length)

    Dumps FIR filter coefficients to file in C language format in forward
    order. Then "cd" to the desired directory PRIOR to execution.
    This will provide for increased C code portability.

    e.g., FIR_dump2c('coeff', 'B', filt1.tf.num, length(filt1.tf.num))

    Arguments: filename    - File to write coefficients, no extension
               varname     - Name to be assigned to coefficient array
               coeffs      - Vector with FIR filter coefficients
               FIR_length  - Length FIR data desired
```

---

[4]This assumes you've used MATLAB commands such as `remez`, SPTool, or FDATool to design your filter. Use `help` to explore these commands; later chapters will discuss them further.

This help output discusses the MATLAB `cd` function, which is an alternative to using the Current Directory field in the MATLAB desktop toolbar. Alternatively, you can allow the m-file `FIR_dump2c` to create the two files (`coeff.h` and `coeff.c`) in the current directory and then move the files to your CCS project using, for example, the Windows Explorer program. Once these files are in your CCS project directory you must add the files to your project. Note that Appendix A includes a basic tutorial to help you get started with the basics of using CCS. In addition to adding the two files (`coeff.h` and `coeff.c`) to your project, the single line of C code,

```
#include coeff.h
```

must also be added to your `ISRs.c` file. An example of a `coeff.h` file is shown below.

Listing 3.5: An example `coeff.h` file.

```
1 /* coeff.h                                    */
  /* FIR filter coefficients                    */
3 /* exported by MATLAB using FIR_DUMP2C        */

5 #define N 30

7 extern float B[];
```

Within the `coeff.h` file, line 5 is used to define the filter order and line 7 allows the B coefficients to be defined in another file. In this case, the coefficients are defined in the file `coeff.c`.

Once you become familiar with these procedures and some of the MATLAB filter design techniques, FIR filters can be designed, implemented, and run in real-time very easily.

Now that you understand the code...

The files necessary to run this application are in the Chapter 3 `ccs\FIRrevB` directory. Go ahead and copy all of the files into a separate directory. Open the project in CCS and "Rebuild All." Once the build is complete, "Load Program" into the DSK and click on "Run." Your FIR filter is now running on the DSK.

3.4.3 Circular Buffered FIR Filtering

As previously stated, shifting all of the values in the x array one element to the right after each dot product operation is a very inefficient use of the DSK's computational resources. The need to perform this shift is based on the assumption that the physical memory is linear. Given linear memory, with the inherent static labeling of each memory location, this shifting of values would seem to be an absolute requirement.

Figure 3.16 shows a linear memory model for the input to the filter x. As expected, to buffer the $N + 1$ elements in the x array, there are memory locations labeled $x[0]$, $x[-1]$, $\cdots$, $x[-N]$, but there is also an $x[-(N+1)]$. While this location was not declared, it does physically exist, and any attempt to access the x array beyond its declared bounds will

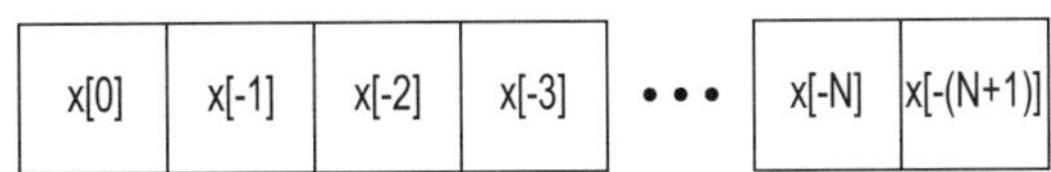

Figure 3.16: The linear memory concept with static memory location labeling.

result in *something* being retrieved and used in any subsequent calculations. The results of this indexing error may be catastrophic (e.g., a run-time error), or more subtle (e.g., the program runs, but gives inaccurate results). Either way, this type of indexing error must be avoided at all costs.

An alternative to linear memory is circular memory. As shown in Figure 3.17, the circular memory concept wraps the next memory location "beyond" the one labeled $x[-N]$ back to the memory location labeled $x[0]$. Since the purpose of this circular memory is to store or buffer x, this concept is routinely referred to as circular buffering.

If instead of using static memory location labels, a pointer can be used to point to and insert the newest sample that just arrived, $x[0]$, into the memory location containing the oldest sample, $x[-N]$, that is no longer needed, then a circular buffer has been created. No shifting of the x values is required since the pointer will always point to the most recent sample value. As the pointer advances, the oldest sample in the buffer is replaced by the most recent sample. This process can be continued indefinitely. The results of inserting the next sample into the buffer is shown in Figure 3.18.

To implement the circular buffer, a pointer must be established to the array `xLeft`. The required code to create this pointer is shown below.

```
float xLeft[N+1], *pLeft = xLeft;
```

The remainder of the circular buffered FIR filter code is shown below. The code comments explain the algorithm.

Listing 3.6: FIR filter using a circular buffer.

```
*pLeft = CodecDataIn.Channel[LEFT];        // store LEFT input value

output = 0;                    // set up for LEFT channel
p = pLeft;                     // save current sample pointer
if(++pLeft > &xLeft[N])        // update pointer, wrap if necessary
    pLeft = xLeft;             // and store
for (i = 0; i <= N; i++) {     // do LEFT channel FIR
    output += *p-- * B[i];     // multiply and accumulate
    if(p < &xLeft[0])          // check for pointer wrap around
        p = &xLeft[N];
}
CodecDataOut.Channel[LEFT] = output; // store filtered value
```

The files necessary to run this application are in the Chapter 3 `ccs\FIRrevD` directory.[5] Go ahead and copy all of the files into a separate directory. Open the project in CCS and "Rebuild All." Once the build is complete, "Load Program" into the DSK and click on "Run." Your FIR filter is now running on the DSK.

3.5   Follow-On Challenges

Consider extending what you have learned.

1. Change the B coefficients associated with the `FIRrevA` code and verify that as additional terms are added to the moving average filter, the LP filter cutoff frequency decreases (see Figure 3.13). As you increase the filter order, don't forget to scale each filter coefficient so that they all sum to 1 (i.e., the D.C. response equals 0 dB).

---

[5]We haven't mentioned the code in `ccs\FIRrevC` for Chapter 3. Consider it a bonus.

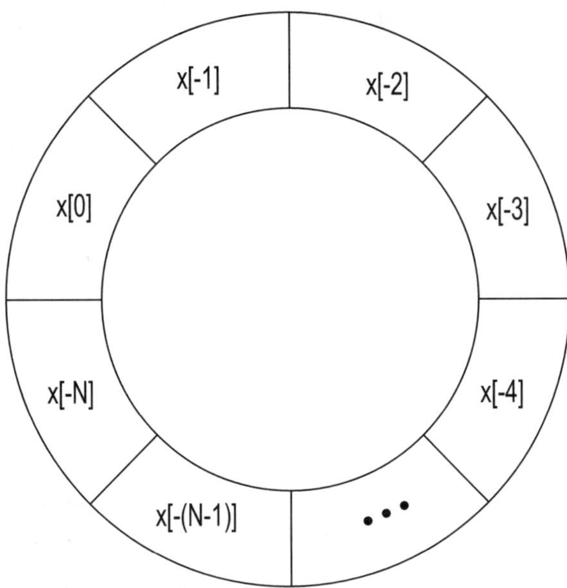

Figure 3.17: The circular buffer concept with static memory location labeling.

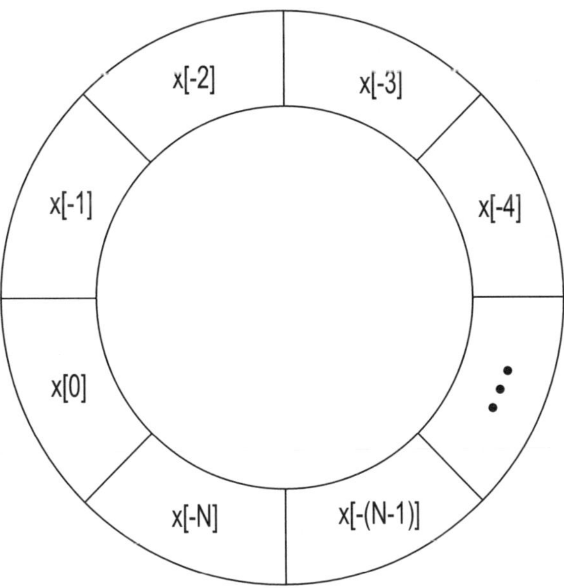

Figure 3.18: The circular buffer concept with dynamic memory location labeling, one sample time after Figure 3.17.

2. There are dozens of different ways to implement an FIR filter. Most of the examples in this chapter have used only the direct-form one (DF-I) techniques. Investigate and implement other forms, e.g., DF-II, DF-II transposed, second order sections (SOS), etc.

3. Routinely, FIR filter coefficients have even or odd symmetry. Develop an algorithm that takes advantage of this symmetry. This will result in lower storage requirements for the filter coefficients.

4. Some FIR filters, for example a Hilbert transforming filter (bandpass filter) symmetrically centered around $\frac{F_s}{4}$, contain a significant number of zeros in the filter coefficients. Develop an algorithm that takes advantage of the fact that you don't need to calculate the terms that involve multiplication by zero.

5. Explore some of the FIR filter design tools that are available in MATLAB (e.g., `remez`, `SPTool`, and `FDATool`). A complete listing of the functional capabilities included in the signal processing toolbox can be found by typing `help signal`. The toolbox functions are grouped by category. You are looking for the `FIR filter design` heading. Use the `FIR_dump2c` function from the text CD-ROM to export the filter coefficients. Implement your design using the `FIRrevB` code.

6. There is a limit to the number of calculations that the DSK can complete before the next input sample arrives. Design and implement an increasingly higher order lowpass filter using the `FIRrevB` code (compiled as a DEBUB build), until the output of the filter sounds either distorts or no longer can be heard at all. These are two of the possible indications that you are no longer meeting your real-time schedule. Is the `FIRrevD` code able to implement a higher order filter compared to the `FIRrevB` code before it also fails?

7. Repeat the previous challenge, but compile your code as a RELEASE build. You should be able to use a higher filter order before the real-time schedule fails, since a greater number of optimizations have been used by the compiler.

Chapter 4

IIR Digital Filters

4.1 Theory

GIVEN the simplicity and stability of the FIR filter, why would you ever want to consider using an IIR filter? This issue has been debated over the years in numerous articles, papers, and book chapters in the DSP literature. The short answer to the FIR versus IIR question can be distilled down to two brief points.

1. There is a tremendous amount of analog filter design knowledge available and, as mentioned in Chapter 3, these analog filters are all IIR in nature. There are times when we should take advantage of this wealth of design information.

2. To meet certain filter design specifications, very high order FIR filters may be needed. Yet we can usually implement a lower (sometimes *much* lower) order IIR filter that can meet such filter specifications.

Discrete IIR filter design takes advantage of decades of filter design discoveries and advances and provides implementations usually requiring considerably less complexity (lower order) than an equivalently performing FIR filter. With this realization that IIR filters can play a vital role in some real-time DSP systems, we provide below a brief reminder of how IIR digital filters are typically designed.

Analog filters are, in most cases, the basis upon which digital IIR filters are created. As an example of why analog filters are IIR, let's examine a simple RC analog filter. A first order analog filter can be built using a single resistor R and capacitor C. If the output of the circuit is taken to be the voltage across the capacitor, then a lowpass (LP) filter has been created. Figure 4.1 shows the schematic diagram of this circuit. Similarly, if the output is taken to be the voltage across the resistor, then the circuit implements a highpass (HP) filter.

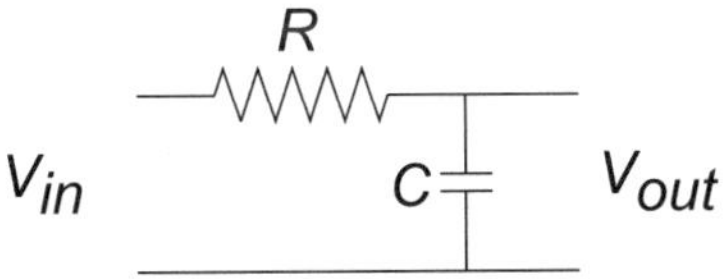

Figure 4.1: Schematic for a continuous-time first order RC filter. This configuration of the resistor and capacitor results in a lowpass filter.

Using voltage division and the fact that the impedance of a capacitor is $Z_c = 1/sC$, the transfer function of this first order RC LP filter is

$$H(s) = \frac{\frac{1}{sC}}{R + \frac{1}{sC}} = \frac{1}{sRC + 1} = \frac{\frac{1}{RC}}{s + \frac{1}{RC}}.$$

Since the impulse response and the transfer function of a system are a Laplace transfer pair,

$$h(t) \xleftrightarrow{\mathcal{L}} H(s)$$

the impulse response of this filter, $h(t)$, can be obtained by taking the inverse Laplace transform of $H(s)$; that is $h(t) = \mathcal{L}^{-1}\{H(s)\}$. This results in

$$h(t) = \frac{1}{RC}\, e^{-\frac{t}{RC}}\, u(t)$$

where $u(t)$ is the unit step function. For this system, as t gets very, very large (i.e., as $t \to \infty$), the impulse response $h(t)$ gets very, very small—but it would still take an *infinite amount of time* for $h(t)$ to actually reach and remain at zero. For this reason, the first order LP filter is an example of an IIR filter, because its impulse response lasts an infinite amount of time. An example of this behavior can be seen in Figure 4.2. In this example, the RC time constant ($\tau = RC$) is 1 ms; an engineering rule of thumb states that after approximately five time constants, the system has reached its final or steady-state value. Notice in Figure 4.2 that the impulse response value at 5 ms *appears* to be approximately zero. But Figure 4.3 plots the same impulse response on a semilog plot all the way out to 500 ms. Looking at this new figure, it now becomes clear that the impulse response of the system *never* actually reaches and remains at a zero value—so it must be categorized as an IIR filter. We hasten to say that practically speaking, the impulse response decays to a low enough value (usually even below the noise floor of our measurement equipment) shortly after five time constants, so the engineering rule of thumb is useful.

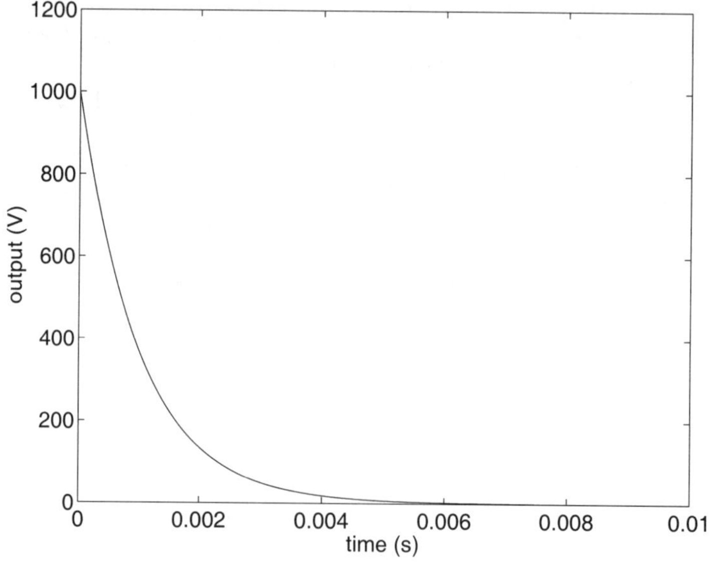

Figure 4.2: Linear plot of the impulse response associated with a first order lowpass filter with ($R = 1000\ \Omega$ and $C = 1\ \mu\mathrm{F}$).

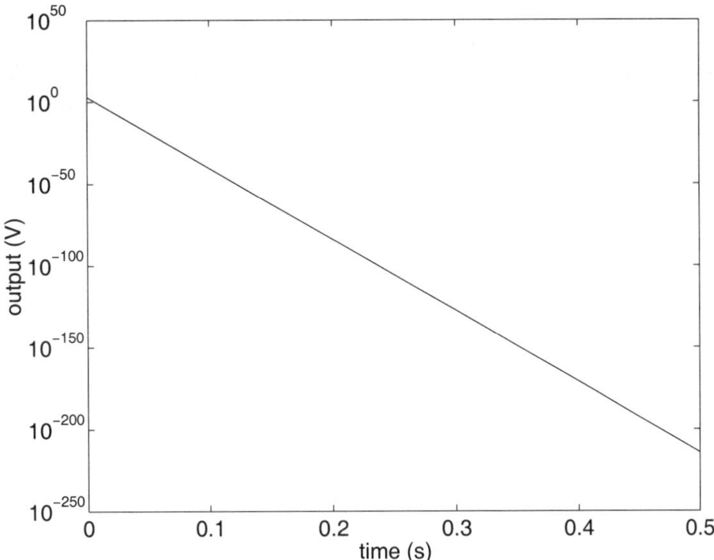

Figure 4.3: Semilog plot of the impulse response associated with a first order lowpass filter with ($R = 1000\ \Omega$ and $C = 1\ \mu$F).

Enough about analog filters—this is a DSP book! IIR digital filters are routinely created by adapting well-proven analog filter designs using one of three major techniques listed below.

Impulse invariance methods These methods are based on the idea that we can design a discrete-time filter with an impulse response that is a scaled and sampled version of an analog (continuous-time) filter's impulse response [2]. Using these techniques the impulse response of the discrete-time filter becomes,

$$h\left[n\right] = T_s h_c\left[nT_s\right]$$

where T_s is the sample period and $h_c[nT_s]$ is the sampled (i.e., discrete-time) version of the continuous-time impulse response associated with the original analog filter.

Bilinear transformation methods These methods take a continuous-time transfer function, $H_c(s)$, and replace its independent variable, s, with a discrete-time, independent transform variable, z, to produce a discrete-time transfer function, $H(z)$. This transformation can be accomplished using the variable substitution,

$$s = \frac{2}{T_s}\left(\frac{1 - z^{-1}}{1 + z^{-1}}\right),$$

which results in the transfer function,

$$H(z) = H_c\left(\frac{2}{T_s}\left(\frac{1 - z^{-1}}{1 + z^{-1}}\right)\right)$$

where T_s is the sample period and H_c is the continuous-time transfer function.

Optimization methods These methods are based on optimizing filter performance using iterative numerical techniques that converge (we hope!) to a design that is close to the stated filter specifications.

In this book we assume you are designing your digital filters with the help of software tools such as MATLAB's SPTool or FDATool. The fact that one of the methods above is how such software arrives at the filter coefficients for you may or may not be of interest to you.

As is almost always the case in engineering design, systems that perform well in some areas routinely perform poorly in other areas. With IIR filters, we are particularly concerned about two issues:

- Stability: Since feedback is always involved in an IIR design, the system may become unstable. For real-time (causal) systems, we can mathematically ensure stability by keeping the poles inside the unit circle (magnitude of the poles < 1) as plotted on the z-plane. As a reminder, the poles are the roots of the denominator polynomial and the zeros are the roots of the numerator polynomial of the system transfer function, $H(z)$. The transfer function normally takes the form,

$$H\left(z\right) = \frac{Y\left(z\right)}{X(z)} = \frac{b_0 + b_1 z^{-1} + b_2 z^{-2} + b_3 z^{-3} + \cdots}{1 + a_1 z^{-1} + a_2 z^{-2} + a_3 z^{-3} + \cdots}.$$

- Phase response: A symmetric FIR filter exhibits a linear phase response, while an IIR filter does not (cannot!) exhibit true linear phase. Different IIR filter design techniques can result in varying approximations to a linear phase response, but can never achieve it completely. Depending on the application, having linear phase (i.e., constant group delay) may be crucial to the proper operation of the DSP-based filtering operation. If so, symmetric FIR filters are recommended.

In summary, IIR filters can take advantage of known analog filter designs and can meet steep requirements (particularly for the magnitude response) at a lower order than FIR filters. But IIR filters are subject to instability if the designer is not careful, and their phase response can never be truly linear. There are times when an IIR filter is exactly what you want for your DSP application, so let's explore them further.

4.2   winDSK6 Demonstration: Notch Filter Application

If you double click on the winDSK6 icon, the winDSK6 application will launch, and a window similar to Figure 4.4 will appear. Before proceeding, ensure the appropriate choices have been selected for "DSK and Host Configuration." These selections are "sticky," such that the next time you run winDSK6 the choices you made will still be selected.

The winDSK6 Notch Filter application implements a second order IIR filter. Clicking on the Notch Filter button will load the program into the attached DSK, and a window similar to Figure 4.5 will appear.

In Chapter 3 we decreased the Q adjustment (determined by the value of variable r in the filter's transfer function) until it reached zero. This put the poles of the filter at the origin of the z-plane and the DSK behaved as if it was running an FIR filter. In this section we will increase the Q adjustment (by increasing $|r|$) which causes the poles to move away from the origin and approach the unit circle. This adjustment (via the slider control at the bottom of the Notch/Bandpass Filter window) is shown in Figure 4.5. Also notice that the "Filter Type" was selected to be "Notch" rather than "Bandpass" for this demonstration; a notch filter may also be called a bandreject filter. As the poles of the notch filter get closer to the unit circle (i.e., as $|r| \to 1.0$) they progressively increase the "steepness" of the notch. But as with all IIR filters, stability is an issue: if $|r| = 1.0$, then the poles are *on* the unit circle and the filter will not be stable (it will tend to break into oscillation).

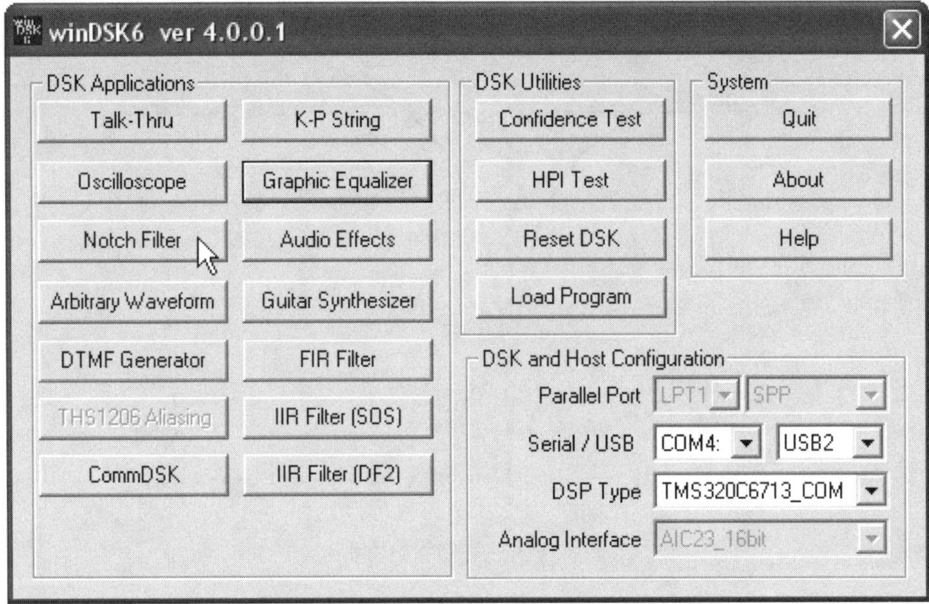

Figure 4.4: winDSK6 ready to load the Notch Filter application.

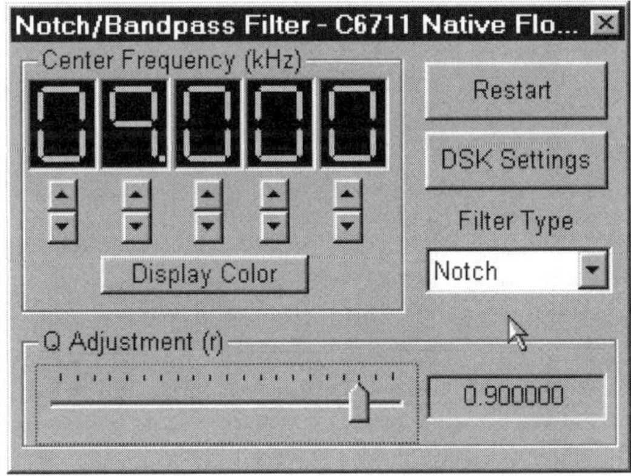

Figure 4.5: winDSK6 running the Notch Filter application with $r = 0.9$.

To show you the effect on the notch filter's magnitude response caused by moving the poles, we keep a constant notch frequency and change only $|r|$. Theoretically, an infinite amount of attenuation is present at the exact notch frequency (in practice it's a very large yet finite amount of attenuation, but it's so large that we can treat it as if it's infinite attenuation). This explains why a properly adjusted (tuned) notch filter can completely remove an interfering tone for all practical purposes. The frequency responses associated with four different settings of $|r|$ for the notch filter are overlaid and shown in Figure 4.6. Comparing the frequency responses shown in this figure demonstrates two ramifications of moving the poles of an IIR filter close to the unit circle.

1. As $|r| \rightarrow 1.0$, the maximum gain of the filter (without scaling) approaches 1 (0 dB). For any value of $|r|$, the maximum gain may always be *forced* to 1 (0 dB) by including a

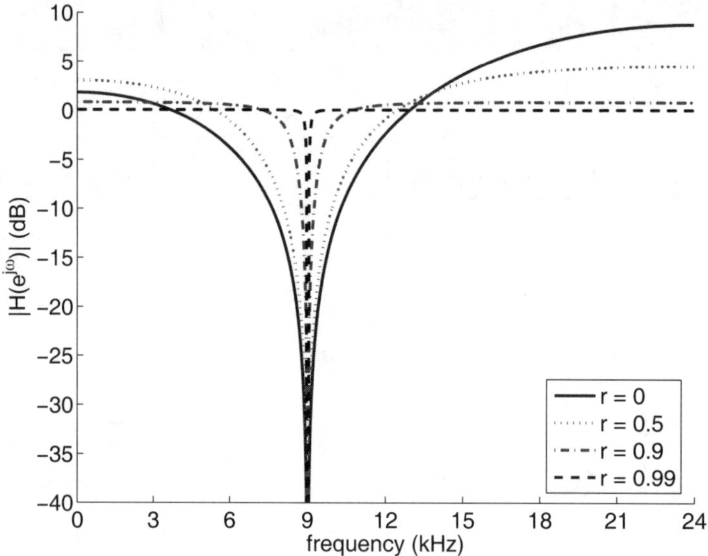

Figure 4.6: The frequency responses of four different notch filters having a notch frequency of 9 kHz and r values of 0, 0.5, 0.9, and 0.99.

multiplicative scale factor that would have to be multiplied by *all* of the b coefficients, but we would prefer to avoid this extra step. Remember, excess gain in a DSP algorithm can cause significant problems if it results in an output value that exceeds the numerical range of the DAC.

2. As $|r| \to 1.0$, the Q of the filter (i.e., steepness of the notch) increases dramatically. At the same time, as the poles approach the unit circle, that portion of the impulse response of the filter with significant values increases in length. This indicates that more time will be needed for the filter to effectively reach a steady-state condition (that's right, there's no free lunch!). And as mentioned earlier, if we ever allow $|r| = 1.0$ (or more precisely $|r| \geq 1.0$) then the filter would be unstable.

To hear the effect of the Notch Filter application, add a sinusoidal signal (tone) to a music signal in a similar fashion as we described in Chapter 3. Most computer sound cards will perform this summing for you. You will need to experiment with your sound card's mixer controls to determine exactly how your particular system responds. Most systems are capable of summing an audio CD signal and another signal by playing the CD on the drive installed in the computer and injecting the second signal from a function generator through the sound card "line input" or "microphone input" connector. The sound card "line output" or "headphone output" is then connected to the signal input of the DSK. As before, the DSK signal output is connected to a set of powered speakers. If a function generator is not available, you may use one of the audio test tones in directory `test_signals` on the CD-ROM that accompanies this book and play them with a second, external CD player. You can also easily create your own audio test tones in MATLAB, then record them on to CD-R or CD-RW media and play them in the same way on an external CD player (this concept will be discussed further in Chapter 5).

When the center frequency of the notch filter equals the frequency of the injected tone, you should hear the sound of the tone disappear from the speakers.

4.3 MATLAB Implementation

4.3.1 Filter Design and Analysis

After designing a filter in MATLAB, the discrete-time difference equation is routinely available in the form of two vectors: B (numerator coefficients) and A (denominator coefficients). Given these two vectors, MATLAB can rapidly analyze and plot your filter's performance using a wide variety of toolbox functions. To find help on the MATLAB signal processing toolbox, type `help signal`. An edited version of the results associated with entering this command are shown below.

```
>> help signal

  Signal Processing Toolbox

  Filter analysis.
     abs          - Magnitude.
     angle        - Phase angle.
     filternorm   - Compute the 2-norm or inf-norm of a digital filter.
     freqs        - Laplace transform frequency response.
     freqspace    - Frequency spacing for frequency response.
     freqz        - Z-transform frequency response.
     fvtool       - Filter Visualization Tool.
     grpdelay     - Group delay.
     impz         - Discrete impulse response.
     phasez       - Digital filter phase response (unwrapped).
     phasedelay   - Phase delay of a digital filter.
     zerophase    - Zero-phase response of a real filter.
     zplane       - Discrete pole-zero plot.
```

Of particular interest are the frequency response, impulse response, pole/zero, and group delay plots.

Creating an Impulse Response Plot

Unlike an FIR filter, the filter coefficients of the IIR filter are *not* the individual terms that make up the impulse response of the system. The impulse response of an IIR filter must be calculated iteratively. The MATLAB command `impz` can greatly simplify this process. For example, if we design our filter using the `butter` command (for Butterworth filters) and wish to examine the impulse response, we will need to use commands similar to:

```
[B, A] = butter(4, 0.25);
impz(B, A, 10, 48000);
```

to create the impulse response plot. In line 1 we design a fourth order Butterworth lowpass filter with a cutoff frequency of $0.25F_s/2$. In the second line of code, we use `impz` to determine the impulse response of the filter, where B and A are the numerator and denominator coefficient vectors (respectively), `10` is the desired number of points of the impulse response to be calculated and plotted, and `48000` is the sample frequency to be used with the filter. Since we included the sample frequency (the fourth input argument), the horizontal axis of the resulting plot will have units of time.

Instead of using the `butter` command, we could have used the command SPTool (Signal Processing Tool), which brings up a graphical user interface (GUI) that, among other things,

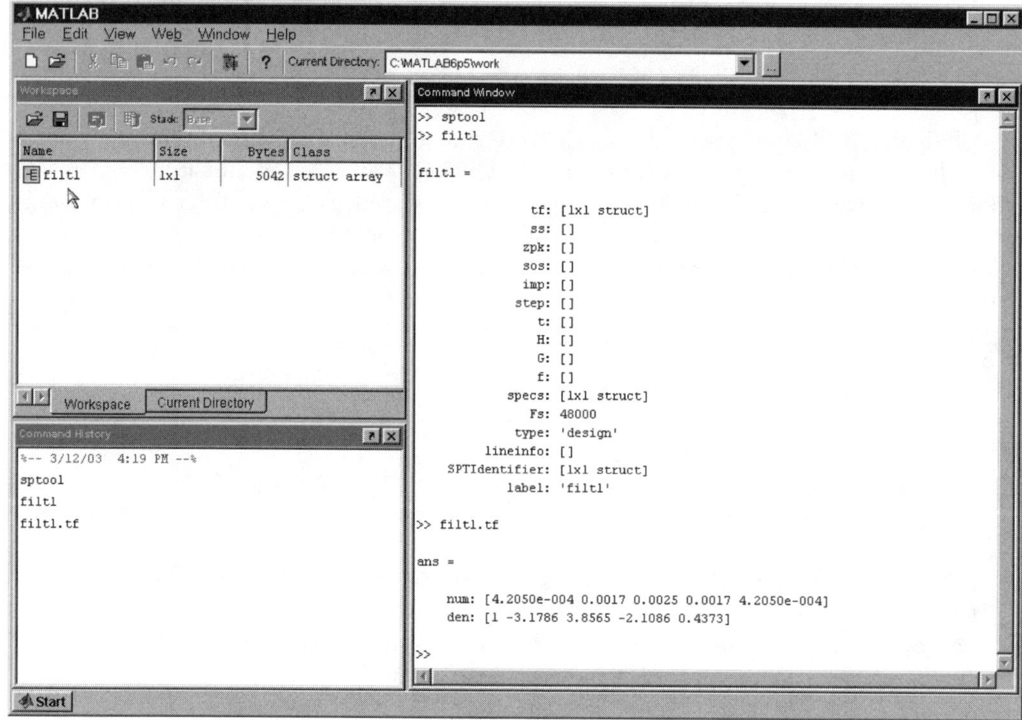

Figure 4.7: Filt1 data structure exported to the MATLAB workspace.

allows us to design digital filters. When you export filters from SPTool to the MATLAB
workspace, the default structure used to hold the filter design is called `filt1`. In that case,
`filt1.tf.num` and `filt1.tf.den` are the filter's numerator and denominator coefficient
vectors instead of B and A that we used above. An example `filt1` data structure is shown
in Figure 4.7.

Alternatively, the three-argument version of the `impz` command is shown below (assuming SPTool was used):

```
impz(filt1.tf.num, filt1.tf.den, 10);
```

Because no sample frequency was specified, the resulting plot will show units of samples
(i.e., sample number n) on the horizontal axis.

Finally, there is a two-argument version of the `impz` command, which is shown below.
If you use this version of the `impz` command, as shown below,

```
impz(filt1.tf.num, filt1.tf.den);
```

the plot will still show sample numbers as horizontal axis units, but the algorithm will
determine for you the number of samples to be evaluated and plotted. An example plot
associated with this simplest version of `impz` is shown in Figure 4.8.

Creating a Frequency Response Plot

Plotting the frequency response can be accomplished using the MATLAB command `freqz`.
An example of using this command follows.

```
freqz(B, A);
```

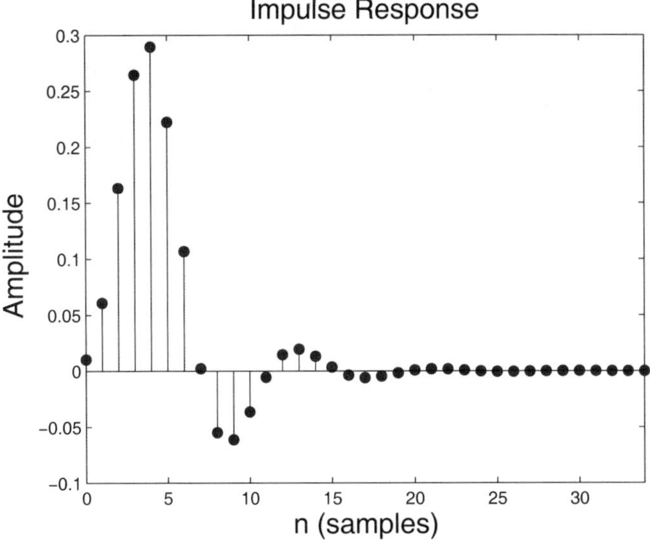

Figure 4.8: The impulse response associated with a fourth order Butterworth lowpass filter having a cutoff frequency of $0.25F_s/2$.

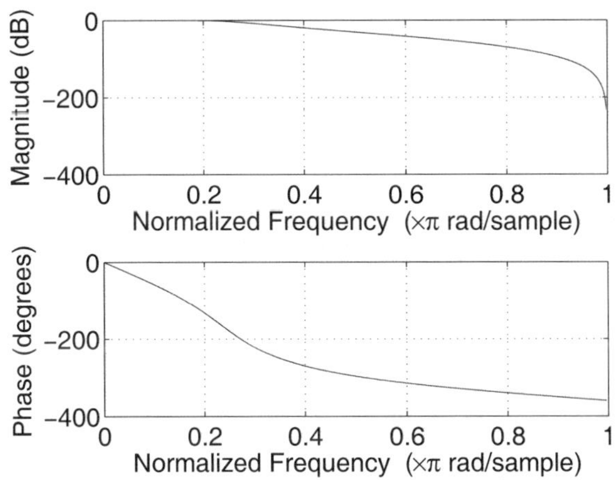

Figure 4.9: The frequency response diagram associated with a fourth order Butterworth lowpass filter having a cutoff frequency of $0.25F_s/2$.

The resulting frequency response plot is shown in Figure 4.9. Without specifying the sample frequency as an additional argument to `freqz`, the frequency axis of the plot is shown as normalized radian frequency in π units, so the far right value shown as 1 on the x-axis is where the normalized radian frequency equals π. This is equivalent to where the frequency equals $F_s/2$. Where the normalized axis equals 0.25 is where $f = 0.25F_s/2$; this is the cutoff frequency (where the filter magnitude response is down by 3 dB) that we specified earlier when we designed the filter. Note the nonlinear phase response of this IIR filter.

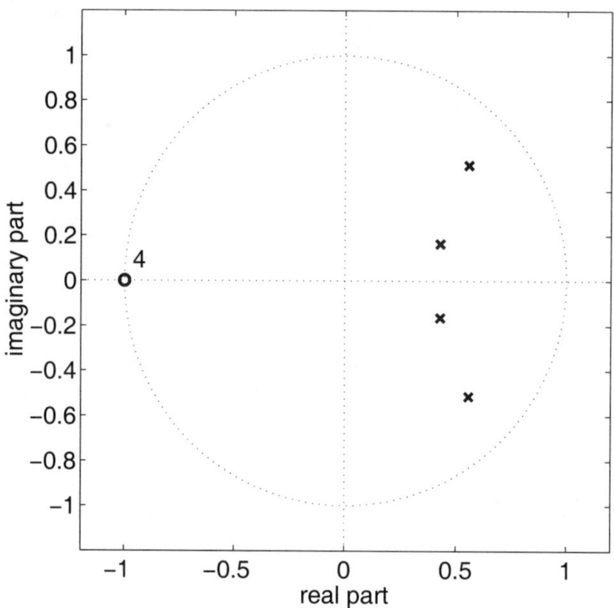

Figure 4.10: The pole/zero diagram associated with a fourth order Butterworth lowpass filter having a cutoff frequency of $0.25F_s/2$.

Creating a Pole/Zero Plot

Plotting the locations of the poles and zeros on the z-plane can be accomplished using the MATLAB command `zplane`. An example using this command follows.

```
zplane(B, A);
```

The resulting pole/zero plot is shown in Figure 4.10. The MATLAB command `zplane` calls another function called `zplaneplot`. This function is actually responsible for creating the pole/zero plot to include plotting the unit circle on the z-plane. Unfortunately, despite repeated feedback to The MathWorks from the authors, MATLAB continues to use only 70 points to form the unit circle. Since MATLAB plots are largely straight lines drawn between points, this leads to a 69-sided polygon. This very *faceted* approach to plotting may be inadequate if your poles and/or zeros are very close to the unit circle. There are two easy solutions to this problem.

- **RECOMMENDED** Use the `ucf` function provided on the CD-ROM (in the Chapter 4 `matlab` directory) to correct the problem. Basically, `ucf` erases the original unit circle by overwriting it with a white line and then plots a more accurate unit circle. The name `ucf` stands for "unit circle fixer," and can be updated to your specifications without causing problems with any MATLAB toolbox functions.

- **NOT RECOMMENDED** Edit line number 91 of the MATLAB m-file `zplaneplot`. As of this printing, this line currently reads:

```
theta = linspace(0, 2*pi, 70);
```

The `linspace` command creates a variable, `theta`, which consists of 70 elements that are uniformly spaced between 0 and 2π. Changing the 70 to a much larger integer, for example, 1000, will result in more points being used to plot the unit circle. More points allows the plotted polygon to much more closely approximate a circle. This approach is not recommended because it modifies a MATLAB toolbox function that was provided to you by The MathWorks. Modifying this and other code that you have paid money to other people to develop and maintain is a bad idea for at least four different reasons.

1. If you believe production code is in error, you should submit a formal request to have the code corrected. This may allow others to benefit from your efforts.

2. If you do modify the MATLAB code, the next update of the toolbox that you receive and install will invariably overwrite the updated function that you have created and all of your modifications will be lost.

3. The chance that you will remember what you have done in the distant future is very small, and you will need to rediscover the entire unit circle plotting problem after installing a new version of the toolbox.

4. You may break the toolbox function so that it no longer works or even worse: you think that the function is working, but it is actually returning inaccurate results.

If you remember these ideas each time that you inevitably encounter a problem with someone else's software, you will have fewer problems in the long run.

Creating a Group Delay Plot

Plotting the group delay can now be accomplished using the MATLAB command `grpdelay`. An example using this command follows.

```
grpdelay(B, A);
```

The resulting plot is shown in Figure 4.11. Note that the group delay is *not* constant, which is due to the nonlinear phase response of this filter.

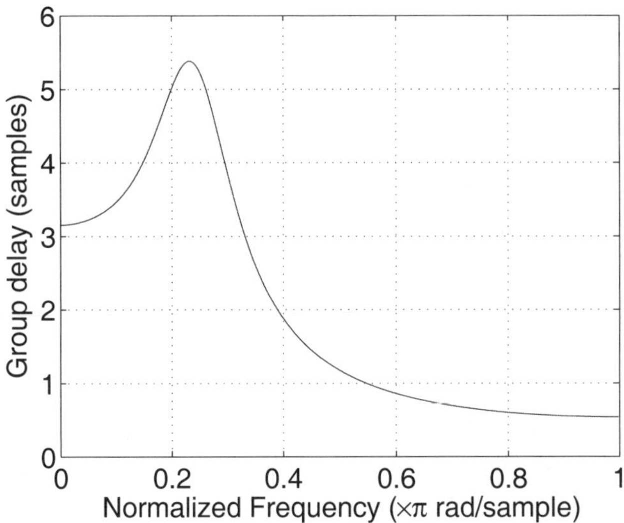

Figure 4.11: The group delay associated with a fourth order Butterworth lowpass filter having a cutoff frequency of $0.25F_s/2$.

Using FDATool and FVTool

The MATLAB environment provides even more tools for designing digital filters. For example, you can use MATLAB's FDATool (Filter Design and Analysis Tool) to design your filter. As shown in Figure 4.12, several software pushbuttons are available that will allow you to not only specify and design your filter, but also to view a number of filter analysis plots. The filter coefficients exported from FDATool are just vectors (by default) rather than being embedded in a `filt` structure as was the case with SPTool.

Finally, regardless of how you design your filter, you can analyze your filter by using MATLAB's FVTool (Filter Visualization Tool):

```
fvtool(B, A);
```

If your filter specifications are in structure form, the MATLAB command would be

```
fvtool(filt1.tf.num, filt1.tf.den);
```

As shown in Figure 4.13, several software pushbuttons are available in FVTool that will allow you to view a number of filter analysis plots, which we've annotated with labels and arrows on the figure. In that figure, the group delay of the filter has been selected.

While all the MATLAB commands and tools were discussed here in the context of IIR filters, they can all just as easily be used for FIR filters. The primary difference is that the "vector" A for all FIR filters is equal to the scalar value of 1.

4.3.2　IIR Filter Notation

IIR filters are more complicated than FIR filters, and there are various choices the designer must make when implementing them. The remainder of this chapter will focus primarily on implementation issues associated with IIR filters. Recall that the difference equation associated with a causal IIR filter is

$$\sum_{k=0}^{M} a[k]y[n-k] = \sum_{k=0}^{N} b[k]x[n-k]$$

or in output variable form,

$$a[0]y[n] = -\sum_{k=1}^{M} a[k]y[n-k] + \sum_{k=0}^{N} b[k]x[n-k].$$

The $a[0]$ term, the coefficient of $y[n]$, is usually normalized to 1. In fact MATLAB normalizes the $a[0]$ coefficient prior to almost all of its calculations. This normalization results in

$$y[n] = -\sum_{k=1}^{M} a[k]y[n-k] + \sum_{k=0}^{N} b[k]x[n-k]$$

where each of the remaining $a[k]$ and $b[k]$ terms are scaled by $a[0]$. We have chosen not to rename this normalized version of the coefficients in the equation above, as this form is how most DSP books depict the difference equation of an IIR filter.

Alternatively, the IIR filter's difference equation can be converted to a transfer function in the z-domain:

$$H(z) = \frac{b_0 + b_1 z^{-1} + b_2 z^{-2} + \cdots + b_N z^{-N}}{1 + a_1 z^{-1} + a_2 z^{-2} + \cdots + a_M z^{-M}}.$$

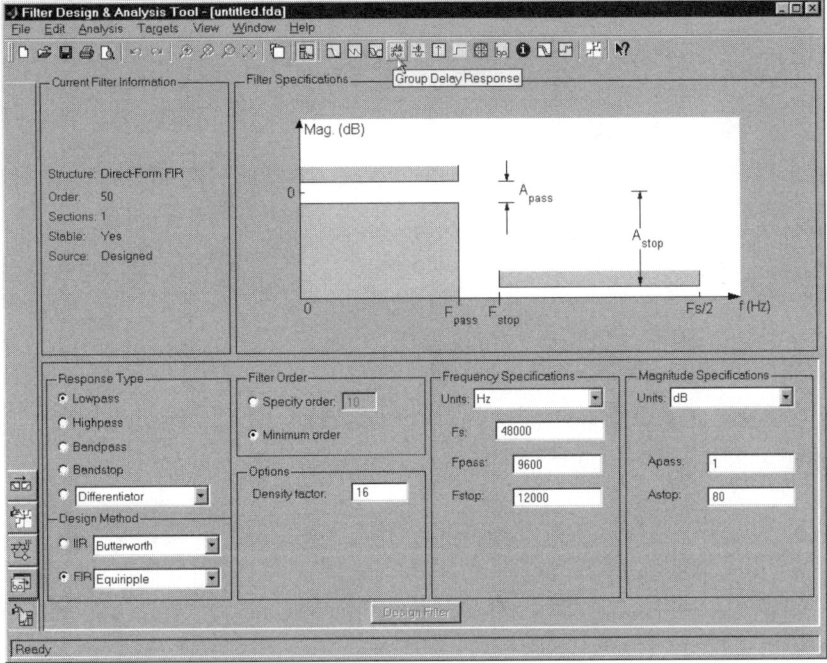

Figure 4.12: The **FDATool** filter design and analysis program GUI.

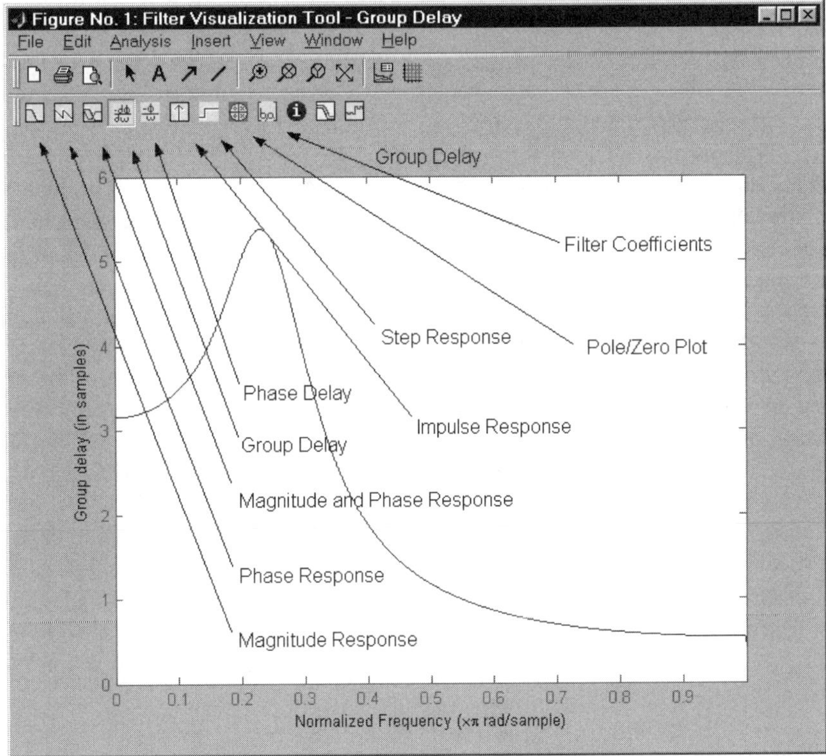

Figure 4.13: The **FVTool** filter viewer program GUI.

If we use a similar notation for filter implementation as we did in Chapter 3, the transfer function becomes,

$$H\left(z\right) = \frac{b\left[0\right] + b\left[1\right]z^{-1} + b\left[2\right]z^{-2} + \cdots + b\left[N\right]z^{-N}}{1 + a\left[1\right]z^{-1} + a\left[2\right]z^{-2} + \cdots + a\left[M\right]z^{-M}}.$$

Notice that the number of a terms $(M+1)$ and the number of b terms $(N+1)$ are often not equal. That is the reason for using M for the order of the denominator and N for the order of the numerator in the transfer function polynomial.

To calculate $y[0]$ (the current output value of the IIR filter), we must perform two operations:

1. the dot product $\mathbf{B}\cdot\mathbf{x}$, where $\mathbf{B} = \{b[0], b[1], \ldots, b[N]\}$ and $\mathbf{x} = \{x[0], x[-1], \ldots, x[-N]\}$ (the current and past values of the input signal), and

2. a dot product of $\mathbf{A}\cdot\mathbf{y}$, where $\mathbf{A} = \{1, a[1], \ldots, a[M]\}$ and $\mathbf{y} = \{y[0], y[-1], \ldots, y[-M]\}$ (the current and past values of the output signal).

Specifically,

$$y[0] = -a[1]y[-1] - a[2]y[-2] - \cdots - a[M]y[-M] + b[0]x[0] + b[1]x[-1] + \cdots + b[N]x[-N].$$

Notice that the $\mathbf{A}\cdot\mathbf{y}$ term is really only a partial dot product because the $a[0]y[0]$ term is not needed and is therefore not calculated.

4.3.3   Block Diagrams

Routinely, engineers use block diagrams to help understand implementation issues and signal flow. One of the standard block diagram forms associated with implementing this IIR filter is shown in Figure 4.14. The blocks containing z^{-1} are delay blocks that store the input into the block for one sample period. The delay blocks may be thought of as synchronous shift registers that have their clocks tied to the ADC and DAC's sample clock, but are typically just memory locations accessed by the DSP CPU. This form is called direct form I (DF-I) and is the most straightforward implementation of the standard difference equation. Alternatively, a single summing node can be used to more accurately implement the difference equation as a single equation. This can be seen in Figure 4.15.

Another variant is shown in Figure 4.16. It is called direct form II (DF-II) and is achieved by reversing the order of the feed forward and feedback terms and combining the delay elements. This form only requires half the DF-I memory elements, and thus would be a more efficient implementation. Figure 4.17 is called a cascade of two second order sections (SOS). Higher order filters can be divided into a number of first or second order terms that can then be multiplied (cascaded) together. Second order terms are preferred over higher order terms since real coefficients can be used to accurately describe the locations of complex conjugate pairs. This often forgotten fact is a result of the fundamental theorem of algebra. Coefficient quantization effects are also far less problematic using cascaded SOS compared to other implementations. MATLAB does provide a wide variety of conversion m-files for various implementations. Of particular interest is `tf2sos` (transfer function to second order section) and `zp2sos` (zero/pole to second order section). A complete listing of the MATLAB conversion routines can be found using `help signal`; look for the Linear system transformations heading.

The final block diagram we will mention is the parallel form, which is shown in Figure 4.18. We have shown only first order numerator (b) terms in this figure because that is routinely the result after the parallel decomposition is complete. MATLAB's signal processing toolbox does not currently have an m-file that converts to parallel form. We have written

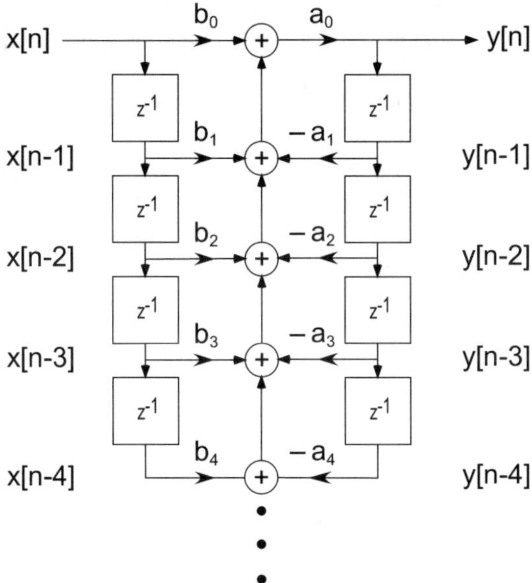

Figure 4.14: Block diagram associated with the DF-I implementation of an IIR filter. Typically, a_0 is normalized to 1.0.

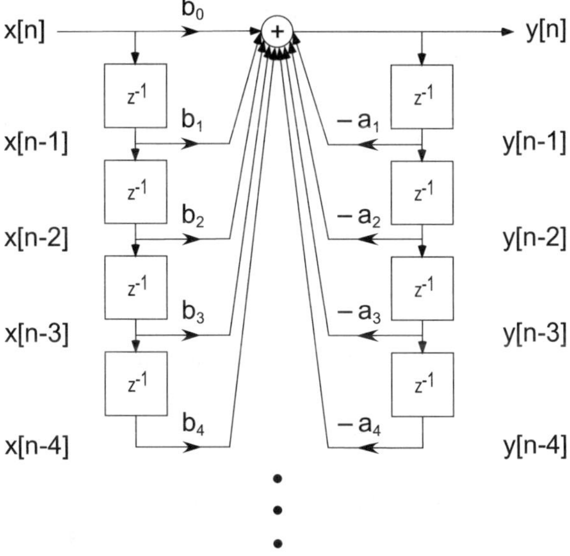

Figure 4.15: Block diagram associated with the direct form I (DF-I) implementation of an IIR filter using only one summing node. Coefficient a_0 is not shown (assumed to be 1.0).

such an m-file and included it in the `matlab` directory for Chapter 4. Conforming to the MATLAB naming convention, our m-file is named `filt2par`. This m-file performs the filt structure to parallel conversion.

There are dozens of implementation block diagrams and we have selected only a few of the most common forms for discussion here. Each form has advantages and disadvantages and we will not dwell on this topic; however, a design example will be used to help explain

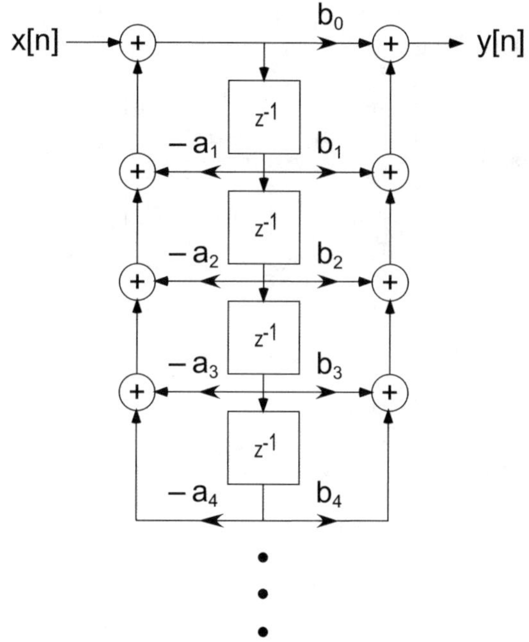

Figure 4.16: Block diagram associated with the DF-II implementation of an IIR filter. Coefficient a_0 is not shown (assumed to be 1.0).

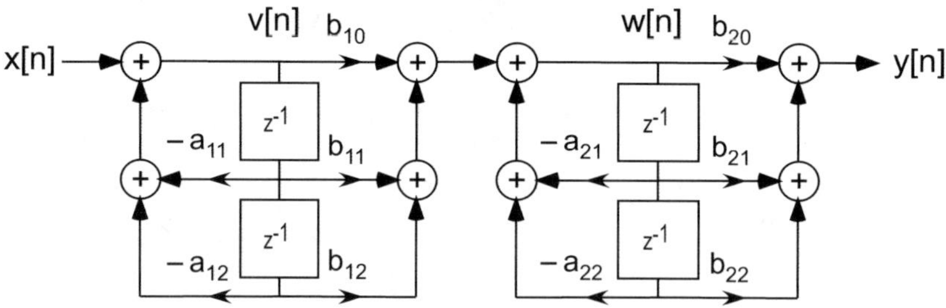

Figure 4.17: Block diagram associated with the second order section (SOS) implementation of an IIR filter.

a few of these issues. The MATLAB code associated with this example can also be found in the `matlab` directory for Chapter 4. The m-file is called `ellipticExample` and as its name implies it designs and implements an elliptic filter. The `ellipticExample` m-file generates the filter coefficients, pole/zero diagrams, and frequency responses associated with DF-I/DF-II, SOS, and parallel implementation. This fourth order filter was carefully selected to cause the filter to become unstable if it is implemented using single precision DF-I or DF-II techniques. Instability in IIR filters is caused by the result of finite precision arithmetic (i.e., coefficient quantization) causing the poles to move outside the unit circle. This can be seen in Figure 4.19 where the smaller circles (zeros) and x's (poles) represent the correct location of the filter's poles and zeros. The larger circles and x's are where the poles and zeros end up when 16-bit direct form arithmetic is used. Because one of the larger-x poles is outside the unit circle, this filter will be unstable if implemented as a direct form I or direct form II using 16-bits to represent each the coefficient. Any attempt to implement such a filter in either MATLAB or real-time code will produce an undesirable result. Simply

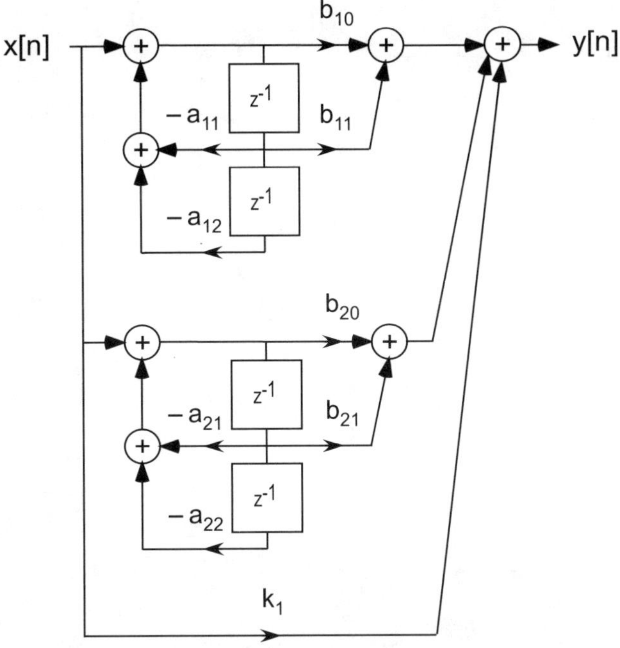

Figure 4.18: Block diagram associated with the parallel implementation of an IIR filter. Coefficient k_1 is an overall gain factor.

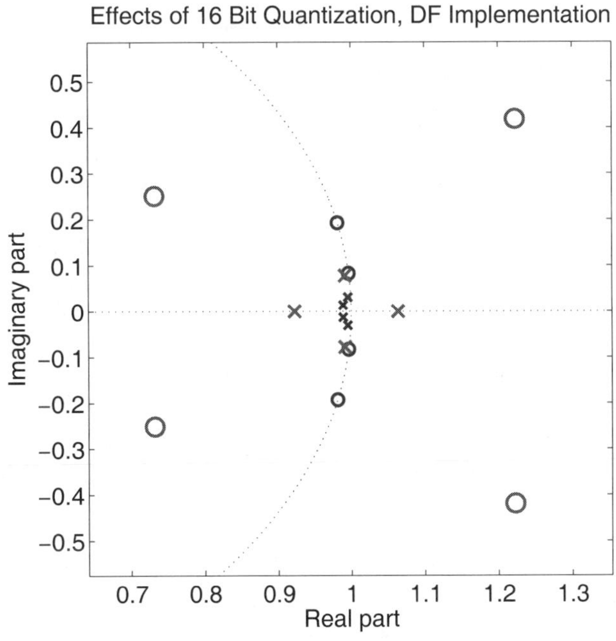

Figure 4.19: Pole/zero plot (zoomed in) associated with a fourth order elliptic filter implementation using direct form techniques.

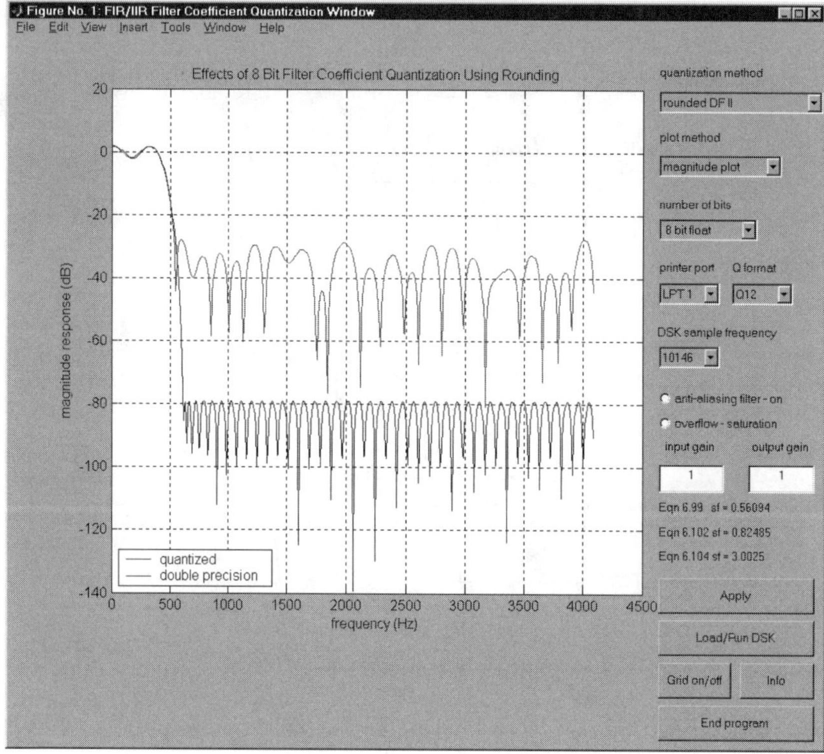

Figure 4.20: The `qfilt` GUI evaluating the performance of a lowpass filter.

changing to an SOS implementation fixes the problem. Of course *zeros* outside the unit circle do not effect stability in any way.

A useful MATLAB-based tool to help evaluate the implementation effects associated with DF-I, DF-II, and SOS is included on the CD. The tool is a collection of MATLAB m-files that are controlled through the `qfilt` GUI. This program is designed to work with a `filt1` data structure and its GUI can be seen in Figure 4.20. This GUI was written not only to evaluate the finite precision arithmetic effects but to implement the quantized coefficient FIR or IIR filter on a TI TMS320C31 DSK. Not having a C31 DSK attached to the host PC will only prevent you from loading and running the displayed filter on the DSK. You will be able to use all the other features of the program. Years after we introduced the `qfilt` program, the MathWorks released a number of tools and toolboxes (e.g., FDATool) to deal with the same finite precision effects. They have also introduced a series of toolboxes that allow some of your work to be run in SIMULINK via CCS on selected TI hardware targets. You may want to investigate these recent tools from The MathWorks on your own.

Should you implement an unstable filter, it may sound similar to audio feedback (getting the microphone too close to the speakers). But most of the time it sounds as if the speakers are not plugged in, since the output of the DSP algorithm grows rapidly to the point that the numbers can no longer be represented in the DSP hardware. Using a CCS watch window to troubleshoot this and other logical programming errors can be a very effective technique. Evaluating and plotting the frequency response of an unstable system, although allowed by MATLAB, is meaningless, since the DFT operation upon which the `freqz` command is based is undefined for unstable systems.

If you run the `ellipticExample` m-file, the filter coefficients will be available in the MATLAB workspace. The resulting transfer functions (rounded) are shown below.

$$H_{DF}(z) = \frac{0.000996 - 0.0039z^{-1} + 0.0059z^{-2} - 0.0039z^{-3} + 0.000996z^{-4}}{1 - 3.97z^{-1} + 5.909z^{-2} - 3.911z^{-3} + 0.971z^{-4}}$$

$$H_{sos}(z) = \frac{0.00101 - 0.00195z^{-1} + 0.00101z^{-2}}{1 - 1.99z^{-1} + z^{-2}} \cdot \frac{1 - 1.98z^{-1} + 0.978z^{-2}}{1 - 1.99z^{-1} + 0.992z^{-2}}$$

$$H_{parallel}(z) = \frac{-0.00385 + 0.00360z^{-1}}{1 - 1.99z^{-1} + 0.992z^{-2}} + \frac{0.00382 - 0.00348z^{-1}}{1 - 1.98z^{-1} + 0.978z^{-2}}$$

In Figures 4.21, 4.22, 4.23, and 4.24, we show the DF-I, DF-II, SOS, and parallel implementation block diagrams, respectively, with the filter coefficients. The coefficient values in the figures are also rounded to three significant digits in the block diagrams for compactness. In each figure the same filter is implemented, but each implementation has its own advantages and disadvantages.

4.3.4 Built-In Approach

As mentioned in Chapter 3, MATLAB has a built-in function called `filter.m`. It can be used to implement FIR filters (using only the numerator (B) coefficients and setting $A = 1$) and IIR filters (using both the numerator (B) and the denominator (A) coefficients). The first few lines of the on-line help associated with the `filter` command are provided below.

```
>> help filter


FILTER One-dimensional digital filter.
   Y = FILTER(B,A,X) filters the data in vector X with the
   filter described by vectors A and B to create the filtered
   data Y.  The filter is a "Direct Form II Transposed"
   implementation of the standard difference equation:

   a(1)*y(n) = b(1)*x(n) + b(2)*x(n-1) + ... + b(nb+1)*x(n-nb)
                         - a(2)*y(n-1) - ... - a(na+1)*y(n-na)
```

This function is useful for quickly implementing a filter with minimal programming on your part.

4.3.5 Creating Your Own Filter Algorithm

In this next MATLAB example, we are trying to implement a first order IIR notch filter (this is *not* the same notch filter implemented in winDSK6). We desire the filter to have a zero at $z = 1$ and a pole at $z = 0.9$. The transfer function associated with this pole/zero diagram is

$$H(z) = \frac{1 - z^{-1}}{1 - 0.9z^{-1}}$$

and the difference equation is

$$y[n] = 0.9y[n-1] + x[n] - x[n-1].$$

We will use a unit impulse as the input to the system. If we calculate an infinite number of output terms, we will have determined the system's impulse response. Manually calculating a few terms from the difference equation is very helpful in understanding this process.

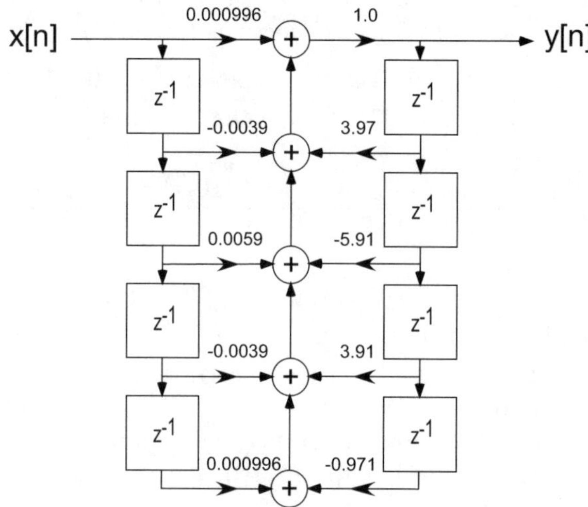

Figure 4.21: Block diagram of direct form I (DF-I) fourth order elliptic filter.

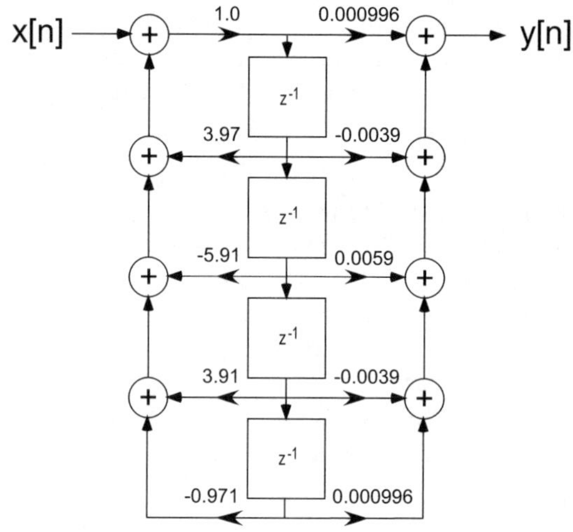

Figure 4.22: Block diagram of a direct form II (DF-II) fourth order elliptic filter.

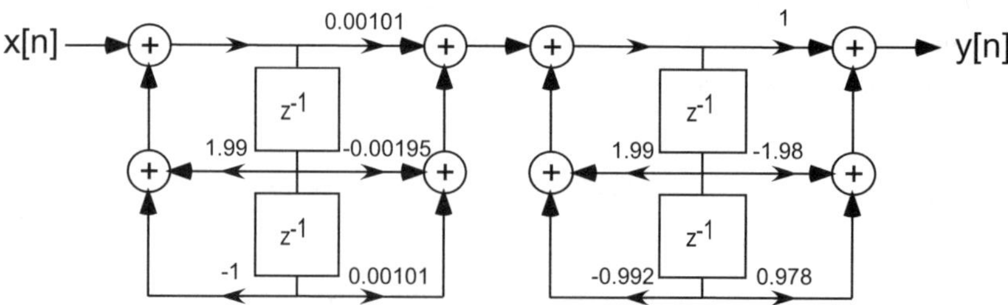

Figure 4.23: Block diagram of a the second order section (SOS) implementation of a fourth order elliptic filter.

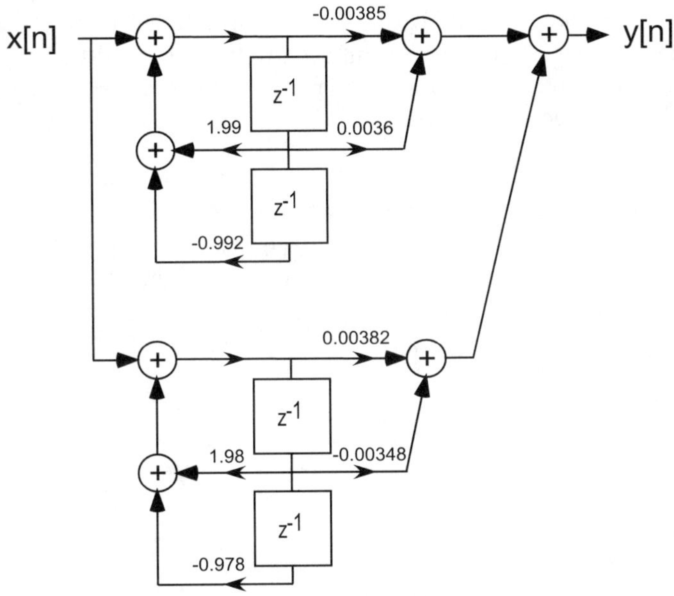

Figure 4.24: Block diagram of a parallel implementation of a fourth order elliptic filter.

1. Label the columns as shown below.

$$n \quad y[n] \quad y[n-1] \quad x[n] \quad x[n-1]$$

2. Fill in the $n = 0$ row information.

n	$y[n]$	$y[n-1]$	$x[n]$	$x[n-1]$
0		0	1	0

3. Calculate the $y[0]$ term.

n	$y[n]$	$y[n-1]$	$x[n]$	$x[n-1]$
0	1	0	1	0

4. Fill in the $n = 1$ row information. Notice the "down and to the right" flow of the stored values.

n	$y[n]$	$y[n-1]$	$x[n]$	$x[n-1]$
0	1	0	1	0
1		1	0	1

5. Calculate the $y[1]$ term.

n	$y[n]$	$y[n-1]$	$x[n]$	$x[n-1]$
0	1	0	1	0
1	-0.1	1	0	1

6. Continue this process until you have calculated all of the terms that you need.

The MATLAB code shown in the listing below will only calculate the $y[1]$ term. This code more closely implements the algorithm required for the real-time process. While it may seem strange to calculate only a single term, you must remember that this is *exactly* how sample-by-sample processing works.

Listing 4.1: Simple MATLAB IIR filter example.

```matlab
% Simulation inputs
x = [0  1];          % input vector x = x[0] x[-1]
y = [1  1];          % output vector y = y[0] y[-1]
B = [1  -1];         % numerator coefficients
A = [1  -0.9];       % denominator coefficients

% Calculated terms
y(1) =  -A(2)*y(2) + B(1)*x(1) + B(2)*x(2);
x(2) = x(1);         % shift x[0] into x[-1]
y(2) = y(1);         % shift y[0] into y[-1]

% Simulation outputs
x                    % notice that x(1) = x(2)
y                    % notice that y(1) = y(2)
```

As in the manual calculations, you should find that the output value is -0.1. In summary, the input (receive) ISR provides a new sample to the algorithm, the algorithm calculates the new output value, the algorithm prepares for the arrival of the next sample, and finally, the algorithm gives the new output value to the output (transmit) ISR, so that it may be converted back into an analog value. Notice that for low order filters, the actual calculation of the output value may be a *single* line of code!

4.4 DSK Implementation in C

4.4.1 Brute Force IIR Filtering

This version of the IIR implementation code, similar to the last MATLAB example, takes a brute force approach. The intention of this approach is understandability, which comes at the expense of efficiency.

The files necessary to run this application are in the `ccs\IIRrevA` directory of Chapter 4. The primary file of interest is the `ISRs.c` interrupt service routine. This file contains the necessary variable declarations and performs the actual IIR filtering operation. To allow for the use of a stereo codec (e.g., the native C6713 codec, the PCM3006-based daughtercard codec for the C6711, etc.), the program implements independent Left and Right channel filters. For clarity, however, only the Left channel will be discussed below. In the code shown below, N is the filter order, the B array holds the filter's numerator coefficients, the A array holds the filter's denominator coefficients, the x array holds the current input value $x[0]$, and past values of x (namely $x[-1]$ for this filter), and the y array contains the current output value of the filter, $y[0]$, and past values of y (namely $y[-1]$ for this filter).

Listing 4.2: Brute force IIR filter declarations.

```c
#define N 1                          // filter order

float B[N+1] = {1.0,  -1.0};         // numerator filter coefficients
float A[N+1] = {1.0,  -0.9};         // denominator filter coefficients
```

```
   float x[N+1];                        // input values
 6 float y[N+1];                        // output values
```

The code shown below performs the actual filtering operation. The four main steps involved in this operation will be discussed following the code listing.

Listing 4.3: Brute force IIR filtering for real-time.

```
   /* I added my routine here */
 2 x[0] = CodecDataIn.Channel[LEFT];  // current input value

 4 y[0] = -A[1]*y[1] + B[0]*x[0] + B[1]x[1];  // calc. the output

 6 x[1] = x[0];  // setup for the next input
   y[1] = y[0];  // setup for the next input
 8
   CodecDataOut.Channel[LEFT] = y[0];  // output the result
10 /* end of my routine */
```

The four real-time steps involved in brute force FIR filtering

An explanation of Listing 4.3 follows.

1. (Line 2): This code receives the next sample from the receive ISR and assigns it to the current input array element, x[0].

2. (Line 4): This code calculates a single value of the difference equation's output, y[0].

3. (Lines 6–7): These 2 lines of code shift the values in the x and y arrays one element to the right. The equivalent operation is,

$$x[0] \rightarrow x[1]$$
$$y[0] \rightarrow y[1].$$

   After the shift to the right is complete, the next incoming sample, $x[0]$ can be written into the $x[0]$ memory location without a loss of information.

4. (Line 9): This line of code completes the filtering operation by transferring the result of the filtering operation, $y[0]$, to the `CodecDataOut.Channel[LEFT]` variable for transfer to the DAC side of the codec via the transmit ISR.

Now that you understand the code...

Go ahead and copy all of the files into a separate directory. Open the project in CCS and "Rebuild All." Once the build is complete, "Load Program" into the DSK and click on "Run." Your IIR HP filter (actually a D.C. blocking filter) is now running on the DSK. Remember this program would typically be used for audio filtering, so a good way to experience the effects of your filter is to listen to unfiltered and filtered music.[1]

---

[1]You may need to adjust the value of $A[1]$ from -0.9 to a value such as -0.7 or even -0.5 to hear the effect well.

4.4.2 More Efficient IIR Filtering

Making the processor physically shift the location of the x and y values to make room for the next sample (lines 6 and 7 above) is very inefficient. For the particular example above that has such a low filter order, it doesn't take much time to do it that way. But for larger order filters this would be a bad idea. Looking back at Section 3.4.3, review how the idea of a circular buffer was implemented for an FIR filter; that same idea using the same technique with pointers, can be applied to IIR filters. Rather than just give you the code, this is left as one of the Follow-On Challenges below.

4.5 Follow-On Challenges

Consider extending what you have learned.

1. We have discussed only the brute force approach to IIR filtering. Similar to the Chapter 3 discussion, investigate and implement a version of the code that will work with MATLAB exported coefficient files (`coeff.c` and `coeff.h`). Use the `IIR_dump2c` function from the text CD to export the filter coefficients.

2. There are dozens of different ways to implement an IIR filter. Most of the examples in this chapter have used only the direct-form one (DF-I) techniques. Investigate and implement other forms, e.g., DF-II, DF-II transposed, lattice structure, parallel form, second order sections, etc.

3. Explore the IIR filter design tools that are available in MATLAB (e.g., `butter`, `cheby1`, `cheby2`, `SPTool`, and `FDATool`). A complete listing of the functional capabilities included in the signal processing toolbox can be found by typing `help signal`. The toolbox functions are grouped by category. You are looking for the `IIR filter design` heading. Use the `IIR_dump2c` function from the text CD to export the filter coefficients. Implement your design using the `IIRrevB` code.

4. Create an IIR filter routine using circular buffers.

Chapter 5

Periodic Signal Generation

5.1 Theory

$\mathbf{M}$ ANY interesting and useful signals can be generated using a DSP. Applications of some DSP-generated signals include, but are not limited to,

alerting signals such as different telephone rings, beeper message alerts, the call waiting tone, and the emergency alert system, which is the replacement for the emergency broadcast system tones;

system signaling such as telephone dialing tones (DTMF) and caller ID tones; and

oscillators such as a sine and/or cosine waveform that are routinely used to generate a wide variety of communications signals.

To keep the length of this chapter reasonable, we will only discuss periodic signal generation. We first review how periodic signals are represented as discrete-time signals, then transition to how such signals can be generated by a DSP.

5.1.1 Periodic Signals in DSP

Periodic signals have a fundamental period that is usually just called the period. During the period the entire signal is defined, and the signal repeats for every period that follows. For a continuous-time signal the fundamental period T_0 is the least amount of time required to completely define the signal; we shall see that the associated fundamental frequency is $f_0 = 1/T_0$. A periodic signal may contain many frequencies but only a single fundamental frequency. The simplest periodic signal is a sinusoid because it contains only a single frequency. Using the sine wave as an example, the concept of the fundamental period means that the sine must satisfy the equation

$$\sin\left(2\pi f_0 t + \phi\right) = \sin\left(2\pi f_0 t + 2\pi f_0 T_0 + \phi\right) = \sin\left(2\pi f_0 (t + T_0) + \phi\right)$$

where f_0 is the frequency (Hz) of the sine, t is the time (s) variable, ϕ is some arbitrary phase (rad), and T_0 is the period.[1] For T_0 to be one full period of 2π radians, $2\pi f_0 T_0 \equiv 2\pi$. This means that, as we mentioned earlier, $f_0 = 1/T_0$. Notice that T_0 must be both positive and real. While this seems to be a trivial discussion, it proves useful when changing from continuous-time to discrete-time representations.

---

[1] If you prefer to work with angular frequency, simply substitute ω_0 for $2\pi f_0$.

For a discrete-time version of our sine wave, we sample every T_s seconds (recall $T_s = 1/F_s$), and thus we replace the variable t with nT_s for $n = 0, 1, 2, \ldots$ for however many samples we obtain. The period N of a discrete-time signal will be expressed in units of *samples* and in the case of our sine wave example must satisfy the equation

$$\sin\left[2\pi f_0 n T_s + \phi\right] = \sin\left[2\pi \frac{f_0}{F_s} n + \phi\right]$$

$$= \sin\left[2\pi \frac{f_0}{F_s} n + 2\pi \frac{f_0}{F_s} N + \phi\right]$$

$$= \sin\left[2\pi \frac{f_0}{F_s}(n + N) + \phi\right]$$

where the value $2\pi(f_0/F_s)$ is the normalized discrete angular frequency (radians/sample). If the discrete-time signal is periodic, then the value at sample n must be equal to the value at sample $n + N$ for some integer N. This implies that $2\pi(f_0/F_s)N \equiv 2\pi k$, where k is another arbitrary integer. Rearranging this equation results in

$$\frac{N}{k} = \frac{F_s}{f_0}.$$

Since both N and k are integers, the ratio F_s/f_0 must be rational for the discrete-time signal to be periodic. In other words, if there are no integer values of N and k that solve this equality, then the sampled version of the signal is *not* periodic. This result of the sampling process is why many continuous-time signals that are periodic do *not* result in periodic discrete-time signals.

The information that defines a discrete-time signal is not necessarily unique; we can define one period of the signal starting at any point. To help understand this concept, Figure 5.1 shows portions of both a continuous and discrete-time 1 kHz sinusoid. Part (a) shows the continuous-time sinusoid with the horizontal axis (time axis) labeling on top of the figure. Additionally, the period ($T = 1$ ms) is also shown. To calculate the period of the sampled, discrete-time version of this signal we must solve

$$\frac{N}{k} = \frac{F_s}{f_0} = \frac{48000}{1000}.$$

The obvious solution is $N = 48$ and $k = 1$. Part (b) shows the first 48 samples (i.e., for $n = 0, 1, 2, \ldots, 47$) of a discrete-time (sampled) version of the 1 kHz sinusoid ($F_s = 48$ kHz) starting at $t = 0$. It is *very* important to realize that the last sample in part (b), where $n = 47$, does *not* equal the value at $n = 0$. Rather, the *next* sample (at $n = 48$) would equal the value at $n = 0$. If a complete discrete-time period is provided, the signal can be "continued" by replicating the information in the selected period. This concept is demonstrated in part (c) where 48 consecutive samples are replicated and concatenated together for two full periods. Repeating this concatenation process will allow you to generate arbitrary length versions of the signal. Parts (d–f) are examples of the same signal where the signal was defined by starting the sampling process at $n = 20$, 30, and 40 samples, respectively, and providing the next $N = 48$ samples of the signal.

5.1.2 Signal Generation

In order to limit the discussion to a reasonable length, we will only discuss the following techniques to generate a sinusoid:

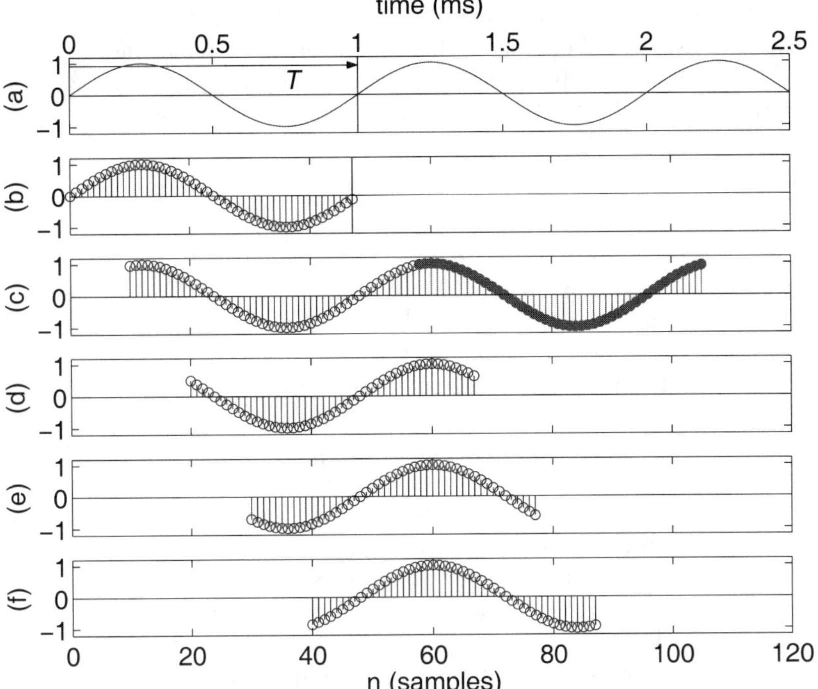

Figure 5.1: Continuous and discrete-time sinusoids. (a): 1 kHz continuous-time sinusoid. (b–f): 1 kHz sinusoid sampled at 48 kHz. (b): one period starting at $n = 0$. (c): demonstration of the periodic nature of a sampled sinusoid. (d–f): sampling/display commencing at $n = 20$, 30 and 40 samples, respectively.

direct digital synthesizer (DDS): these techniques can use a phase accumulator with a sin() or cos() trigonometric function call or use a table lookup system.

special cases: this includes sine and cosines with $f = F_s/2$, $f = F_s/4$, and other frequencies that result in reasonable values for N.

digital resonator: this technique uses an impulse excited, second order, IIR filter, where the complex conjugate pole-pairs are placed *on* the unit circle.

impulse modulator (IM): this technique is based upon using scaled impulses to periodically excite an FIR filter. Impulse modulation is commonly used in digital communications transmitters and is discussed further in Chapter 13.

We will discuss the theory of these signal generation techniques before proceeding to examples.

Direct Digital Synthesizer Case

You are probably familiar with plotting deterministic waveforms such as

$$w(t) = A\sin(2\pi f t)$$

in many of your math, physics, and engineering classes. In the equation above, A is the waveform's amplitude, f is the desired output frequency, and t represents the time variable.

The DDS idea, as it is implemented in real-time hardware, starts by converting $w(t)$ to a discrete-time process. This conversion is accomplished by replacing t by nT_s, where n is an integer and T_s is the sample period. Therefore, $w(t)$ becomes $w[nT_s]$, which is equal to $A\sin[2\pi f n T_s]$. Remembering that $T_s = 1/F_s$, where F_s is the sample frequency and rearranging the argument of the sine function, we arrive at

$$w[n] = A\sin[2\pi f n T_s] = A\sin\left[n\left(2\pi\frac{f}{F_s}\right)\right] = A\sin\left[n\phi_{inc}\right]$$

where we have used the common notation of using $w[n]$ in place of $w[nT_s]$ since T_s is tacitly assumed. The value $\phi_{inc} = 2\pi f/F_s$ is called the *phase increment*. A phase accumulator can be used to add the phase increment to the previous value of the phase accumulator every sample period. The phase accumulator is kept in the interval 0 to 2π by a modulus operator. Since real-time processes can run for an indefinite amount of time, a modulus operation is required to prevent an overflow of the phase accumulator. Finally, the sin() of the phase accumulator's value can be calculated and the value provided as the system's output. The block diagram for this process is shown in Figure 5.2. Notice that since ϕ_{inc} is added to the phase accumulator each time the input ISR is called (which is every $T_s = 1/F_s$ seconds or 48,000 times/sec in this example), the value n never appears in the algorithm.

The argument of the sin() operator, $n\phi_{inc}$, is a linearly increasing function whose slope depends on the desired output frequency. To illustrate this point, Figure 5.3 plots the accumulated phase as a function of time for 4 different frequencies. Figure 5.4 plots the sampled ($F_s = 48$ kHz) version of Figure 5.3 for only the 1000 Hz case. Figure 5.5 expands a portion of Figure 5.4 and adds additional labeling. To prevent aliasing, a minimum of two samples per period are required. This limit requires that $\phi_{inc} \leq \pi$. Finally, Figure 5.6 demonstrates the effect of the modulus operation on the phase accumulator's value.

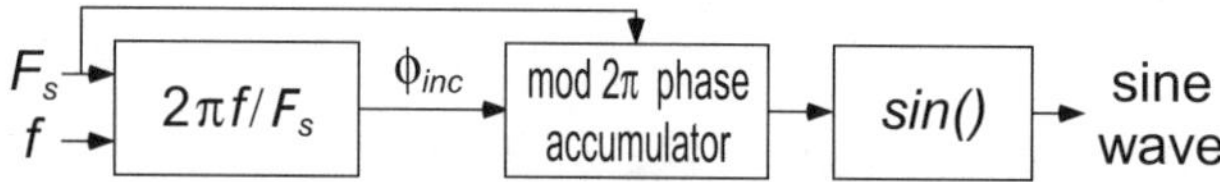

Figure 5.2: Block diagram associated with sinusoid generation.

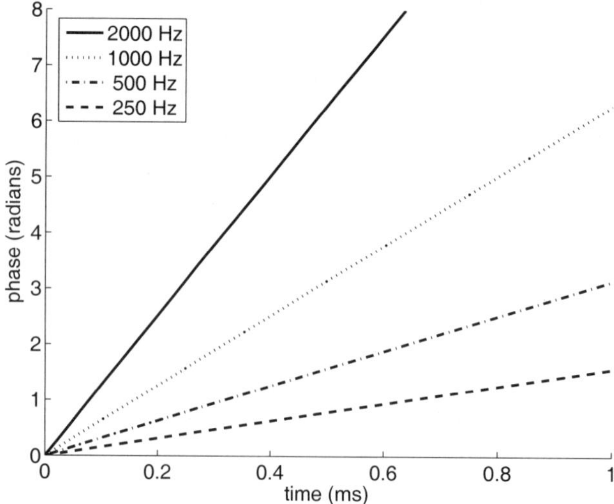

Figure 5.3: Accumulated phase for four different frequencies.

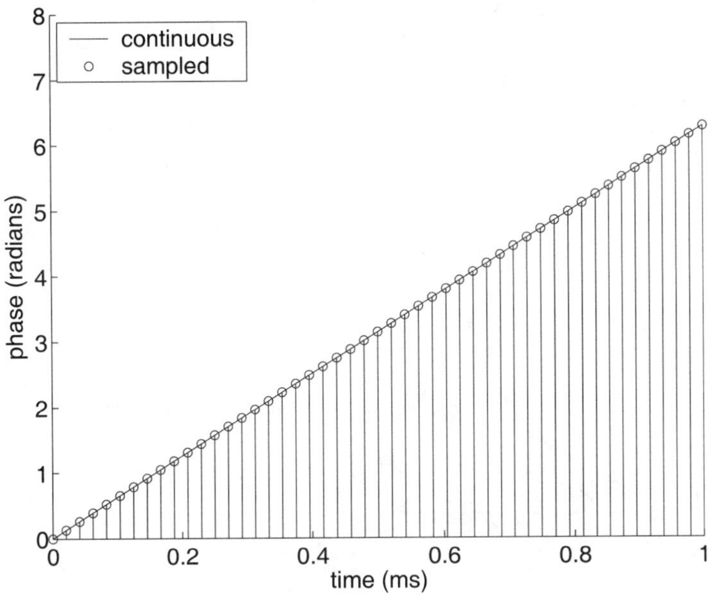

Figure 5.4: Accumulated phase for a 1000 Hz sinusoid.

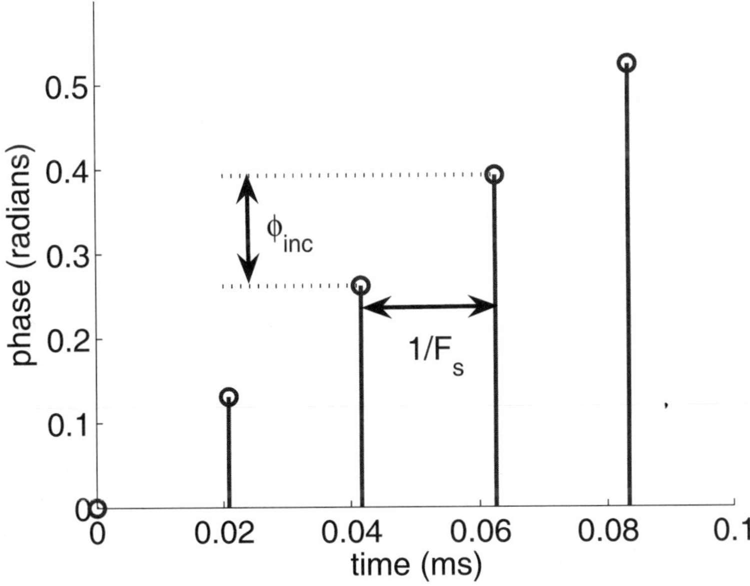

Figure 5.5: Accumulated phase for a 1000 Hz sinusoid (zoomed in from Figure 5.4).

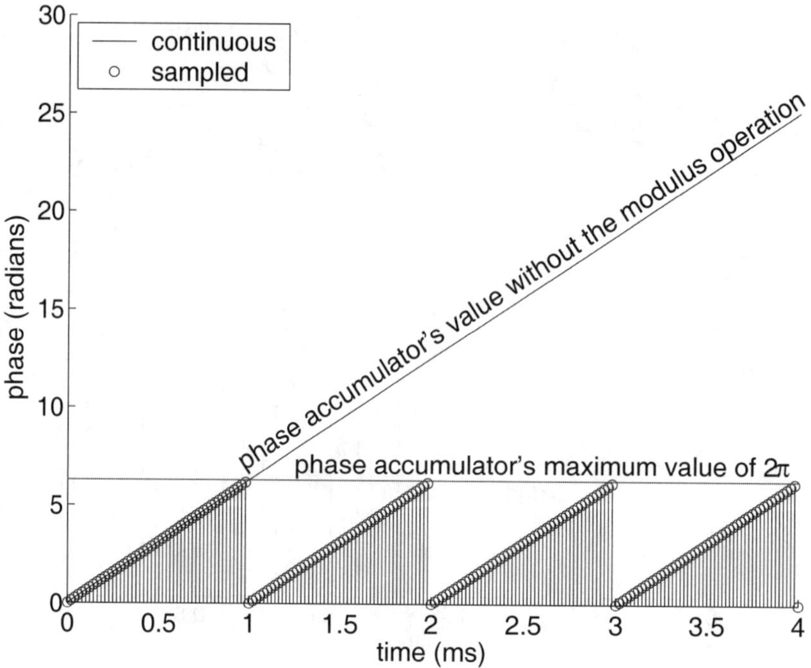

Figure 5.6: Accumulated phase for a 1000 Hz sinusoid with modulus 2π applied.

Special Cases

If the characteristics (frequency and phase) of the signal that you need to generate will not change with time, then you may not need a phase accumulator at all. A few special cases follow.

1. Sine and cosine with $f = \frac{F_s}{2}$. Substituting $f = \frac{F_s}{2}$ into the ϕ_{inc} equations results in

$$\phi_{inc} = 2\pi \left(\frac{f}{F_s} \right)\bigg|_{f=\frac{F_s}{2}} = \pi.$$

This is the aliasing limit for ϕ_{inc} and results in,

$$w\left[n\right] = A\sin\left(n\phi_{inc}\right)\big|_{(\phi_{inc}=\pi)} = A\sin\left(n\pi\right) = 0.$$

A signal generator that always has an output $= 0$ is of little use. However, the cosine version results in

$$w\left[n\right] = A\cos\left(n\phi_{inc}\right)\big|_{(\phi_{inc}=\pi)} = A\cos\left(n\pi\right) = A, -A, A, -A\ldots$$

This implies that a cosine waveform of frequency $F_s/2$ can be created by simply generating $A, -A, \ldots$ for however long you need the signal. The CPU resources required to generate these or other alternating values are inconsequential. Notice that for $f = F_s/2$, $N = 2$ (2 samples per period).

2. Sine and cosine with $f = \frac{F_s}{4}$. Substituting $f = \frac{F_s}{4}$ into the ϕ_{inc} equations results in

$$\phi_{inc} = 2\pi \left(\frac{f}{F_s} \right)\bigg|_{f=\frac{F_s}{4}} = \frac{\pi}{2}.$$

Table 5.1: Some special case frequencies of direct digital synthesis (DDS). The frequency values shown in the right column assume $F_s = 48$ kHz.

F_s ratio	N	frequency (Hz)
$F_s/2$	2	24,000
$F_s/3$	3	16,000
$F_s/4$	4	12,000
$F_s/5$	5	9,600
$F_s/6$	6	8,000
$F_s/8$	8	6,000
$F_s/10$	10	4,800
$F_s/12$	12	4,000
$F_s/15$	15	3,200
$F_s/16$	16	3,000
$F_s/20$	20	2,400
$\vdots$	$\vdots$	$\vdots$
F_s/N	N	$48,000/N$

This yields

$$w[n] = A \sin(n\phi_{inc})\big|_{\left(\phi_{inc}=\frac{\pi}{2}\right)} = A \sin\left(n\frac{\pi}{2}\right) = 0, A, 0, -A, \ldots$$

and

$$w[n] = A \cos(n\phi_{inc})\big|_{\left(\phi_{inc}=\frac{\pi}{2}\right)} = A \cos\left(n\frac{\pi}{2}\right) = A, 0, -A, 0, \ldots$$

This implies that a sine or cosine waveform of frequency $\frac{F_s}{4}$ can be created by simply generating $0, A, 0, -A, \ldots$ or $A, 0, -A, 0, \ldots$ respectively. As in the $\frac{F_s}{2}$ case, the CPU resources required to generate these or other repeating values are inconsequential. Notice that for $f = \frac{F_s}{4}$, $N = 4$ (4 samples per period).

3. Sines and cosines with other frequencies that result in reasonable values for N. Assuming from our discussion in Section 5.1 that $k = 1$, we simply divide F_s by the desired frequency f to find N. Table 5.1 shows several possible frequencies and the corresponding values of N.

All of the table entries in the right column are based on $F_s = 48$ kHz. For example, to generate a 4,800 Hz cosine waveform, we would only need to calculate the first 10 values of the sequence. These values are based upon $N = 10$ and $\phi_{inc} = \frac{\pi}{5}$. Specifically, we would need to evaluate

$$w[n] = A \cos\left(\frac{\pi}{5}n\right), \quad \text{for } n = 0, 1, \ldots, 9.$$

These values can be calculated once by the real-time program (i.e., in `StartUp.c`) or off line using tools such as a handheld calculator, spreadsheet program, or MATLAB. Continuously repeating all 10 values (in the proper order), with one value every sample time of $T_s = \frac{1}{48,000}$ seconds, will result in the desired 4,800 Hz signal.

Digital Resonator

The digital resonator technique is based on the idea that if you refer to any z-transform table you will find an entry similar to,

$$[r^n \sin(\omega_0 n)] \, u[n] \xleftrightarrow{\ Z\ } \frac{r \sin(\omega_0) z^{-1}}{1 - [2r\cos(\omega_0)] z^{-1} + r^2 z^{-2}}.$$

Letting $r = 1$ (equivalent to placing the poles on the unit circle), this equation simplifies to

$$[\sin(\omega_0 n)] \, u[n] \xleftrightarrow{\ Z\ } \frac{\sin(\omega_0) z^{-1}}{1 - [2\cos(\omega_0)] z^{-1} + z^{-2}}.$$

This transform pair implies that if you excite this system with an impulse, the system's output will be a sine wave. The system's difference equation can be determined from the transfer function

$$H(z) = \frac{Y(z)}{X(z)} = \frac{\sin(\omega_0) z^{-1}}{1 - [2\cos(\omega_0)] z^{-1} + z^{-2}}.$$

Cross multiplying, taking the inverse z-transform, and rearranging the terms into the standard form results in the difference equation

$$\boxed{\; y[n] = \sin(\omega_0)x[n-1] + 2\cos(\omega_0)y[n-1] - y[n-2] \;}.$$

Thus to create a sine wave of digital frequency ω_0, we need to excite this second order IIR filter with an impulse. To find out where the poles and zeros are located, we convert the transfer function to positive powers of z then factor the transfer function. This leads to

$$\frac{\sin(\omega_0) z^{-1}}{1 - [2\cos(\omega_0)] z^{-1} + z^{-2}} = \frac{\sin(\omega_0) z^{-1}}{1 - [2\cos(\omega_0)] z^{-1} + z^{-2}} \cdot \frac{z^2}{z^2} = \frac{\sin(\omega_0) z}{z^2 - [2\cos(\omega_0)] z + 1}.$$

The numerator term reveals a single zero at the origin. For the denominator, we apply the quadratic equation

$$\frac{2\cos(\omega_0) \pm \sqrt{(2\cos(\omega_0))^2 - 4(1)(1)}}{2(1)} = \cos(\omega_0) \pm \sqrt{\cos^2(\omega_0) - 1}$$

which, using the trigonometric identity

$$\sin^2(\omega_0) + \cos^2(\omega_0) = 1 \;\;\therefore\;\; \cos^2(\omega_0) - 1 = -\sin^2(\omega_0)$$

can be simplified to

$$\cos(\omega_0) \pm \sqrt{-\sin^2(\omega_0)} = \boxed{\; \cos(\omega_0) \pm j\sin(\omega_0) = e^{\pm j\omega_0} \;}.$$

This result, shown above in both rectangular and polar forms, should be recognized as indicating that the complex conjugate poles are located *on* the unit circle at the frequency $\pm\omega_0$. This is at best a marginally stable system (in that it oscillates at a constant frequency of ω_0) and some authors would call this system unstable. Actually it is "intentionally unstable." Oscillators and resonators are definitely on a fine line between systems that are clearly stable and those that are clearly unstable. While the system is unstable in the sense that the output does not change regardless of the input, it is stable in the sense that the output frequency remains the same.

We will now use a unit impulse as the input to the system and calculate the first few output terms. Manually calculating a few terms for the difference equation is very helpful, not only in understanding this process, but it will greatly assist our real-time algorithm development. Remember that the difference equation is

$$y[n] = \sin(\omega_0)x[n-1] + 2\cos(\omega_0)y[n-1] - y[n-2].$$

1. Label the columns as shown below.

n	$y[n]$	$y[n-1]$	$y[n-2]$	$x[n]$	$x[n-1]$

2. Fill in the $n = 0$ *at rest* row information.

n	$y[n]$	$y[n-1]$	$y[n-2]$	$x[n]$	$x[n-1]$
0		0	0	1	0

3. Calculate the $y[0]$ term.

n	$y[n]$	$y[n-1]$	$y[n-2]$	$x[n]$	$x[n-1]$
0	0	0	0	1	0

4. Fill in the $n = 1$ row information. Notice the "down and to the right" flow of the stored y values and the stored x values.

n	$y[n]$	$y[n-1]$	$y[n-2]$	$x[n]$	$x[n-1]$
1		0	0	0	1

5. Calculate the $y[1]$ term.

n	$y[n]$	$y[n-1]$	$y[n-2]$	$x[n]$	$x[n-1]$
1	$\sin(\omega_0)$	0	0	0	1

6. Fill in the $n = 2$ row information.

n	$y[n]$	$y[n-1]$	$y[n-2]$	$x[n]$	$x[n-1]$
2		$\sin(\omega_0)$	0	0	0

We stop here with the $n = 2$ initial conditions loaded and ready to calculate $y[2]$. This is an excellent place for us to pause and pick the idea up again for our real-time C digital resonator discussed later in this chapter. The difference equation for the rest of time (i.e., where $n \geq 2$) now simplifies to

$$y[n] = 2\cos(\omega_0)y[n-1] - y[n-2].$$

since, from this point on, all of the $x[n-1]$ terms will equal zero.

Note that the impulse modulator (IM) technique is commonly used in digital communications transmitters. We therefore postpone its discussion until Chapter 16, which includes a digital transmitter project.

5.2   winDSK6 Demonstration

If you double click on the winDSK6 icon, the winDSK6 application will launch, and a window similar to Figure 5.7 will appear. Before proceeding, be sure the selections in "DSK and Host Configuration" are correct.

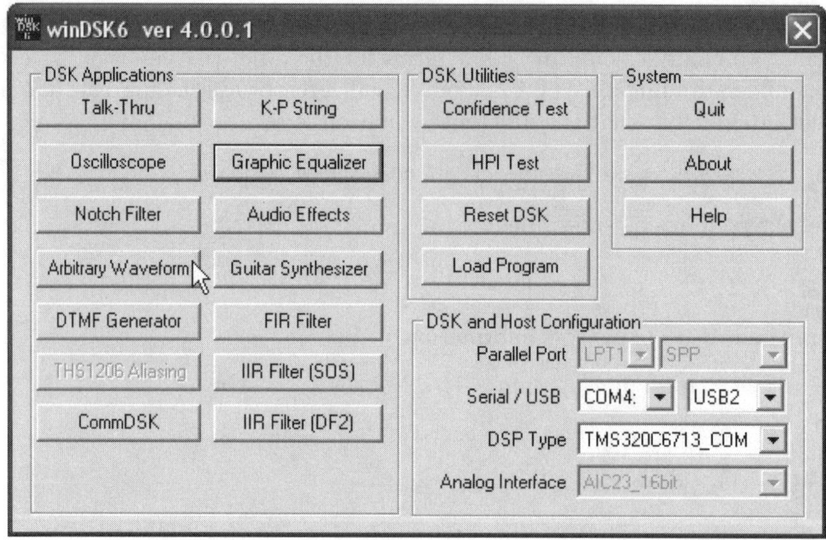

Figure 5.7: winDSK6 ready to load the Arbitrary Waveform application.

5.2.1   Arbitrary Waveform

Clicking on the winDSK6 Arbitrary Waveform button will load that program into the attached DSK, and a window similar to Figure 5.8 will appear. The arbitrary waveform program generates sine, square, and triangle waves at frequencies between 1 Hz and the upper limit of the CODEC in use. For multichannel CODECs, each output channel is capable of simultaneous independent operation. The displays show the settings for the currently selected channel, as indicated by the channel number display. The frequency displays will turn red if the selected frequency exceeds the capabilities of the CODEC, and the DSK fre-

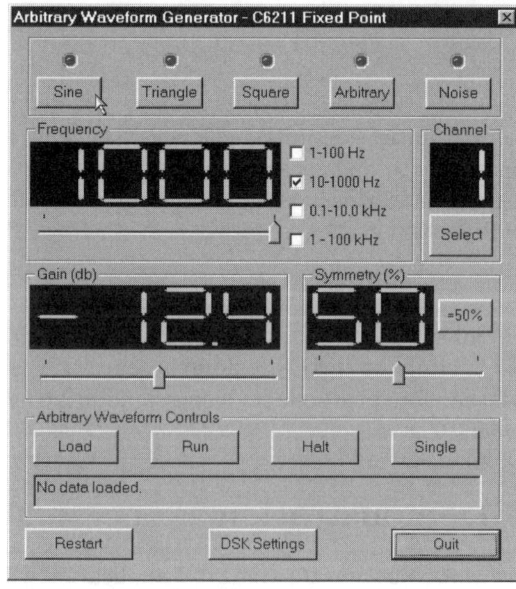

Figure 5.8: winDSK6 running the Arbitrary Waveform application.

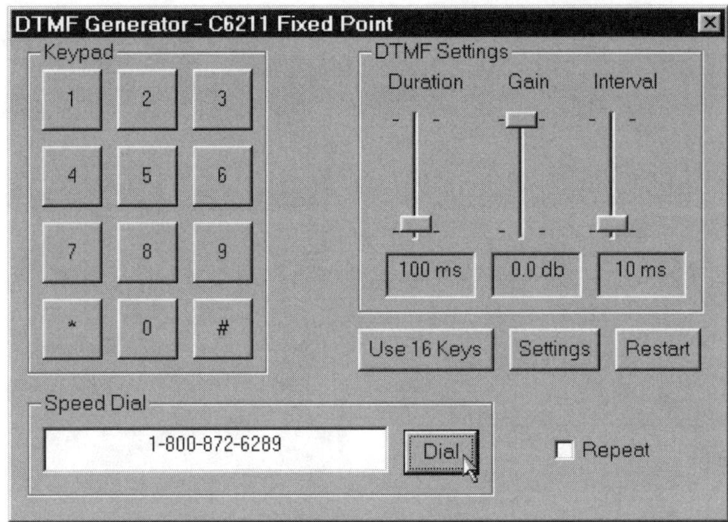

Figure 5.9: winDSK6 running the DTMF application (12-keys).

quency will not be updated. As an arbitrary waveform generator, the program can load up to 2,000,000 sample values (depending on the DSK version) per channel from a text file. In this mode the sample values in the file are repeatedly used as the system output. The values will be automatically scaled to fit within the ADC range. A sample waveform file called `chirp.asc` is included in the winDSK6 installation. This file contains a 2,500 sample chirp waveform that can be played through the application.[2] The arbitrary waveform generator can also function as a noise generator. Finally, one-shot operation is also supported.

Selecting the arbitrary waveform generator in the "sine" mode will run a program in the DSK that is most similar to the examples given in the preceding discussion regarding periodic signal generation. Of course square and triangle waves are also periodic signals.

5.2.2 DTMF

Clicking on the winDSK6 DTMF button will load and run that program into the attached DSK, and a 12-keypad window similar to Figure 5.9 will appear by default. Clicking on the "16-keys" button will add a fourth column to the keypad display as shown in Figure 5.10.

This application generates standard Dual-Tone, Multiple-Frequency (DTMF) signals as defined by telephone companies. These are signals that consist of two sinusoids of different frequencies that added together. Any time you dial a modern telephone, DTMF tones are generated that correspond to the buttons you pressed on the telephone's keypad (or correspond to a telephone number that was stored as an autodial selection). The DTMF standard specifies that tones must persist for at least 40 ms and have at least 50 ms of "quiet time" between tones. Additionally, DTMF tones must not occur at a faster rate than 10 characters/sec [51].

A speed-dial feature is available on the DTMF application of winDSK6 by clicking the "Dial" button; it provides automatic generation of DTMF sequences based upon the number you type into the "Speed Dial" entry window. For this option, only the standard 12-key tone pairs are generated by the characters 0-9, #, *. Any other characters are ignored. The

---

[2]A *chirp* is typically a short duration signal in which the frequency sweeps monotonically (up or down) with time. Chirps can have linear sweeps or logarithmic sweeps, and are used in a variety of radar, sonar, and communications applications.

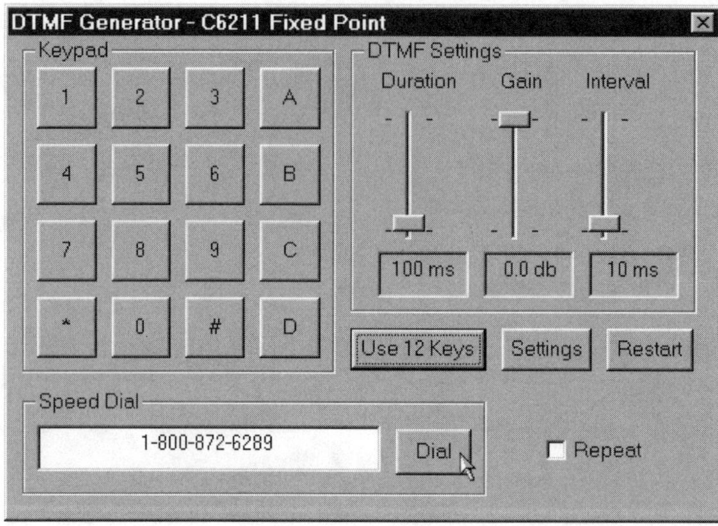

Figure 5.10: winDSK6 running the DTMF application (16-keys).

duration and volume (i.e., gain) of a tone, as well as the interval of silence between tones, may be adjusted by using the slider controls in the upper right of the DTMF application window.

If you're using a stereo codec on the DSK, both channels are driven with the same signal. As mentioned above, a 12-key or 16-key keypad can be selected. In 16-key mode, all 16 standardized tone pairs can be generated. The two frequencies that are generated for any given key press can be determined by inspecting Figure 5.11.

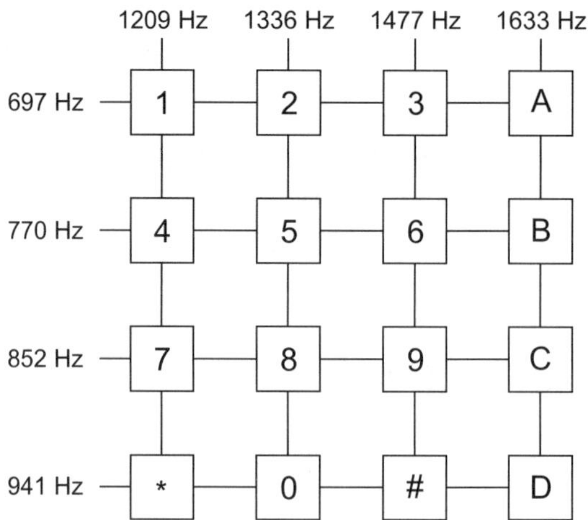

Figure 5.11: The DTMF frequencies.

5.3 MATLAB Implementation

MATLAB has a number of ways of generating sinusoids. However, we will focus on three of the techniques that can help prepare us for the realities of real-time signal generation using DSP hardware.

5.3.1 Direct Digital Synthesizer Technique

In this technique, MATLAB is used to implement the phase accumulator process. A listing demonstrating this technique is shown below.

Listing 5.1: MATLAB implementation of phase accumulator signal generation.

```
   % Simulation inputs
 2 A = 32000;                       % signal's amplitude
   f = 1000;                        % signal's frequency
 4 phaseAccumulator = 0;            % signal's initial phase
   Fs = 48000;                      % system's sample frequency
 6 numberOfTerms = 50;             % calculate this number of terms


 8 % Calculated and output terms
   phaseIncrement = 2*pi*f/Fs; % calculate the phase increment
10
   for i = 1:numberOfTerms
12     % ISR's algorithm begins here
           phaseAccumulator = phaseAccumulator + phaseIncrement;
14         phaseAccumulator = mod(phaseAccumulator, 2*pi);
           output = A*sin(phaseAccumulator)
16     % ISR's algorithm ends here
   end
```

A few items need to be discussed concerning this listing.

1. Variable initialization section (lines 2–6). Remembering that the sine and cosine functions are constrained to being between ± 1 requires an amplitude scale factor, A, or the DAC's output will only use the least significant bit (LSB).

2. For a constant output frequency, the calculation of the phase increment (line 9) need only be accomplished once. The calculated value of the phase increment *must* be $\leq \pi$ or signal aliasing will occur.

3. The actual algorithm to generate the sinusoidal signal requires only three lines of code (lines 13–15), inside a "for" loop that simulates the execution of an ISR in C that is called each time a new sample arrives. These lines of code accomplish the following three tasks each time the "ISR" is called:

   (a) line 13: add the phase increment's value to the phase accumulator.

   (b) line 14: perform a modulus 2π operation to keep the phase accumulator in the range 0 to 2π.

   (c) line 15: calculate the system's output value by scaling the sine of the phase accumulator's value by A.

   These three lines of code could be combined, but that would result in less understandable code.

5.3.2   Table Lookup Technique

This section demonstrates how MATLAB can be used to implement the table lookup technique, which is a very efficient method to generate a discrete-time signal. In this technique, we repeatedly cycle through a stored vector of predefined signal values. We will again simulate the execution of an ISR at the sample frequency by using a "for" loop. A new value of the signal is read from the table each time the "ISR" is called.

Listing 5.2: MATLAB implementation of the table lookup-based signal generation.

```matlab
1  % Simulation inputs
   signal = [32000 0 -32000 0];   % cosine signal values (Fs/4 case)
3  index = 1;                     % used to lookup the signal value
   numberOfTerms = 20;            % calculate this number of terms
5
   % Calculated and output terms
7  N = length(signal);           % signal period

9  for i = 1:numberOfTerms
       % ISR's algorithm begins here
11         if (index >= (N + 1))
               index = 1;
13         end
           output = signal(index)
15         index = index + 1;
       % ISR's algorithm ends here
17 end
```

A few items need to be discussed concerning this listing.

1. Variable initialization section (lines 2–4). These lines of code establish the variable `signal` that stores the required values of the output signal and the integer variable `index` that is used to access the different storage locations of `signal`.

2. Period determination (line 7). This line of code determines the period of the signal based upon the length on the variable `signal`.

3. The actual algorithm to generate the sinusoidal signal requires only five lines of code (lines 11–15). These lines of code accomplish the following three tasks each time the ISR is called:

   (a) line 11–13: performs a modulus N operation to keep `index` in the range 1 to N. Remember, that unlike C/C++, MATLAB array indices start at 1 instead of 0.

   (b) line 14: calculates the system's output value by selecting the appropriate `index` of `signal`.

   (c) line 15: increments the integer variable `index`.

5.4   DSK Implementation in C

Note that the examples in this section may require you to change the ISR file in a project to change the operation of the code. **Important:** you must have only *one* of these ISR files loaded as part of your project at any given time. To switch from using one ISR file to another, right click the current ISR file in the left project window and select "Remove from project." At the top of the Code Composer Studio window click "Project," "Add Files to Project," and select the new ISR file. Then click "Rebuild All" (or "Incremental Build"). Once the build is complete, "Load Program" (or "Reload Program") into the DSK and click on "Run." You will then be using the new ISR file.

5.4.1 Direct Digital Synthesizer Technique

This version of the direct digital synthesizer technique is very similar to the DDS MATLAB example. The intention of this first approach is understandability, which often comes at the expense of efficiency.

The files necessary to run this application are in the `ccs\sigGen` directory of Chapter 5. The primary file of interest is the `sinGenerator_ISRs.c`, which contains interrupt service routines; ensure this is the only ISR file included in the project. This file contains the necessary variable declarations and performs the actual sinusoid generation. However, as with all the Code Composer Studio projects we include with this text, you should make a habit of inspecting other files in the project, such as `StartUp.c`, to be sure you understand the full workings of the program.

If you're using one of the stereo codecs on your DSK, the program could implement two independent sinusoid generators for the Left and Right channels. For clarity, this example program will contain only a single phase accumulator, but that phase will be used to generate a sine wave for the Left channel and a cosine wave for the Right channel.

In the code shown below, `A`, `fDesired`, and `phase` (lines 1-3) are the signal's amplitude, frequency, and phase, respectively. Remember, that a 16-bit DAC has a range of $+32,767$ to $-32,768$. The variable `phase` sets not only the signal's initial phase, but will also serve as the phase accumulator. Having π (`pi` on line 5), and the system's sample frequency (`fs` on line 8), defined allows us to calculate the phase increment, which is declared on line 6.

Listing 5.3: Variable declaration associated with sinusoidal signal generation.

```
1  float  A = 32000;          /* signal's amplitude */
   float  fDesired = 1000;    /* signal's frequency */
3  float  phase = 0;          /* signal's initial phase */

5  float  pi = 3.1415927;     /* value of pi */
   float  phaseIncrement;     /* incremental phase */
7
   int  fs = 48000;           /* sample frequency */
```

The code shown below performs the actual signal generation operation. The four main steps involved in this operation will be discussed below the code listing.

Listing 5.4: Algorithm associated with sinusoidal signal generation.

```
   /* algorithm begins here */
2  phaseIncrement = 2*pi*fDesired/fs;
   phase += phaseIncrement;            // calculate the next phase
4
   if (phase > 2*pi) phase -= 2*pi;    // modulus 2*pi operation
6
   CodecDataOut.Channel[ LEFT] = A*sinf(phase); // scaled L output
8  CodecDataOut.Channel[RIGHT] = A*cosf(phase); // scaled R output
   /* algorithm ends here */
```

The four real-time steps involved in DDS-based signal generation

An explanation of Listing 5.4 follows.

1. (Line 2): This code calculates the phase increment each time the ISR is called. This will allow us to *change* the signal's frequency if desired during program execution.

2. (Line 3): This code adds the phase increment to the current value of the phase.

3. (Line 5): This code performs the equivalent of a modulus 2π operation. To prevent signal aliasing, the phase *increment* must be $\leq \pi$. With a maximum increment value of π, the modulus operation can be simplified to just a test and a subtraction of 2π. Subtracting 2π "starts over" by one full period. This method is far more efficient than using the modulus operation.

4. (Lines 7–8): These two lines of code calculate the sine and cosine values, scale these values by A, and write the results to the DAC.

Now that you understand the code...

Go ahead and copy all of the files into a separate directory. Open the project in CCS and "Rebuild All." Once the build is complete, "Load Program" into the DSK and click on "Run." Your 1 kHz sine generator is now running on the DSK.

5.4.2   Table Lookup Technique

This version of the table lookup technique is very similar to the table lookup MATLAB example. The files necessary to run this application are in the same place as before, the `ccs\sigGen` directory of Chapter 5. The primary file of interest this time is `sinGenerator_ISRs1.c`, which contains the interrupt service routines; remove the previous ISR file from your project and add this one. This file contains the necessary variable declarations and performs the actual sinusoid generation.

To allow for the use of a stereo codec (e.g., the native C6713 codec, the PCM3006-based daughtercard codec for the C6711, etc.), the program implements independent Left and Right channel sinusoid generators. For clarity, this example program will only generate $f = F_s/4 = 12$ kHz, but will output a sine wave to the Left channel and a cosine wave to the right channel.

In the code shown below, N (line 1) is the signal's period, `signalCos` (line 3) and `signalSin` (line 4) store the table values for the cosine and sine waveforms, respectively, and `index` (line 5) is a integer used to cycle through the different values stored in the table.

Listing 5.5: Variable declaration associated with sinusoidal signal generation.

```
#define N 4             // signal period for f = Fs/4

int signalCos[N] = {32000, 0, -32000, 0}; // cos waveform
int signalSin[N] = {0, 32000, 0, -32000}; // sin waveform
int index = 0;          /* signal's indexing variable */
```

The code shown below performs the actual signal generation operation. The three main steps involved in this operation will be discussed below the code listing.

Listing 5.6: Algorithm associated with sinusoidal signal generation.

```
/* algorithm begins here */
if (index == N) index = 0;
```

```
3
  CodecDataOut.Channel[ LEFT] = signalCos[index]; // cos output
5 CodecDataOut.Channel[RIGHT] = signalSin[index]; // sin output

7 index++;
  /* algorithm ends here */
```

The three real-time steps involved in table lookup-based signal generation

An explanation of Listing 5.6 follows.

1. (Line 2): This code performs a modulus operation and keeps the variable `index` between 0 and 3.

2. (Line 4–5): This code outputs the next value from the cosine and sine tables. Notice that the signal's amplitude is included in the `signalCos` and `signalSin` array values.

3. (Line 7): This code increments (increases by 1) the value of `index`.

Now that you understand the code...

Go ahead and copy all of the files into a separate directory. Open the project in CCS and "Rebuild All." Once the build is complete, "Load Program" into the DSK and click on "Run." Your 12 kHz ($\frac{F_s}{4}$) cosine and sine generators are now running on the DSK.

5.4.3 Table Lookup Technique with Table Creation

This version of the table lookup technique adds a table creation routine. The files necessary to run this application are in the `ccs\sigGenTable` directory of Chapter 5. In the previous examples, we only looked at the ISR file in detail. For this example, there are two files of interest: `StartUp.c` and `tableBasedSinGenerator_ISRs.c`. The file `StartUp.c` contains code not tied to any interrupt, and so is the appropriate place for the code that runs just once to create the table values. The file `tableBasedSinGenerator_ISRs.c` contains the interrupt service routines. These files contain routines to generate and fill the table as well as the necessary variable declarations and routines to perform the actual sinusoid generation.

To allow for the use of a stereo codec (e.g., the native C6713 codec, the PCM3006-based daughtercard codec for the C6711, etc.), the program could implement independent Left and Right channel sinusoid generators. For clarity, this example program will only generate a 6 kHz sine wave, which will be heard on both the Left and Right channels.

In the code shown below, `NumTableEntries` (line 2) defines the size of the table, `desiredFreq` (line 4) is the desired output frequency, and `SineTable` (line 5) is the array (i.e., table) that will be filled with sine values. The `FillSineTable` function (lines 7–13) is only called once by the `StartUp.c` file. This function call occurs just after the DSK finishes initializing. This one-time calculation prevents repeated calls of the computationally expensive trigonometric function `sinf()`.

Listing 5.7: Variable declaration and table creation associated with table-based sinusoidal signal generation.

```
  /* declared at file scope for visibility */
2 #define NumTableEntries 100
```

```
 4  float desiredFreq = 6000.0;
    float SineTable[NumTableEntries];
 6
    void FillSineTable()
 8  {
       int i;
10
       for(i = 0; i < NumTableEntries; i++)  // fill table values
12         SineTable[i]=sinf(i*(float)(6.283185307/NumTableEntries));
    }
```

The code shown below performs the actual signal generation operation. The four main steps involved in this operation will be discussed below the code listing.

Listing 5.8: Algorithm associated with table-based sinusoidal signal generation.

```
 1  /* ISR's algorithm begins here */
    index += desiredFreq;             // calculate the next phase
 3  if (index >= GetSampleFreq())   // keep phase between 0-2*pi
        index -= GetSampleFreq();
 5
    sine=SineTable[(int)(index/GetSampleFreq()*NumTableEntries)];
 7  CodecData.Channel[LEFT]  = 32767*sine; // scale the result
    CodecData.Channel[RIGHT] = CodecData.Channel[LEFT];
 9  /* ISR's algorithm ends here */
```

The four real-time steps involved in table lookup-based signal generation with table creation

An explanation of Listing 5.8 follows.

1. (Line 2): This code is the equivalent operation of adding the phase increment to the phase accumulator.

2. (Lines 3–4): This code performs a modulus operation and maintains `index` between 0 and the sample frequency (typically, $F_s = 48$ kHz). Note the use of the function `GetSampleFreq()` in these lines and line 6. This is a very simple function call which returns a float value equal to the sample frequency you chose for the DSK. Using a function call instead of hard-coding a number makes the code a bit more portable.

3. (Line 6): This code calculates the next floating point value of the table index, converts this number to an integer, and then uses this to access the required table value's location in the `SineTable` array.

4. (Lines 7–8): This code scales the table's output (`sine`) and outputs the result to both the Left and Right channels.

Now that you understand the code ...

Go ahead and copy all of the files into a separate directory. Open the project in CCS and "Rebuild All." Once the build is complete, "Load Program" into the DSK and click on "Run." Your 6 kHz sine generator is now running on the DSK.

5.4.4 Digital Resonator Technique

The digital resonator technique implements a second order IIR filter with very special initial conditions stored in the $y[n-1]$ and $y[n-2]$ memory locations. You may want to refer back to the theoretical discussion of digital resonators given earlier in the chapter. The intention of the approach we use is understandability, which may come at the expense of efficiency.

The files necessary to run this application are in the same `ccs\sigGen` directory of Chapter 5 that contained the files for the first two C examples of this chapter. The primary file of interest this time is `resonator_ISRs.c`, which contains the interrupt service routines; remove any other ISR fie from the project and add this one. This file contains the necessary variable declarations and performs the actual sinusoid generation.

If you're using one of the stereo codecs on your DSK, the program could implement independent Left and Right channel sinusoid generators. For clarity, this example program will generate only one sine wave but will output the sine wave to both the Left and Right channels.

In the code shown below, `fDesired` and `A` (lines 1–2) are the signal's frequency and amplitude, respectively. Remember that a 16-bit DAC has a range of $+32,767$ to $-32,768$. Having π (`pi` on line 4), and the system's sample frequency (`fs` on line 8), defined will allow us to calculate `theta`, the digital frequency, which is declared on line 5. Finally, `y[3]` declares the storage for the current and past output values. Only three terms are needed to implement the second order difference equation.

Listing 5.9: Variable declaration associated with a digital resonator.

```
1  float  fDesired = 1000;    // your desired signal frequency
   float  A = 32000;          // your desired signal amplitude
3
   float  pi = 3.1415927;     // value of pi
5  float  theta;             // the digital frequency
   float  y[3] = {0, 1, 0};  // the last 3 output values.
7
   int  fs = 48000;          // sample frequency
```

The code shown below performs the actual signal generation operation using the digital resonator techniques. The four main steps involved in this operation will be discussed below the code listing.

Listing 5.10: Algorithm associated with a digital resonator.

```
   /* algorithm begins here */
2  theta = 2*pi*fDesired/fs;     // calculate digital frequency

4  y[0] = 2*cosf(theta)*y[1] - y[2];   // calculate the output
   y[2] = y[1];                          // prepare for the next ISR
6  y[1] = y[0];                          // prepare for the next ISR

8  CodecDataOut.Channel[ LEFT] = A*sinf(theta)*y[0];  // scale
   CodecDataOut.Channel[RIGHT] = CodecDataOut.Channel[LEFT];
10 /* algorithm ends here */
```

The four real-time steps involved in digital resonator-based signal generation

An explanation of Listing 5.10 follows.

1. (Line 2): This code calculates the digital frequency. This term is needed as an input argument for both a scale factor and a filter coefficient.

2. (Line 4): This code calculates the current output value by implementing the system's difference equation.

3. (Lines 5–6): These two lines of code shift the values in the y arrays one element to the right. The equivalent operation is,

$$y[1] \to y[2]$$
$$y[0] \to y[1].$$

4. (Lines 8–9): These two lines of code scale the filter's output to achieve an amplitude of A. The resulting value is then sent to both the Left and Right output channels.

Now that you understand the code...

Go ahead and copy all of the files into a separate directory. Open the project in CCS and "Rebuild All." Once the build is complete, "Load Program" into the DSK and click on "Run." Your 1 kHz sine generator is now running on the DSK.

5.5 Follow-On Challenges

Consider extending what you have learned.

1. The trigonometric function call-based technique discussed in this chapter uses single precision variables (floats). Design and implement a double precision variable (doubles) version of the real-time program. What are the advantages and disadvantages of this higher precision technique?

2. Create a DTMF signal generator that creates the tones associated with your phone number.

   (a) Consider starting the process by embedding the phone number in your ISR. You can always add more sophisticated coding techniques later.

   (b) Ensure that your system will work with any phone number.

   (c) Ensure that your system will handle parentheses, space, and dash characters.

   (d) Ensure that your system allows for user defined tone durations and tone spacing.

   (e) Ensure that your system has the ability to repeat dial.

3. The table lookup technique actually only needs to define 1/4 of the table since the sine and cosine functions are symmetric functions (only 1/4 of a 0 to 2π table is unique). Design and implement a real-time program that takes advantage of this symmetry.

Chapter 6

Frame-Based DSP

6.1 Theory

THE discussion up until this point in the book has usually assumed that the real-time processing is accomplished on a sample-by-sample basis. That is, an input sample $x(t)$ was converted to digital form $x[n]$ by the ADC part of the codec and transferred to the DSK's CPU for whatever processing was desired. This processed sample $y[n]$ was then transferred to the DAC part of the codec, converted back to analog form $y(t)$, and sent to an output device (e.g., a speaker). Processing the data in this way has two distinct advantages. First, this approach makes the DSP algorithms easier to understand and easier to program. Second, sample-by-sample processing also minimizes the system latency by acting on each sample as soon as it is available. However, sample-by-sample processing has serious drawbacks, and is in fact not commonly used for commercial code.

6.1.1 Drawbacks of Sample-Based DSP

One of the implications of real-time sample-based DSP is that all processing must be completed in the time between samples. For many complex DSP algorithms this becomes difficult if not impossible, especially for fast sampling rates. For example, if we are using one of the PCM-300x stereo codecs with a sample frequency of $F_s = 48$ kHz, we have only $T_s = 1/(48 \times 10^3) = 20.83$ μs to process $both$ the left and the right channel samples (assuming stereo operation), or 10.42 μs per sample. Given the 225 MHz clock frequency of the C6713 DSK (4.444 ns per clock cycle), this means we have approximately 2,345 clock cycles per sample.[1] Note that we must also include in this time period all the "overhead" such as that associated with codec transfers, memory access, instruction and data cache latency, and other unavoidable factors when we assess the available clock cycles. While this many clock cycles may seem to be plenty, a complex algorithm with many memory transfers could easily exceed this number. Of course if we are performing *non-real-time* DSP with previously stored data, then this limitation does not exist.

Another implication of real-time sample-based DSP, which seems fairly obvious, is that there is only one sample available for processing at any given time. Certain classes of DSP algorithms such as implementations that use the FFT (Fast Fourier Transform) require some contiguous range of samples to be available at any given time, which is clearly impossible with sample-based DSP. A second implication of real-time sample-based DSP is that the processor must respond to each interrupt from the devices that are data sources and sinks

---

[1] The C6711 DSK runs at a clock frequency of 150 MHz, or 6.667 ns per clock cycle.

(such as a codec) in order to perform the required data transfers. Doing this means that the current processing is interrupted, the state of the processor preserved, control is transferred to the appropriate interrupt service routine and it is executed, then the processor state must be restored and execution restarted at the interrupted point. During this process, additional inefficiencies also occur, such as pipeline flushes and cache misses. This overhead represents lost processing time, and can significantly reduce the overall performance of the DSP. To remove this burden from the processor, specialized hardware components referred to as direct memory access (DMA) controllers are normally included as peripheral elements on the DSP device itself. Once the DMA controller is programmed to respond to a device that is sourcing or sinking data, it will automatically perform the required transfers to or from a memory buffer without processor intervention. When a buffer has been filled or emptied, the DMA controller then interrupts the DSP. This frees up the DSP from the mundane task of repetitive data transfers, and allows its resources to be focused on the computationally-intense processing once a buffer of data is available. In order to make this process efficient, the buffers will typically be designed to contain hundreds or thousands of samples. For situations where one or both of these situations occur, we need an alternative method of processing signals, which we shall call frame-based DSP.

6.1.2   What Is a Frame?

A frame is the name we will use to describe a group of consecutive samples. Some other texts may use the term "block" or "packet" instead of "frame," but they mean essentially the same thing. In order to implement frame-based DSP, we must collect N samples, and at that point initiate the processing of the frame. See Figure 6.1 for a pictorial comparison of sample-based versus frame-based processing.

How many samples constitute a frame? While frame sizes are common where the number of samples is some power of two (i.e., $N = 2^n$), there is no particular number that *must* be

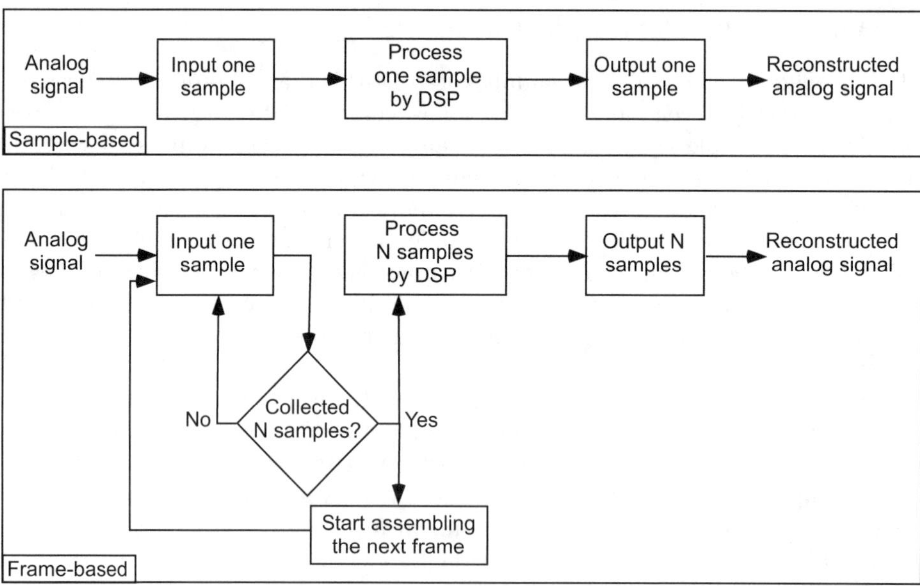

Figure 6.1: A comparison of a generic sample-based (top) versus frame-based (bottom) processing system. Note some systems may not require the input or output conversions from/to analog as shown here.

used. The frame size is selected based upon several factors such as the DSP algorithm to be used, the speed and efficiency of the ADC, the overhead required for memory transfers, other hardware limitations, and the performance of the DSP system. This last consideration is driven by the fact that whatever result is obtained by the DSP, a new "updated" result cannot be obtained any faster than it takes to sample an entire frame of data.

For example, suppose we are graphically displaying the spectrum of a signal based upon the FFT of that signal. The FFT requires a frame of data to be available at one time. If we assume a sample frequency of $F_s = 48$ kHz (thus $T_s = 1/(48 \times 10^3) = 20.83$ μs) and a frame size of 2048 samples, then the spectrum display cannot be updated any faster than 2048×20.83 μs $= 42.66$ ms, which equates to 23.44 display updates per second. In the United States, standard television video is updated 30 frames per second, and movies in theaters are updated 24 frames per second, so this might be satisfactory.[2] Note the implication in this example is that we now have 42.66 ms or over 9.6 *million* clock cycles of the C6713 to process the data! If we double the frame size, we get twice as much time to process the data, but we can only update the spectrum display half as fast, which may not be acceptable to the user. Thus as the frame size increases we get more time to process the data, but the response time of the system output gets slower. Frame size is one of many engineering design tradeoffs that need to be made. If the system output frames are sent to the DAC, they will be converted to an analog signal on a sample-by-sample basis, at the given sample frequency F_s. For proper operation, the next frame for output must be available by the time the last sample of the current frame is converted. If, for example, we were performing audio processing, this ensures that from the listener's perspective there would be no "gap" in the music and the output would sound no different than if sample-based processing were being used. Of course there is a time lag, or latency, equal to the frame period but it is imperceptible to the listener.

Most real-time DSP, such as a CD player, the telephone system (both land-line and wireless cellular), internet communications, and digital television (such as HDTV) implement a form of frame-based processing. For example, CD players use a data frame that is made up of six sample periods (six left channel samples, six right channel samples, alternating) [52]. Each sample is two bytes (16-bits), so the initial frame size is 24 bytes in length.[3]

6.2 winDSK6 Demonstration

Most of the functions available in winDSK6 are implemented as sample-based programs to keep the code simple. However, one exception is the Oscilloscope function, which must transfer information to be displayed in real-time on the video screen of the PC via the I/O port interface (e.g., the USB port). Since there is significant timing overhead in both I/O port transfers and video screen rendering, frame-based processing is used. In fact, the overhead incurred just in writing to the video screen would make sample-by-sample video transfers impractical. The actual frame size used in this part of winDSK6 is 512 samples per channel.

If you haven't yet tried the Oscilloscope function of winDSK6, try it now. The primary user interface window for this function is shown in Figure 6.2. Note that this function can provide both time domain (i.e., waveform) or frequency domain (i.e., spectrum) displays. In some DSP texts "time domain" is called "sample domain." A time domain example is shown in Figure 6.3. When the spectrum display is selected, frequency domain values are

---

[2] Actually, standard television uses two interlaced fields per frame and movie projectors use a light source chopper to increase the apparent update rate so the human visual system will not perceive a flickering image.

[3] Various DSP steps including error correction and modulation expand this to 73.5 bytes (588 bits) per frame that is actually stored on the CD.

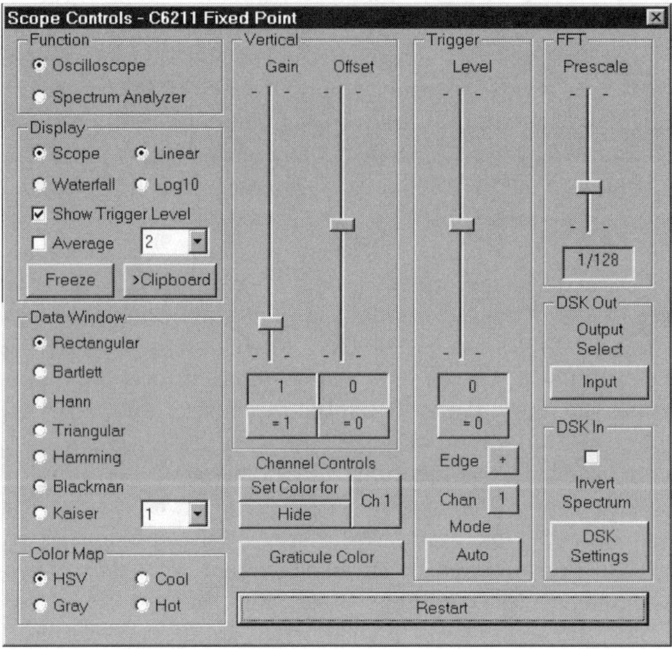

Figure 6.2: The primary user interface window for the Oscilloscope function of winDSK6.

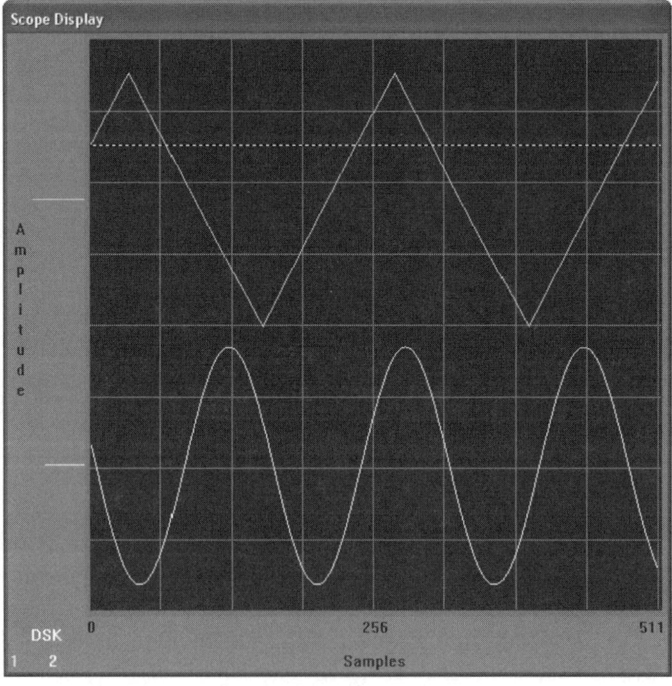

Figure 6.3: An example of a two-channel time domain display using the Oscilloscope function of winDSK6.

calculated by performing the FFT on each 512 sample block of data. Using the **Log10** units option often provides better results for the spectrum display. Try the "waterfall" option as well, which adds a moving time-axis to the display. A waterfall spectral display is often called a *spectrogram*, which may seem familiar to you if you have tried the non-real-time **spectrogram** function (or the related **specgramdemo**) in MATLAB. In winDSK6 the most recent data is shown at the top of the waterfall display.[4]

6.3 MATLAB Implementation

Frame-based processing in MATLAB was demonstrated previously in Section 2.4.2, where frames of 500 samples each were transferred from the DSK to the PC and manipulated using MATLAB. Another MATLAB-related application where frame-based processing is typically used is SIMULINK, which was demonstrated in Section 2.4.1. In fact, for those MATLAB Toolboxes that generate code for a C6x target DSP from SIMULINK models, we have found that the actual code generated uses a double buffering scheme for both input and output streams, or four buffers total. The triple-buffered frame-based approach we explain below is actually a more efficient technique as it only requires three buffers rather than four.

6.4 DSK Implementation in C

Important: As mentioned in Chapter 1, some of the code listings in this book (and particularly in this section) include lines that, despite our best efforts, are too long to fit within the book's margins and still allow us to use meaningful variable and function names. So we remind you that in all program listings where a line wrap occurred in the listing due only to page margins, the characters "[+]" show up to identify the beginning of the wrapped part of the line. The line numbers shown at the left edge of the listings do not increment for the wrapped part of a line.

Note that for a real-time process, the collection of samples by the ADC never stops as long as the system is in operation. When N samples have been collected they are transferred for processing by the DSP and the collection of a new frame begins without interruption. As shown in Figure 6.1, frame-based processing requires some means to determine when we have acquired N samples. A common technique is to incorporate a counter and a flag in the input ISR.

6.4.1 Triple Buffering

For real-time frame-based processing, we need at least three memory buffers at any given time: one for filling a frame with new samples, one for processing a frame by the DSP, and one for sending a frame of processed samples to the output. Let's call them buffer A, buffer B, and buffer C, then follow a single frame through the system to illustrate the basic process. A brute force method would fill the input frame using buffer A to store samples from the ADC, then copy the contents of buffer A to buffer B for processing (freeing up buffer A for the next frame), then after processing the samples copy the contents of buffer B to buffer C, whereupon the contents of buffer C are sent to the DAC and so on... but this technique is *extremely* inefficient. All the memory transfers involved incur considerable overhead.

The most efficient way to implement frame-based processing in real-time is to use a technique known as "triple buffering." There is a related method called "ping pong buffers,"

---

[4]This is similar to the readout form of sonar systems on U.S. nuclear submarines.

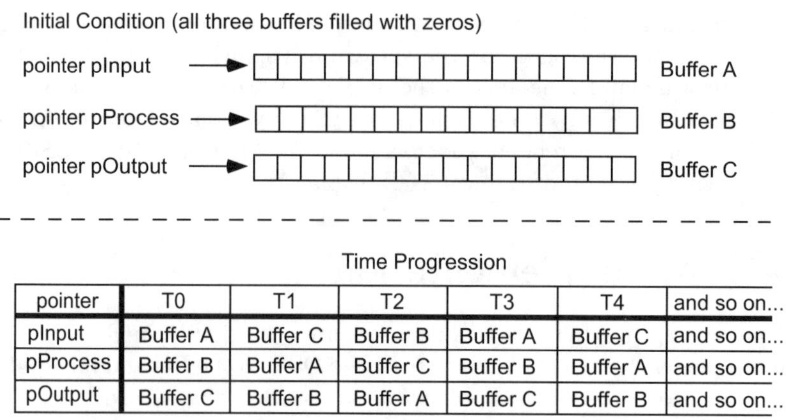

Time Progression

pointer	T0	T1	T2	T3	T4	and so on...
pInput	Buffer A	Buffer C	Buffer B	Buffer A	Buffer C	and so on...
pProcess	Buffer B	Buffer A	Buffer C	Buffer B	Buffer A	and so on...
pOutput	Buffer C	Buffer B	Buffer A	Buffer C	Buffer B	and so on...

Notes: 1. Each time block is the amount of time needed to fill one frame with samples.
2. Time T0: Buffer A is filling, Buffers B and C are still filled with zeros.
3. Time T1: Buffer C is filling, Buffer A is being processed, Buffer B is all zeros.
4. Time T2: the first actual output appears when Buffer A is sent to the DAC.
5. The same pattern repeats as shown above for as long as the program runs.

Figure 6.4: A pictorial representation of triple buffering.

which uses four buffers (two for input, two for output), but we focus here on the more efficient triple buffering method. In this technique, no copying of buffer contents is needed. All we do is define three pointers that will be used for the addresses of the input, processing, and output buffer memory locations. When the input buffer fills up, we just change the *pointers* instead of physically copying any of the buffer contents. The best way to visualize the process is with a picture; Figure 6.4 shows the typical sequence in which the pointers are updated. We could think of this as a variation on circular buffering (first discussed in Chapter 3), which rotates through *frames* instead of samples.

The simplest way to implement triple buffering is to use an input ISR and an output ISR, just as we programmed sample-based processing. The input ISR brings samples in from the ADC and fills the input buffer, the output ISR sends samples from the output buffer to the DAC, and the main program runs whatever algorithm is needed on the processing buffer when it isn't being interrupted by an ISR. One of the two ISRs (typically the input ISR) must be responsible for keeping track of when a full frame has been gathered, and set a flag so the main program can update the pointer assignments for the buffers. This ISR is also typically used to check if the buffer processing is "finished" in time, or else erroneous output may occur without the user's knowledge.

A minor variation on this scheme would be to have a single ISR that performs both input and output of the data and tracks the status of the frame filling and processing as described above. This may be slightly more efficient, but less general in its approach. To keep our code easy to understand, we separate the input and output functions into separate ISRs.

6.4.2   A Frame-Based DSP Example

We will keep the DSP algorithm very simple so we don't obscure the main point: to introduce you to frame-based processing. A more realistic example will be shown in a later chapter. Since the primary purpose of this section is to show you how to input, process, and output data in "blocks" of samples called frames, the actual processing we do doesn't matter as long as it doesn't exceed the real-time schedule. Suppose we simply make the left channel output be the sum $(L + R)$ of the left and right channel inputs and make the right channel output be the difference $(L - R)$ of the left and right channel inputs. An example of this simple program implemented in frame-based code is given in the **ccs\Frame** directory of Chapter 6. Be sure to inspect the full code listings in this directory to understand the complete working of the program.[5] Here we simply point out some highlights. The main program (**main.c**) is very basic, as shown below.

Listing 6.1: Main program for simple frame-based processing using ISRs.

```
#include "..\Common_Code\DSK_Config.h"
#include "frames.h"

int main() {
    // initialize all buffers to 0
    ZeroBuffers();

    // initialize DSK for selected codec
    DSK_Init(CodecType, TimerDivider);

    // main loop here, process buffer when ready
    while(1) {
        if(IsBufferReady()) // process buffers in background
            ProcessBuffer();
    }
}
```

While this is similar in many respects to the main program used in previous examples that implemented sample-based processing, there are a few significant differences. For example, in line 6 we ensure that the contents of all three buffers are set to zero before the program proceeds.[6] In line 12 we enter a continuous **while** loop similar to what we did for sample-based processing, only in this instance the **while** loop isn't empty: we first test to see if the buffer is full in line 13, and when it is full we process the samples in the full buffer in line 14.

The real "meat" of the program is in the file **ISRs.c**, which contains the interrupt service routines. The first part of this file contains various declarations as shown below.

Listing 6.2: Declarations from the "ISRs.c" file.

```
// Data is received from codec as 2 16-bit words (left/right)
// packed into one 32-bit word.  The union allows the data to be
// accessed as a single word for transfers or as separate
// left and right channel data.
#define LEFT  0
#define RIGHT 1
volatile union {
    unsigned int UINT;
    short Channel[2];
    } CodecData;

// frame buffer declarations
```

---

[5]Be sure to look also in the **DSK_Support.c** file in the **common_code** directory for initialization functions shared by multiple programs.

[6]While many modern compilers set the contents of a newly created buffer or variable to zero, it may not be wise to rely on that feature.

```
   #define  BUFFER_LENGTH          96000 // buffer length in samples
14 #define  NUM_CHANNELS           2     // supports stereo audio
   #define  NUM_BUFFERS            3     // don't change
16 #define  INITIAL_FILL_INDEX     0     // start filling this buffer
   #define  INITIAL_DUMP_INDEX     2     // start dumping this buffer

18

   #pragma DATA_SECTION (buffer, "CEO"); // put buffers in SDRAM
20 volatile float buffer[NUM_BUFFERS][NUM_CHANNELS][BUFFER_LENGTH];
   // 3 buffers are in use: one being filled by McBSP_Rx_ISR,
22 // one being operated on, and one being emptied by McBSP_Tx_ISR
   // fill_index   --> buffer being filled by the ADC
24 // dump_index --> buffer being written to the DAC
   // ready_index --> buffer ready for processing
26 short buffer_ready = 0, over_run = 0, ready_index = 2;
```

Note that lines 7 to 10 continue our previous technique of efficiently bringing in both the left
and right 16-bit samples as a single 32-bit integer, but still allowing separate manipulation
of the left and right channels by declaring `CodecData` as a union. Lines 13 and 14 specify
the dimensions of the buffers to be used for frames: in this example, each frame will be
192,000 samples long, consisting of 96,000 samples from the left channel and 96,000 samples
from the right channel; line 15 specifies that there will be three identical buffers of this size.
Lines 16 and 17 specify which of the three buffers will start out as the input buffer and
which will start out as the output buffer. Line 20 is the actual declaration that allocates
memory space for all three buffers, and the compiler pragma on line 19 ensures that the
external SDRAM memory space is used for these buffers.[7] Line 26 establishes variables that
will indicate when a bufffer is full, whether or not there has been a buffer over run (i.e., the
DSP operation on the current processing buffer didn't complete before the current input
buffer needed to become the new processing buffer), and an index value used to determine
the next buffer to start filling with input samples.

The interrupt service routine which transfers input samples from the ADC to the proper
input buffer is called `McBSP_Rx_ISR`, and is shown below.

Listing 6.3: The input interrupt service routine from the "ISRs.c" file.

```
   McBSP_Rx_ISR() {
2      static short fill_index = INITIAL_FILL_INDEX; // index of
           [+]buffer to fill
       static int sample_count = 0; // current sample count in
           [+]buffer
4      volatile McBSP *port;

6      if(CodecType == TLC320AD535)
           port = McBSP0_Base;        // McBSP0 used with TLC320AD535
8      else
           port = McBSP1_Base;        // McBSP1 used with codec
               [+]daughtercards
10     CodecData.UINT = port->drr;    // get input data from serial
           [+]port

12     // store in buffer
```

---

[7]Unless the buffers are rather small, they typically won't fit in the RAM space on the DSP chip itself.
We'll see this again later.

```
       buffer[fill_index][LEFT][sample_count] = CodecData.Channel[
           [+]LEFT];
14     buffer[fill_index][RIGHT][sample_count] = CodecData.Channel[
           [+]RIGHT];

16     /* update sample count and swap buffers when filled */
       if(++sample_count >= BUFFER_LENGTH) {
18         sample_count = 0;
           ready_index = fill_index;
20         if(++fill_index >= NUM_BUFFERS)
               fill_index = 0;
22         if(buffer_ready == 1) // buffer isn't processed in time
               over_run = 1;
24         buffer_ready = 1;
       }
26 }
```

The variable `fill_index` in line 2 is used to select which buffer is the current input buffer,
and the variable `sample_count` in line 3 is used as a counter to determine when the buffer
is filled. As we saw in previous chapters, declaring these variables as `static` allows them
to keep their values between calls of the ISR. Keep in mind that this ISR gets called
many times (96,000 times in this implementation) before the input buffer is filled, and
the processing routine `ProcessBuffer` has all this time (minus the brief times devoted to
any ISRs) to do its work. The code from lines 6 to 10 is essentially the same as used in
previous ISR code to set up the proper input port and codec type, and to bring the new
sample in from the ADC. Lines 13 and 14 transfer the new sample into the appropriate
location in the current input buffer. Lines 17 to 25 contain the logic that changes to the
next buffer when the current input buffer is full and determines if the processing buffer is
ready to be switched as well. Note that if the processing buffer is *not* ready (indicated by
`buffer_ready`[8] being equal to 1 in line 22) the program will still switch the buffers, but
the `over_run` flag indicates that this error condition has occurred. A buffer overrun means
you have not met the real-time schedule; this means the `ProcessBuffer` function must be
made faster somehow.

Now that we've seen how the input buffer is filled, just what does the `ProcessBuffer`
function look like? As stated earlier, we show a very simple example of forming the sum
and difference of the left and right channels just to illustrate how the buffer values are
manipulated.

Listing 6.4: Abbreviated version of `ProcessBuffer` from the "ISRs.c" file.

```
ProcessBuffer() {
2      int i;
       float *pL = buffer[ready_index][LEFT];
4      float *pR = buffer[ready_index][RIGHT];
       float temp;

6
/* addition and subtraction */
8      for(i=0;i < BUFFER_LENGTH;i++){
           temp = *pL;
10         *pL = temp + *pR;  // left = L+R
```

---

[8]The `buffer_ready` flag is set to 0 at the end of the `ProcessBuffer` function, indicating that processing
is completed.

```
            *pR = temp - *pR; // right = L-R
12          pL++;
            pR++;
14      }

16  buffer_ready = 0;
}
```

Two things in particular should be noticed about the `ProcessBuffer` function. First, all
the values in the buffer (96,0000 samples of the left channel and 96,000 samples of the right
channel) are processed before this function is completed (although it is interrupted many
times by ISRs). Second, when the processing is completed, the variable `buffer_ready` is
set to 0, so that the rest of the program will know that this is the case.

How do the processed samples get sent to the output? When the buffers change, the
current processing buffer becomes the new output buffer, which is sent to the DAC by the
ISR `McBSP_Tx_ISR` shown below.

Listing 6.5: The output interrupt service routine from the "ISRs.c" file.

```
1  interrupt void McBSP_Tx_ISR()
   {
3      static short dump_index = INITIAL_DUMP_INDEX; // index of
           [+]buffer to dump
       static int sample_count = 0; // current sample count in
           [+]buffer
5      volatile McBSP *port;

7      // bound output data before packing
       // use saturation of SPINT to limit to 16-bits
9      CodecData.Channel[ LEFT] = _spint(buffer[dump_index][LEFT][
           [+]sample_count] * 65536) >> 16;
       CodecData.Channel[RIGHT] = _spint(buffer[dump_index][RIGHT][
           [+]sample_count] * 65536) >> 16;
11
       // pack output data without bounding
13 //  CodecData.channel[ LEFT]=buffer[dump_index][LEFT][
       [+]sample_count];
   //  CodecData.channel[RIGHT]=buffer[dump_index][RIGHT][
       [+]sample_count];
15
       // update sample count and swap buffers when filled
17     if(++sample_count >= BUFFER_LENGTH) {
           sample_count = 0;
19         if(++dump_index >= NUM_BUFFERS)
               dump_index = 0;
21     }
         // now, send the data to the codec
23     if(CodecType == TLC320AD535) {
           port = McBSP0_Base;            // McBSP0 used with
               [+]TLC320AD535
25         CodecData.UINT &= 0xfffffffe;// mask off LSB to prevent
               [+]codec reprogramming
           }
```

```
27      else {
            port = McBSP1_Base;        // McBSP1 used with codec
                [+]daughtercards
29      }

31      port->dxr = CodecData.UINT; // send output data to serial
            [+]port
}
```

In lines 9 and 10, the processed sample is bounded to the allowable range of a signed 16-bit number by the compiler intrinsic function _spint, which is faster than a double comparison with the largest positive value and the smallest negative value as would otherwise have to be performed. Lines 9 and 10 also transfer the processed sample from the buffer to the appropriate `CodecData` variable. If for some reason you didn't care to bound the processed sample, you could comment out lines 9 and 10 and uncomment lines 13 and 14. Lines 17 to 21 contain the logic that changes to the next buffer when the current output buffer is empty. Note that the static variable `sample_count` used here is local to this function, and is different from the static variable `sample_count` used in the input ISR routine shown in Listing 6.3. The code from lines 23 to 31 is essentially the same as used in previous ISR code to set up the proper output port and codec type, and to send the processed sample out to the DAC. Thus the input and output of samples are coordinated by the associated ISRs, and the actual DSP algorithm is implemented within the `ProcessBuffer` function.

Note that each time an ISR interrupts the CPU, some time is "lost" that might have been used for the main algorithm. If we can reduce the number and duration of these interruptions, we can gain even more programming efficiency and gain time for our main algorithm. One very elegant method of achieving this goal, briefly mentioned earlier in this chapter, is to take advantage of something called Direct Memory Access.

6.4.3 Using Direct Memory Access

Direct Memory Access (DMA) is a mechanism that transfers data from one memory location to another without any intervention or work required by the CPU. In essence, the DMA hardware contains a controller unit that can perform memory transfer operations, either one location at a time or in blocks, independent of the CPU. Once the DMA hardware has been configured with the initial source and destination memory locations, and the number of transfers to perform, it can run on its own without any need to "bother" the CPU. Since transferring data from a memory buffer to the codec (or vice versa) is essentially a memory transfer, we can cut the CPU interruptions dramatically by delegating these tasks to the DMA hardware instead of using the CPU to perform such memory transfers.

In order to implement triple buffering with DMA, our program must first initialize and configure the DMA hardware. The C6x DSK documentation calls this hardware "EDMA," where the "E" stands for *enhanced* because it has more capabilities than typical DMA hardware. For a complete description of the C6x EDMA, see the *TMS320C6000 Peripherals Reference Guide* [53]. For our purposes here, we shall only use a subset of the EDMA capabilities.

One consideration we can't ignore when using DMA is the need to keep the input and output synchronized. While we gain the advantage of not needing the CPU to perform memory transfers, we must realize that the CPU will thus be unaware of when and how fast these transfers are taking place unless we include some type of code to ensure synchronization. Without such code, the three buffers may get "out of step" with each other, leading to unpredictable behavior. Luckily, the EDMA hardware has the ability to configure and monitor "events" that behave in many ways as interrupts do with the CPU. This gives us a flexible method of keeping the three buffers synchronized, and enables us to swap buffer pointers as needed to implement the triple buffering scheme, all with minimal interruption of the CPU. Thus the CPU's time can be devoted almost entirely to whatever DSP algorithm is being performed. This technique represents one of the most efficient ways possible to implement a real-time DSP program.

In the following example program, we have modified our previous example in only one significant aspect: we now get the input from the ADC side of the codec and send the output to the DAC side of the codec using the EDMA hardware instead of having the CPU do it inside of ISR functions. This EDMA program is given in the `ccs\Frame_EDMA` subdirectory for Chapter 6. Be sure to inspect the full code listings in this directory to understand the complete working of the program. Here we simply point out some highlights. The main program (`main.c`) is once again very basic, as shown below, with only a few minor differences from Listing 6.1 (see lines 10 and 13).

Listing 6.6: Main program for frame-based processing using EDMA.

```
#include "..\Common_Code\DSK_Config.h"
#include "frames.h"

int main()
{
    // initialize all buffers to 0
    ZeroBuffers();

    // initialize EDMA controller
    EDMA_Init();

    // initialize DSK for selected codec using EDMA
    DSK_Init_EDMA(CodecType, TimerDivider);

    // main loop here, process buffer when ready
    while(1) {
        if(IsBufferReady()) // process buffers in background
            ProcessBuffer();
    }
}
```

The functions `ZeroBuffers` and `IsBufferReady` are unchanged from the non-EDMA versions presented earlier. However, the initialization routines required, the single ISR routine (triggered by an EDMA "event"), and the `ProcessBuffer` function are all different from what we have presented before, so we need to examine these. Because we're changing the way the interrupts are assigned, we'll also use a different file that assigns the interrupt vectors; that is, instead of `vectors.asm` we'll use `vectors_EDMA.asm`. We'll also define a different buffer size just for variety. Note that, as with our previous Code Composer Studio projects, the EDMA version of the DSK initialization functions can be found in the file `DSK_Support.c`, which is in the `common_code` directory. This is also where the interrupt vector files are located. Other files for the EDMA version are located in the `ccs\Frame_EDMA` subdirectory for Chapter 6.

We begin by listing the declarations in the EDMA version of the file `ISRs.c`, as they differ somewhat from the non-EDMA version.

Listing 6.7: Declarations from EDMA version of the "ISRs.c" file.

```c
#include "..\Common_Code\DSK_Config.h"
#include "math.h"
#include"frames.h"

// frame buffer declarations
#define BUFFER_COUNT    1024    // buffer length in samples (L+R)
#define BUFFER_LENGTH   BUFFER_COUNT*2 // two shorts each time
#define NUM_BUFFERS     3  // don't change this!

#pragma DATA_SECTION (buffer , "CE0"); // put buffers in SDRAM
short buffer[NUM_BUFFERS][BUFFER_LENGTH];
// 3 buffers used at all times, one being filled from the McBSP,
// one being operated on, and one being emptied to the McBSP
// ready_index --> buffer ready for processing
volatile short buffer_ready = 0, over_run = 0, ready_index = 0;
```

Note that in this EDMA version of the declarations the data buffers in line 11 are declared as type `short`, which for the DSK is a 16-bit integer. In all our past code we have used buffers of type `float` to take advantage of the ease and flexibility that floating point numbers provide. We will continue to do so here, and the values in these buffers will be converted to type `float` before processing. Why the extra conversion step? Recall that the samples coming in from the ADC (or going out to the DAC) of the codec are integers. In non-EDMA code, the transfer of values to or from the codec and the buffer is performed by the CPU, which takes care of the necessary conversion between floating point and integer data types. However, in the EDMA code the transfer to or from the codec and the buffer is performed without any participation of the CPU, so in effect this is a "brainless" transfer that cannot do any data type conversions. Such conversions require the CPU, so we move the necessary conversion from integer to floating point (and back to integer again) into the `ProcessBuffer` function as will be shown subsequently.

The frame size specified in line 6 is 1024 samples per frame (1024 left channel samples and 1024 right channel samples). Previously, we efficiently moved both the left and right channel samples (each 16-bits) in a single 32-bit transfer operation, but by declaring a **union** we were able to individually manipulate the two channels. We accomplish a similar feat here by observing that the data type **integer** (32-bits long for the DSK) is twice the length of a **short**, so you will soon see that the EDMA hardware is initialized to transfer integers.

We now discuss `EDMA_ISR`, which is the function that actually implements the triple buffering scheme (i.e., enabling the pointer addresses to change as necessary). In the non-EDMA program, this task was performed by the function `MCBSP_Rx_ISR` in Listing 6.3. `EDMA_ISR` is contained in the `ISRs.c` file located in the `ccs\Frame_EDMA` subdirectory for Chapter 6 and is shown below. This code is simpler and faster than the code in Listing 6.3.

Listing 6.8: Function for implementing triple buffering using the EDMA hardware.

```c
interrupt void EDMA_ISR() {
    *(unsigned volatile int *)CIPR=0xf000; // clear McBSP events
    if(++ready_index >= NUM_BUFFERS) // update buffer index
        ready_index = 0;
    if(buffer_ready == 1) // if buffer isn't processed in time
        over_run = 1;
    buffer_ready = 1; // mark buffer as ready for processing
}
```

Note that while in the non-EDMA version the ISR was triggered by an interrupt that was generated for every sample, in the EDMA version the ISR-like "event" only occurs when the entire frame of samples has been transferred. This event triggers an interrupt which causes interrupt service routine EDMA_ISR to run. Thus in this example EDMA_ISR is called 1024 times *less often* than MCBSP_Rx_ISR would be in the non-EDMA program for a frame size of 1024. But the use of the EDMA event allow us to keep all the buffers synchronized just as well.

While we still use an interrupt service routine for the EDMA version shown here, EDMA_ISR is a very short and fast routine that minimizes the interruption of the CPU. It is important to reiterate here that in order to map the vector for the proper interrupt (INT8, triggered by an EDMA event) to the proper address of the EDMA_ISR function, the Code Composer Studio project for this program must include the file vectors_EDMA.asm (provided for you in the common_code directory) *not* the file vectors.asm as was the case in our previous programs that used CPU ISRs.

Because the DSK initialization is slightly different for this EDMA program compared to previous programs, we show the initialization function below.

Listing 6.9: Function for initializing the DSK when using EDMA.

```c
void DSK_Init_EDMA(int Codec, unsigned int Divider)
{
    CSR=0x100;                          // disable all interrupts
    IER=0;


    if(Codec == DSK6713_16bit) {
        Init_AIC23();          // initialize codec using McBSP0
    }
    else {
        if(Codec == TLC320AD535) {    // using onboard codec
            SampleFreq = 8000.0;
            Init_McBSP(McBSP0_Base, AD535_mode);
        }
        else {                          // using codec daughtercard
            SampleFreq = SetTimer1Period(Divider);
            Init_McBSP(McBSP1_Base, (Codec == TI_PCM3003_16bit)?
                [+]PCM0_Slave_Inverted_mode:PCM0_Master_mode);
        }
    }


    IER |= 0x0102;                      // enable EDMA interrupt (INT8)
    ICR = 0xffff;                       // clear all pending interrupts
    CSR |= 1;                           // set GIE
}
```

The only difference in this version compared to the non-EDMA version used before is in line 20, where we enable interrupt 8 for the EDMA hardware (rather than enabling interrupts 11 and 12 that we used before for the McBSP interrupts). You may want to verify for yourself that vectors_EDMA.asm maps interrupt 8 to the EDMA_ISR function.

The next function we need to discuss, EDMA_Init, is quite a bit more involved. It is contained in the ISRs.c file located in the ccs\Frame_EDMA subdirectory for Chapter 6. and is shown next.

Listing 6.10: Function for initializing the EDMA hardware.

```c
void EDMA_Init() {
    EDMA_params* param;

    // McBSP tx event params
    param = (EDMA_params*)((CodecType == TLC320AD535)?
        EVENTC_PARAMS:EVENTE_PARAMS);
    param->options = 0x211E0002;
    param->source = (unsigned int)(&buffer[2][0]);
    param->count = (0 << 16) + (BUFFER_COUNT);
    param->dest = (CodecType == TLC320AD535)?0x30000000:0
        x34000000;
    param->reload_link = (BUFFER_COUNT << 16) + (EVENTN_PARAMS &
        0xFFFF);

    // set up first tx link param
    param = (EDMA_params*)EVENTN_PARAMS;
    param->options = 0x211E0002;
    param->source = (unsigned int)(&buffer[0][0]);
    param->count = (0 << 16) + (BUFFER_COUNT);
    param->dest = (CodecType == TLC320AD535)?0x30000000:0
        x34000000;
    param->reload_link = (BUFFER_COUNT << 16) + (EVENTO_PARAMS &
        0xFFFF);

    // set up second tx link param
    param = (EDMA_params*)EVENTO_PARAMS;
    param->options = 0x211E0002;
    param->source = (unsigned int)(&buffer[1][0]);
    param->count = (0 << 16) + (BUFFER_COUNT);
    param->dest = (CodecType == TLC320AD535)?0x30000000:0
        x34000000;
    param->reload_link = (BUFFER_COUNT << 16) + (EVENTP_PARAMS &
        0xFFFF);

    // set up third tx link param
    param = (EDMA_params*)EVENTP_PARAMS;
    param->options = 0x211E0002;
    param->source = (unsigned int)(&buffer[2][0]);
    param->count = (0 << 16) + (BUFFER_COUNT);
    param->dest = (CodecType == TLC320AD535)?0x30000000:0
        x34000000;
    param->reload_link = (BUFFER_COUNT << 16) + (EVENTN_PARAMS &
        0xFFFF);


    // McBSP rx event params
    param = (EDMA_params*)((CodecType == TLC320AD535)?
        EVENTD_PARAMS:EVENTF_PARAMS);
    param->options = 0x203F0002;
    param->source = (CodecType == TLC320AD535)?0x30000000:0
```

```
           [+]x34000000;
41    param->count = (0 << 16) + (BUFFER_COUNT);
      param->dest = (unsigned int)(&buffer[1][0]);
43    param->reload_link = (BUFFER_COUNT << 16) + (EVENTQ_PARAMS &
           [+]0xFFFF);

45    // set up first rx link param
      param = (EDMA_params*)EVENTQ_PARAMS;
47    param->options = 0x203F0002;
      param->source = (CodecType == TLC320AD535)?0x30000000:0
           [+]x34000000;
49    param->count = (0 << 16) + (BUFFER_COUNT);
      param->dest = (unsigned int)(&buffer[2][0]);
51    param->reload_link = (BUFFER_COUNT << 16) + (EVENTR_PARAMS &
           [+]0xFFFF);

53    // set up second rx link param
      param = (EDMA_params*)EVENTR_PARAMS;
55    param->options = 0x203F0002;
      param->source = (CodecType == TLC320AD535)?0x30000000:0
           [+]x34000000;
57    param->count = (0 << 16) + (BUFFER_COUNT);
      param->dest = (unsigned int)(&buffer[0][0]);
59    param->reload_link = (BUFFER_COUNT << 16) + (EVENTS_PARAMS &
           [+]0xFFFF);

61    // set up third rx link param
      param = (EDMA_params*)EVENTS_PARAMS;
63    param->options = 0x203F0002;
      param->source = (CodecType == TLC320AD535)?0x30000000:0
           [+]x34000000;
65    param->count = (0 << 16) + (BUFFER_COUNT);
      param->dest = (unsigned int)(&buffer[1][0]);
67    param->reload_link = (BUFFER_COUNT << 16) + (EVENTQ_PARAMS &
           [+]0xFFFF);

69    *(unsigned volatile int *)ECR = 0xf000; // clear all McBSP
           [+]events
      *(unsigned volatile int *)EER = (CodecType == TLC320AD535)?0
           [+]x3000:0xC000;
71    *(unsigned volatile int *)CIER = 0x8000; // interrupt on rx
           [+]reload only
}
```

To fully understand what this function does, you should read the TI documentation that
describes how to use EDMA, *TMS320C6000 Peripherals Reference Guide* [53], and look at
the header file `c6x11dsk.h` in `common_code` that contains the various defines for the DSK.
If you're a bit rusty regarding structures and pointers in C, now might be a good time to
refresh yourself! The function `Init_EDMA` runs only once, and is where we will specify things
like the source address, the destination address, and the number of elements for each DMA
transfer we need to define, and set up links so that a triple buffering scheme is implemented.

In the interests of brevity, only a synopsis of the transmit (i.e., output) function is given here. The receive (i.e., input) function operates similarly.

Keep in mind we are still dealing with three buffers as before. The function `Init_EDMA` sets up two EDMA channels, one to service the McBSP transmitter and one for the McBSP receiver. Each channel of the EDMA has an "event" dedicated to it (as mentioned previously, this relationship is quite similar to that between the CPU and the interrupt lines). Each EDMA transfer is controlled by values set in a parameter RAM block; setting up these RAM blocks is the primary purpose of `Init_EDMA`. EDMA channel 12 is dedicated to McBSP0 transmit, and channel 14 is dedicated to McBSP1 transmit, so the first block of code (lines 5 to 10) selects either event 12 (called `EVENTC`) or event 14 (called `EVENTE`) based on the codec being used. The parameter values given configure the event for the number of elements (which themselves are each 32-bit integers) specified by `BUFFER_COUNT` (defined in `ISRs.c`), and transfers them to buffer[2],[9] the initial buffer designated for output. The code for EVENTC or EVENTE (depending upon the codec in use) is such that when it finishes that transfer, it automatically reconfigures the channel[10] with the information stored at `EVENTN_PARAMS` in the parameter RAM (this is accomplished by the `reload_link` field in line 10). `EVENTN` is set up in the next code block starting at line 13, which configures the EDMA to use buffer[0] and reconfigures the channel with the information stored at `EVENTO_PARAMS` when it finishes. `EVENTO_PARAMS` (starting on line 21) will then cause the EDMA transfer to use buffer[1], and reconfigures the channel with the information stored at `EVENTP_PARAMS` when it finishes. `EVENTP_PARAMS` (starting on line 29) will then cause the EDMA transfer to use buffer[2], and reconfigures the channel with the information stored at `EVENTN_PARAMS` when it finishes, which is effectively a loop back to the original buffer; this cycle of using the three buffers in sequence continues for as long as the program runs. This is illustrates an important point: for zero CPU overhead, we can get automatic n-way (in this case 3-way) buffering. The receive channel (see lines 37 to 67) operates in an identical fashion, but it begins with a different buffer (buffer[1]) so we can keep the input and output transfers working on different buffers.

The last few lines of `Init_EDMA` cannot be ignored. Line 69 clears all McBSP events (although we are only using two events—transmit and receive—we might as well clear them all). Line 70 sets the appropriate value in the event enable register according to the type of codec being used. Finally, line 71 is a critical line in that it sets the channel interrupt enable register so that an EDMA "event" occurs when the particular EDMA channel serving the McBSP receive operation finishes transferring a frame, which means that an event is only triggered each time the input frame buffer is full.

Finally, we now show the function `ProcessBuffer` as modified for the EDMA version of the program. This is the function where the actual DSP algorithm is implemented. The first thing you will notice is that this version of `ProcessBuffer` is much longer than the non-EDMA version. This is due to the need to perform type conversions on the data in the buffers.

Listing 6.11: An abbreviated version of the `ProcessBuffer` function from the EDMA version of the "ISRs.c" file.

```
void ProcessBuffer()
{
    short *pBuf = buffer[ready_index];
    static float Left[BUFFER_COUNT], Right[BUFFER_COUNT];
    float *pL = Left, *pR = Right;
```

---

[9] Recall that `&buffer[2][0]` is the address of the first element of `buffer[2]`.

[10] The actual channel in use depends upon the codec being used. Channel 12 would be for the onboard codec and channel 14 would be for a daughtercard codec.

```c
    int i;
    float temp;

    *(volatile int *)IO_PORT = 0; // set STS0/1 low - for time
        measurement

    for(i = 0;i < BUFFER_COUNT;i++) { // extract data to float
        buffers
        *pR++ = *pBuf++;
        *pL++ = *pBuf++;
    }

    pL = Left; // reinitialize pointers
    pR = Right;

/* addition and subtraction */
    for(i=0;i < BUFFER_COUNT;i++){
        temp = *pL;
        *pL = temp + *pR; // left = L+R
        *pR = temp - *pR; // right = L-R
        pL++;
        pR++;
    }

    pBuf = buffer[ready_index];
    pL = Left;
    pR = Right;

//  for(i = 0;i < BUFFER_COUNT;i++) { // pack into buffer without
    bounding
//        *pBuf++ = *pR++;
//        *pBuf++ = *pL++;
//  }

    for(i = 0;i < BUFFER_COUNT;i++) { // pack into buffer after
        bounding
        *pBuf++ = _spint(*pR++ * 65536) >> 16;
        *pBuf++ = _spint(*pL++ * 65536) >> 16;
    }

    pBuf = buffer[ready_index];
    if(CodecType == TLC320AD535) { // mask off LSB to prevent
        codec reprogramming
        for(i = 0;i < BUFFER_COUNT;i++) {
            *(unsigned int *)pBuf &= 0xfffffffe;
            pBuf += 2;
        }
    }

    *(volatile int *)IO_PORT = -1; // set STS0/1 high - for time
        measurement
```

```
        buffer_ready = 0; // signal we are done
52 }
```

You can see that the actual DSP algorithm, which is the same simple addition and subtraction example used before, is implemented in lines 20 to 26.

In line 3, a pointer is declared that is set to the address of the frame buffer of shorts that was previously filled using EDMA and is now designated to be ready for processing. Line 4 declares two buffers of floats that will be used to contain the left and right frame data, after conversion to floating point. Line 5 declares pointers that will be used to manipulate individual samples (in floating point format) in the frames. Line 9 and line 50 are used only for testing and evaluation purposes during code development. The actual conversion from **short** to **float** occurs in lines 11 to 14; the C compiler ensures that when the CPU transfers the integer values pointed to by *pBuf into the floating point locations pointed to by *pR for the right channel and *pL for the left channel, that the appropriate conversion takes place. Note the transfer and conversion alternates first right channel then left channel, which is the order in which the data was stored in the EDMA buffer in the first place. We could also have used a "cast" from **short** to **float** here, but there is no advantage and we prefer this method. While it may seem inefficient to transfer from one buffer to another in this way, it is in fact very fast. The CPU savings realized by the EDMA transfers far outweigh the CPU time required for this buffer transfer.

Lines 16 and 17 set the appropriate pointers back to the first element of the arrays of floating point values for the left and right channels, and the actual DSP algorithm can then commence. Similarly, lines 28 to 30 reset the appropriate pointers after the DSP algorithm has completed. To convert the now "processed" floating point values back into integer values suitable for EDMA transfer to the codec, the user can select either lines 32 to 35 or lines 37 to 40, depending upon whether one wishes to use bounds checking (a good idea) via the intrinsic _spint function. Lines 43 to 48 exist to ensure compatibility with the onboard TLC320AD535 codec of the C6711 DSK: recall that a non-zero LSb (least significant bit) transmitted to this codec signals that the user wishes to reconfigure the codec, which is *not* what we want to happen when we are just sending output data. Thus line 45 ensures all the LSbs are set to zero if this codec is being used.

6.5   Summary of Frame-Based Processing

This chapter has introduced you to a new way of thinking about how to implement DSP programs. While the sample-by-sample processing of earlier chapters is typically easier to understand, there are many reasons to consider frame-based processing. First and foremost, the speed and efficiency of frame-based processing—especially when using EDMA transfers—cannot be matched using sample-by-sample processing. Secondly, certain DSP algorithms require that a contiguous block of samples be available at once, which is not possible with sample-by-sample processing.

The disadvantages of frame-based processing are somewhat greater code complexity, and a time latency equal to NT_s where N is the frame size and $T_s = 1/F_s$ is the time between samples. While the latency is unavoidable, code complexity should not deter you. Most users can adapt the example code given in this chapter to create a "skeleton" of a frame-based program where they only need to adjust the frame size and the operations performed in the **ProcessBuffer** function to produce their own customized program. Keep in mind that nearly every DSP text discusses theory on a sample-by-sample basis, but that most production real-time DSP code is written on a frame-by-frame basis. Having an understanding of frame-based processing is therefore highly recommended.

The next chapter will introduce frame-based DSP programs that are more complicated (and more useful) than the simple $L + R$ and $L - R$ used in this chapter.

6.6   Follow-On Challenges

Consider extending what you have learned.

1. Replace the addition and subtraction of the left and right channels in `ProcessBuffer` with some other operation of your choosing. Try this with both the interrupt-driven and the EDMA versions of the program.

2. Making no other changes to the non-EDMA version of the triple buffer program, increase the size of the buffer until the process can no longer "keep up" with the real-time schedule.

3. Making no other changes to the EDMA version of the triple buffer program, increase the size of the buffer until the process can no longer "keep up" with the real-time schedule. Is there a difference with what you found for the non-EDMA version?

4. If you are (or are willing to become) familiar with the profiler capability of Code Composer Studio, you can use this to measure the increase in program speed of the EDMA version compared to the non-EDMA version of the program.

5. Some algorithms implement an "overlap and add" process where some portion of the processed frame is combined with some portion of the prior frame. How would you implement such a process with your own code?

Chapter 7

Digital Filters Using Frames

7.1 Theory

AS discussed in Chapter 6, using frame-based processing greatly increases the efficiency of DSP programs. The CPU performing the signal processing algorithm is only interrupted at the end of a frame rather than at every sample. In this chapter, we will show how frames can be used for time-domain digital filtering similar to the filters we discussed in Chapter 3.

Recall that a time-domain implementation of a digital filter involves convolution, which is an iterative calculation of the filter's difference equation. The only change in this chapter is that we will be dealing with our samples a frame at a time rather than a sample at a time. For simplicity, we will restrict this discussion to FIR filters, but IIR filters can be implemented with frames in the same manner.

7.2 winDSK6 Demonstration

The winDSK6 program does not provide an equivalent function; all the filtering in winDSK6 is accomplished sample by sample to keep the code very simple.

7.3 MATLAB Implementation

Frame-based filtering in MATLAB can be accomplished using resources such as the Data Acquistion Toolbox or SIMULINK, both available from The MathWorks, or by using the MATLAB-to-DSK interface software included with this book. See Section 2.4.1 and Section 2.4.2 for examples, and Appendix E for more details.

7.4 DSK Implementation in C

To demonstrate frame-based digital filtering on the DSK, we provide a C program that implements a filter similar to the FIR filters shown in Chapter 3. Before discussing the C code, however, it's important to understand the process required to implement an FIR filter in a frame-based manner.

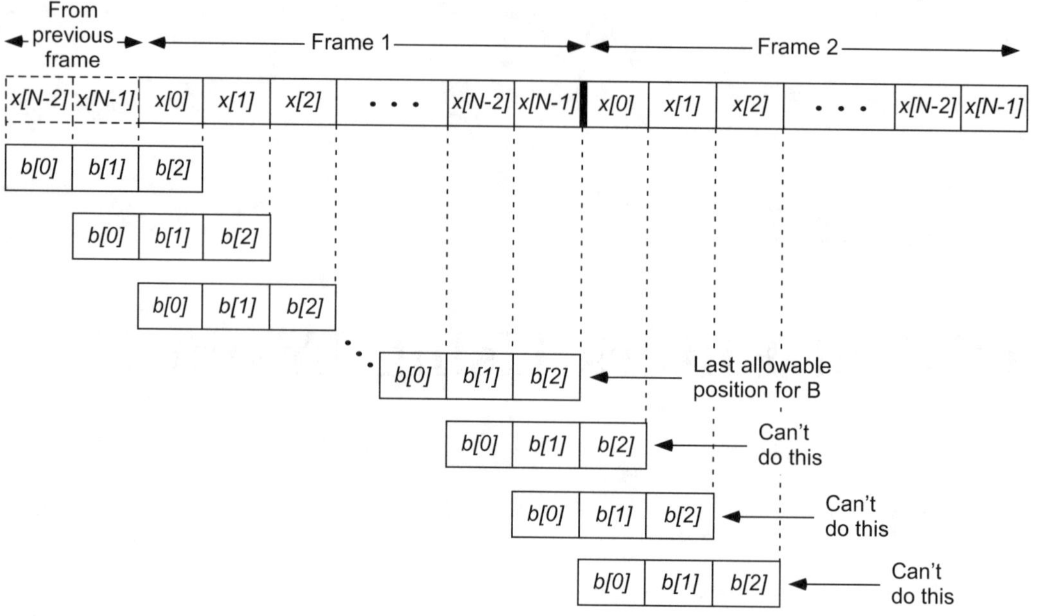

Figure 7.1: Implementing a second order FIR filter with a frame-based approach. Note that the b coefficients can't "slide" past the edge of Frame 1 without some programming "tricks."

7.4.1   Understanding the FIR Process for Frames

Using triple-buffered EDMA-transferred frames for I/O will make the program far more efficient and faster than the sample-based versions discussed in Chapter 3. However, the program will have available to it a fixed-size frame of samples at any given time, which complicates the filter convolution at the "edges" of the frame. Recall from Chapter 3 that the output $y[n]$ of a digital filter is calculated by performing the discrete-time convolution of the input $x[n]$ with the filter's impulse response $h[n]$. When written as the difference equation for an FIR filter, the numerator values $b[n]$ are equivalent to the impulse response $h[n]$. Making this substitution (b for h) and remembering that an FIR filter of order K has $K + 1$ coefficients, the convolution sum becomes the general form of the FIR difference equation, namely,

$$y[n] = \sum_{k=0}^{K} b[k]x[n - k] \quad \text{for} \quad n = 0, 1, 2, \ldots, N - 1$$

where N represents the length of $x[n]$ and $K + 1$ is the length of $b[n]$.[1] This equation tells us that to calculate the values for an entire frame's worth of filter output, we will need to "slide" the filter coefficients "across" the entire frame of filter input, multiplying and summing point-by-point as indicated by the equation above. This process is depicted in Figure 7.1, where it can be seen that there is a potential problem at the "edges" of the frame.

There is only a single frame of input data available at any given time. In Figure 7.1, the values in Frame 2 aren't available yet, and the values from the previous frame shown to

---

[1] There is nothing magic about the letters K, N, M, and so on. Sometimes N is used to represent the length of the input data as we do here, another time N might represent the order of a filter. The context of the discussion should make clear what the letters represent.

the left of Frame 1 would be "gone" by now unless we use some programming "tricks" to keep those values around. If we ignore the "edges" of the frame we would be ignoring the initial and final conditions of the filter for that frame of data, and we would not correctly implement the filter. For audio applications, the result of ignoring these "edge" problems would be heard as a distinctive "clicking" or "popping" noise in the output occurring at the frame rate.

7.4.2 How to Avoid the "Edge" Problems

How do we fix this problem? We create a buffer large enough to contain both the frame of current input data and also have room to hold the necessary edge values from the previous frame. In Figure 7.1 this would mean an array that includes both the values labeled "From previous frame" and the values labeled Frame 1. We "slide" the filter starting from the left most element (the first "From previous frame" element) until the right edge of the filter reaches the end of Frame 1, indicated in the figure by a thick, dark line. We can't go any further than this because the data for Frame 2 isn't available yet. Before the current frame (Frame 1) is transferred out and the next frame (Frame 2) comes in to overwrite it, we copy the right edge values of Frame 1 into the locations labeled "From previous frame." The next frame values are stored in the remaining locations, so we have effectively "saved" the values from the previous frame. In this way the edge effects are eliminated, as will be shown in the actual C code.

7.4.3 Explanation of the C Code

As an illustrative example, we will show you a program that implements a simple low-pass FIR filter that provides a similar output result to that of the FIR filters discussed in Chapter 3. But the method used to obtain our output will be to use frame-based processing, written in such a way that we avoid the "edge" problems discussed above. Before you read the program listing, you may want to glance back at Figure 7.1.

The C program to implement this as frame-based code can be found in the `ccs\` `FiltFrame` directory of Chapter 7. Be sure to inspect the full code listings in this directory to understand the complete working of the program.[2] Here we simply point out some highlights. The discussion which follows assumes you have read and are familiar with the explanation of frame-based processing given in Chapter 6.

Realistically, you would use MATLAB or some other filter design program to determine your filter coefficients. To easily use FIR filter coefficients generated by the filter design tools in MATLAB, you can use the script file named `fir_dump2c.m` that is located in the `MatlabExports` directory for Appendix E (there are other script files in the same directory for IIR filters). As discussed in Chapter 3, this script creates two files needed by your C program, typically named `coeff.h` and `coeff.c`. These files define N, representing the order of the filter, and $B[N+1]$, the array of filter coefficients.

The program we provide makes use of the EDMA capabilities of the DSK. This C code is almost identical to the EDMA version of the frame-based C code described in Chapter 6. The only differences are the need to include `coeff.h` and `coeff.c` in your project, and the changes described below to the `ProcessBuffer()` routine in the `ISRs.c` file. The contents of `coeff.h` and `coeff.c` were discussed in Chapter 3. The code for the `ProcessBuffer()` routine is shown in Listing 7.1.

---

[2]Look also in the `DSK_Support.c` file in the `common_code` directory for initialization functions shared by multiple programs.

Listing 7.1: `ProcessBuffer()` routine for implementing a frame-based FIR filter.

```
void ProcessBuffer()
{
    short *pBuf = buffer[ready_index];
    // extra buffer room of N for convolution "edge effects"
    // N is filter order from coeff.h
    static float Left[BUFFER_COUNT+N]={0};
    static float Right[BUFFER_COUNT+N]={0};
    float *pL = Left, *pR = Right;
    float yLeft, yRight;
    int i, j, k;

    // offset pointers to start filling after N elements
    pR += N;
    pL += N;

    // extract integer data to float buffers
    // order is important here: must go right first then left
    for(i = 0;i < BUFFER_COUNT;i++) {
        *pR++ = *pBuf++;
        *pL++ = *pBuf++;
    }

    // reinitialize pointer before FOR loop
    pBuf = buffer[ready_index];

////////////////////////////////////////////////////////////////////
// Implement FIR filter. Ensure COEFF.C is part of project.
////////////////////////////////////////////////////////////////////
    for(i=0;i < BUFFER_COUNT;i++){
        yLeft  = 0;                 // initialize the L output value
        yRight = 0;                 // initialize the R output value

        for(j=0,k=i;j <= N;j++,k++){
            yLeft  += Left[k] * B[j];     // perform the L dot-product
            yRight += Right[k] * B[j];    // perform the R dot-product
        }

        // pack into buffer after bounding (first right then left)
        *pBuf++ = _spint(yRight * 65536) >> 16;
        *pBuf++ = _spint(yLeft * 65536) >> 16;
    }

    // save end values at end of buffer array for next pass
    //   by placing at beginning of buffer array
    for(i=BUFFER_COUNT,j=0;i < BUFFER_COUNT+N;i++,j++){
        Left[j]=Left[i];
        Right[j]=Right[i];
    }

//////////// end of FIR routine ///////////////
```

```
52    // reinitialize pointer
      pBuf = buffer[ready_index];
54
      // if on-board code of 6711 is being used
56    // mask off LSB to prevent codec reprogramming
      if(CodecType == TLC320AD535) {
58        for(i = 0;i < BUFFER_COUNT;i++) {
              *(unsigned int *)pBuf &= 0xfffffffe;
60            pBuf += 2;
          }
62    }

64    buffer_ready = 0; // signal we are done
}
```

The key parts of the code that differ the EDMA code in Chapter 6 are:

- (lines 6–7) declaring the arrays to be `BUFFER_COUNT+N`, which leaves enough room for the edge values that need to be saved;

- (lines 13–14) advancing the pointers so that the incoming data doesn't overwrite the edge values from the previous frame;

- (lines 33–36) using different array index values to implement the convolution properly across the frame; and

- (lines 45–48) copying the edge values of the current frame to the beginning of the buffer so they will be available when the next frame comes in.

Now that you understand the code...

Go ahead and copy all of the files into a separate directory. If you wish to design your own FIR filter using MATLAB rather than using the provided filter (a simple low-pass filter), use the script file `fir_dump2c.m` that can be found in the `MatlabExports` directory of Appendix E and was first discussed in Chapter 3. Copy any new versions of `coeff.h` and `coeff.c` that you create into your project directory before proceeding. When ready, open the project in CCS and "Rebuild All." Once the build is complete, "Load Program" into the DSK and click on "Run." Your frame-based FIR filter is now running on the DSK.

7.5 Follow-On Challenges

Consider extending what you have learned.

1. Compare the sample-based FIR code to the frame-based FIR code. Which do you predict could handle the higher order filter in real time? Try to get the sample-based FIR code to break the real time schedule by using a large-order filter; try the same filter with the frame-based code.

2. Many FIR filters exhibit some form of symmetry in the coefficients to ensure linear phase response. For example, $b[0] = b[N-1]$, $b[1] = b[N-2]$, $b[2] = b[N-3]$, and so on. Modify the code provided in this chapter to take advantage of this symmetry.

3. In addition to symmetry, some FIR filters also exhibit a regular pattern of zero-valued coefficients (such as every other value). Modify the code provided in this chapter to take advantage of this fact.

4. This chapter demonstrated an FIR filter using frames. Implement an IIR filter using frames.

Chapter 8

The Fast Fourier Transform

8.1   Theory

THE Fast Fourier Transform (FFT) is just what the name says it is: a fast way for computers to calculate the Fourier transform, specifically the Discrete Fourier Transform (DFT). The introduction of the FFT algorithm by Cooley and Tukey in 1965 revolutionized signal processing and has had an enormous effect on engineering and applied science in general [1]. This chapter is by no means a treatise on the FFT. Despite the fact that we will only cover a few of the major points, this is still a long chapter, but we encourage you to read it completely. For more details about the FFT, see [2, 4, 54].

Having a fast way to calculate the DFT has many advantages. We will see in Chapter 9 how it can be used for practical spectral analysis. In this chapter, we investigate how the FFT can be used for efficiently implementing digital filters.

8.1.1   Defining the FFT

Before proceeding with a discussion of this application of the FFT, we first need to briefly discuss the DFT and how the FFT is faster. The N-point DFT for $k = 0, 1, \ldots, N - 1$ is defined as

$$X[k] = \sum_{n=0}^{N-1} x[n]e^{-j2\pi kn/N} \quad = \quad \sum_{n=0}^{N-1} x[n]W_N^{kn}$$

where the "twiddle factor" notation $W_N = \exp\left(-j\frac{2\pi}{N}\right)$ is often used for a more compact notation. This equation is identical for the FFT; we simply use an efficient programming method to implement it. The inverse DFT or FFT is defined by

$$x[n] = \frac{1}{N} \sum_{n=0}^{N-1} X[k]e^{j2\pi kn/N} \quad = \quad \frac{1}{N} \sum_{n=0}^{N-1} X[k]W_N^{-kn}$$

Some books may show the $1/N$ factor as part of the FFT definition rather than the inverse FFT definition, and some even show a $1/\sqrt{N}$ factor in *both* definitions, but the notation of $1/N$ for the inverse FFT shown above is commonly used.

8.1.2   The Twiddle Factors

The only difference between the FFT and the inverse FFT (IFFT) is the division by N and the negative power of the twiddle factors; thus the same basic algorithm can be used for

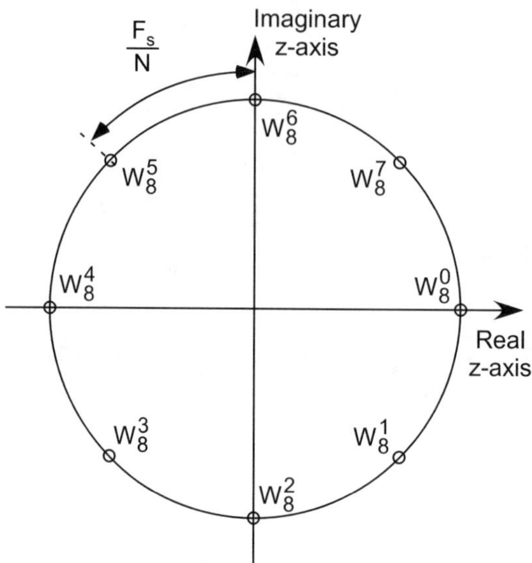

Figure 8.1: The placement of twiddle factor points for the DFT or FFT.

both. Note that the integer powers of W_N form a periodic sequence of numbers, having a period of N. That is, $W_N^n = W_N^{n+N}$. Since in the DFT and FFT definition both k and n in W_N^{kn} are always integers, the same set of twiddle factors appear over and over again. For example, in a 2-point FFT the only two twiddle factors for any of the positive k values are

$$W_2^0 = \exp\left(-j\frac{2\pi(0)}{2}\right) = 1$$

$$W_2^1 = \exp\left(-j\frac{2\pi(1)}{2}\right) = -1.$$

Likewise, for a 4-point FFT the only four twiddle factors for all the positive values of k are $W_4^0 = 1$, $W_4^1 = -j$, $W_4^2 = -1$, $W_4^3 = j$. To visualize this, we can plot W_N^n on a circle (analogous to the unit circle on the z-plane). Figure 8.1 shows where all the twiddle factor points would be for an 8-point FFT given any positive value of k. This can be thought of as a vector rotating clockwise along the unit circle, and the spacing of the twiddle factors in real-world frequency is always F_s/N, where F_s is the sample frequency.[1] For W_N^{-kn} used in the IFFT the vector would be rotating *counter*clockwise, and $W_8^{-1} = W_8^7$, $W_8^{-2} = W_8^6$, ..., $W_8^{-7} = W_8^1$. We take advantage of the periodicity of the twiddle factors to implement the FFT.

8.1.3 The FFT Process

The main idea behind the FFT developed by Cooley and Tukey is that when the length N of the DFT is not a prime number, the calculation can be decomposed into a number of shorter length DFTs. The total number of multiplications and additions required for all the shorter length DFTs is fewer than the number required for the single full-length DFT. Each of these shorter length DFTs can then be further decomposed into a number of even shorter DFTs, and so on, until the final DFTs are of a length that is a prime factor of N.

---

[1]Note that F_s/N is the best frequency resolution possible for the FFT or DFT.

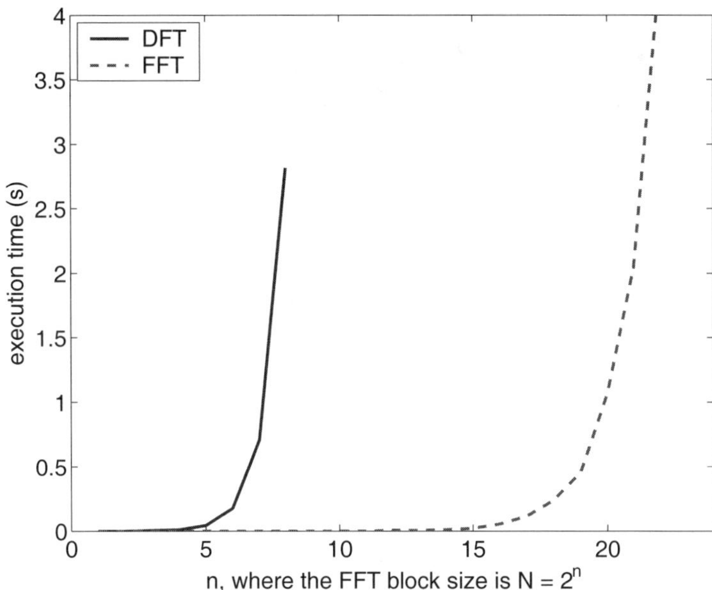

Figure 8.2: Relative calculation time for the "brute force" DFT versus a commercially available radix-2 FFT routine. The data plotted is averaged empirical timing results from a desktop workstation.

Then these shortest DFTs are calculated in the normal fashion. The opposite order can be used, starting with the smallest DFTs and building (recombining) into larger ones for an equivalent savings. For the radix-2 FFT (a common type), the length of the FFT must be a power of 2. If $N = 8$ for example, the FFT would decompose this into two 4-point FFTs, which are each decomposed into two 2-point FFTs (for a total of four 2-point FFTs), and the calculation is complete. Or it could perform four 2-point DFTs, using these results to compute two 4-point DFTs for the final result. The former method is called *decimation-in-frequency*, and the latter method is called *decimation-in-time*. With appropriate summing and reuse of intermediate results, as well as taking advantage of the periodicity of the twiddle factors, the FFT can be performed at a tremendous increase in efficiency compared to the "brute force" DFT. The number of complex mathematical operations required to calculate an N-point DFT is proportional to N^2; for the same N-point radix-2 FFT it would be proportional to $N \log_2 N$. Thus for a 512-point example, the DFT would require on the order of 262,144 complex mathematical operations, while the FFT would require only 4,608 operations—which is over fifty times faster than the DFT! For larger DFTs the difference is even more dramatic. The empirical speed-up provided by the FFT on a typical computer is shown graphically in Figure 8.2. Note from the figure that the FFT can perform the DFT on $N = 2^{20} = 1,048,576$ data points in the same amount of time that the "brute force" DFT would take for $N = 2^7 = 128$ data points. That's a speed-up factor of 8,192, which shows why the FFT is so widely used.

The traditional way to show the FFT algorithm is through the use of "butterfly diagrams," which depict the decomposition, intermediate summing, and application of the twiddle factors. These are shown in Figures 8.3, 8.4, and 8.5 for a 2-point, 4-point, and 8-point decimation-in-time FFT. It is *very* beneficial to manually trace your way through these small butterfly diagrams, until you get a feel for what is going on.

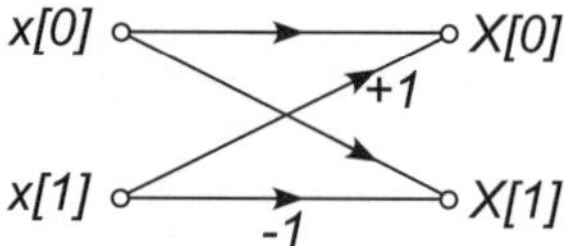

Figure 8.3: Butterfly diagram of a decimation-in-time radix-2 FFT for $N = 2$. Any branches not marked have a gain of $+1$.

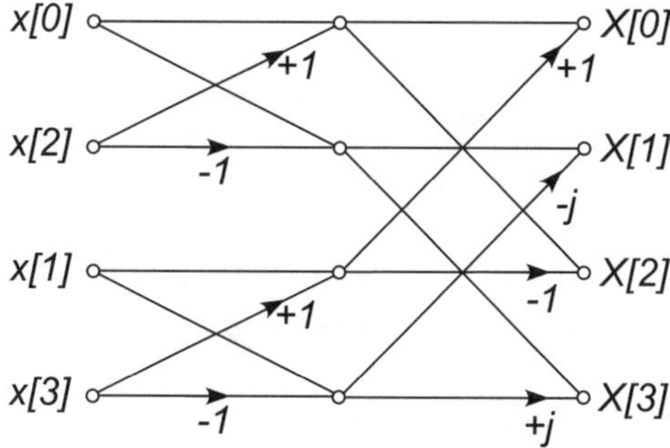

Figure 8.4: Butterfly diagram of a decimation-in-time radix-2 FFT for $N = 4$. Any branches not marked have a gain of $+1$.

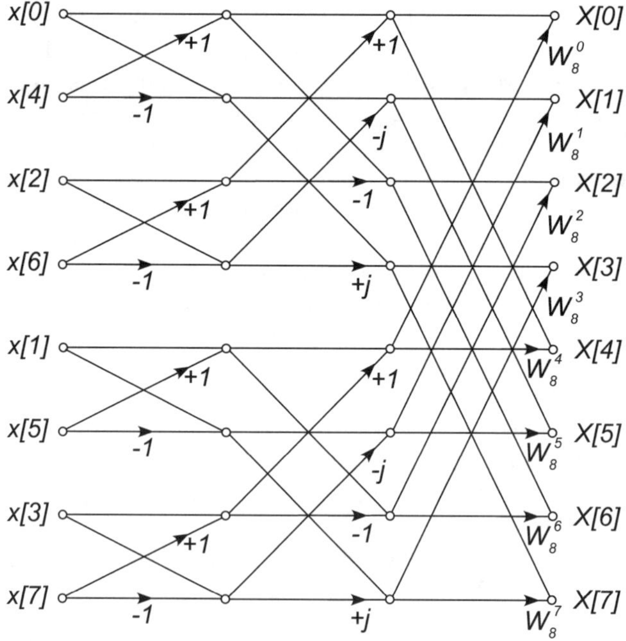

Figure 8.5: Butterfly diagram of a decimation-in-time radix-2 FFT for $N = 8$. Any branches not marked have a gain of $+1$.

8.1.4 Bit-Reversed Addressing

The ordering of the input values on the left side of the butterfly figures may seem strange to you. To account for the most efficient way to perform the butterfly operation by computer for a decimation-in-time FFT, the order of the input values must be rearranged into what is known as *bit-reversed* addressing. You know that to address (or index) N values for an $N = 2^n$ FFT input array requires n bits if the index is expressed in binary from. For bit-reversed addressing, the binary index number of the data array element at the input is reversed left to right. For example, in Figure 8.5, you might expect the second input element to be $x[1]$. With the index expressed as a 3-bit binary number this would be $x[001]$. Reversing the bits of the index yields $x[100]$, which in decimal notation is $x[4]$; this is the actual second input element in bit-reversed addressing. The dual of a decimation-in-time FFT is a decimation-in-frequency FFT. The only difference in the butterfly for a decimation-in-frequency FFT is that the order of the butterfly sections are reversed, the twiddle factors swap positions, and the output values rather than the input values appear in bit-reversed order. There is no intrinsic advantage of one over the other and the implementation choice is typically arbitrary.

When we use bit-reversed addressing on either the input of output side of a butterfly, we can perform what is called an "in-place" calculation, which means the same array that holds the input data is used to hold the output data. The bit-reversed addressing ensures that no input data element gets overwritten by an output value until it is no longer need for any more calculations of the FFT.

8.1.5 Using the FFT for Filtering

As the order of a filter increases the time required to calculate the output value associated with each input sample also increases. As we saw in Chapter 7, frame-based filtering helps increase the overall efficiency of the filtering operation by reducing the time required to pass samples to and from the DSP CPU. Yet we are still calculating a time domain convolution. If we can also take advantage of the FFT to perform the *equivalent* of convolution, then we can save even more time and thus implement even longer (higher order) filters in real-time. Using the FFT in this way is generally referred to as *fast convolution*.

Frequency-domain techniques such as this extend the concepts first introduced in Chapter 3. We know the filtering equation is really just the convolution integral

$$y\left(t\right) = \int_0^\infty h\left(\tau\right)x\left(t - \tau\right) d\tau.$$

If we take the Fourier transform of the convolution integral we obtain

$$Y\left(j\omega\right) = H\left(j\omega\right)X\left(j\omega\right).$$

Notice that the original convolution operation in the time domain has been converted into a multiplication operation in the frequency domain. Similarly, the discrete-time version of the convolution integral (the convolution sum) is

$$y[n] = \sum_{m=0}^{M-1} h[m]x[n - m] \quad \text{for} \quad n = 0, 1, 2, \ldots, N - 1$$

where, in this context, M is the *length* of the filter (thus the filter order is $M - 1$) and N

is the length of the data.[2] The FFT of the convolution sum is

$$Y[k] = H[k]X[k].$$

Again notice that the convolution operation in the discrete-time domain has been converted into a multiplication operation in the discrete-frequency domain.

We might expect that to calculate the digital filter's output back in the time domain, we can use the IFFT and calculate

$$\begin{aligned}
y[k] &= \text{IFFT}\{Y[k]\} \\
&= \text{IFFT}\{H[k]X[k]\} \\
&= \text{IFFT}\{\text{FFT}\{h[n]\}\text{FFT}\{x[n]\}\}.
\end{aligned}$$

Using this approach we would take the FFT of our filter's impulse response, $h[n]$, and multiply the result by the FFT of the input signal, $x[n]$. We would then take the inverse FFT of the product. As we shall see below, this approach requires a slight modification to avoid the effects of circular convolution but otherwise will work splendidly. As the lengths of $h[n]$ and $x[n]$ increase, there comes a point beyond which the frequency-domain transform technique of fast convolution will require fewer mathematical operations than the traditional time-domain convolution approach. For a constant coefficient filter (non-time varying), additional savings can be gained by realizing that the transform of the filter's impulse response need only be calculated once.

8.1.6　Avoiding Circular Convolution

Remembering that for discrete-time systems the transform-based approach results in *circular* convolution instead of *linear* convolution, we must zero pad both $h[n]$ and $x[n]$. When two sequences $h[n]$ and $x[n]$ are properly padded, the circular convolution of the two that is due to multiplication in the frequency domain provides the exact same result as would the linear convolution of $h[n]$ and $x[n]$ in the time domain. Without such padding, circular convolution is *not* equivalent to linear convolution. This can be seen in the MATLAB listing shown below.

Listing 8.1: A MATLAB listing that compares linear and circular convolution.

```
% Simulation inputs
h = [1 2 3 2 1];        % impulse response declaration
x = [1 3 -2 4 -3];      % input term declaration

% Calculated and output terms
y = conv(h, x)
yLength = length(y)
circularConvolutionResult = ifft(fft(h).*fft(x))
circularConvolutionResultLength=length(circularConvolutionResult)
```

In this listing, line 2 declares the impulse response, $h[n]$, of a fourth order filter and line 3 declares the input sequence, $x[n]$, which will be "processed" by the filter. Line 6 performs the linear convolution with MATLAB's built-in `conv()` command and line 7 determines the length of the convolution result. Line 8 performs the *circular* convolution (by using the FFT, point-by-point multiplication, and the IFFT) and line 9 determines the length of that resulting sequence. The MATLAB command window results are similar to:

---

[2]It is common practice in DSP texts to use letters such as N as the filter order in some contexts and as the filter length in other contexts. From the nature of the discussion, it should be clear to the reader which definition is being used.

```
y = 1   5   7   11   6   5   -3   -2   -3
yLength = 9
circularConvolutionResult = 6.0000   2.0000   5.0000   8.0000   6.0000
circularConvolutionResultLength = 5.
```

At this point, two very important observations are required.

1. The outputs of the two processes, y and `circularConvolutionResult` are not the same sequence.

2. The resulting sequences, y and `circularConvolutionResult` are not the same length.

As stated earlier, zero padding (sometimes just called padding), can turn circular convolution into the equivalent of linear convolution. To accomplish this task we must

1. Ensure that the padded lengths of $h[n]$ and $x[n]$ are the same.

2. Adjust the padded lengths of $h[n]$ and $x[n]$ to be at least equal to the length of the resulting linear convolution of the sequences $h[n]$ and $x[n]$, given by

$$N + M - 1$$

where N is the unpadded length of $x[n]$ and M is the unpadded length of $h[n]$. In the previous MATLAB code listing output, $N = 5$ and $M = 5$. Therefore,

$$N + M - 1 = 5 + 5 - 1 = 9.$$

Notice that this result, 9, is the length of the linear convolution, yLength calculated previously by MATLAB. So, to convert the circular convolution to the equivalent of linear convolution we must pad $h[n]$ and $x[n]$ to at least a length of 9. The updated MATLAB code listing to accomplish this task is shown below.

Listing 8.2: A MATLAB listing that demonstrates how to convert circular convolution into the equivalent of linear convolution.

```matlab
% Simulation inputs
format short g          % set format to short g
h = [1 2 3 2 1];       % impulse response declaration
x = [1 3 -2 4 -3];     % input term declaration


hZeroPad = [h zeros(1, 4)];
xZeroPad = [x zeros(1, 4)];


% Calculated and output terms
y = conv(h, x)
yLength = length(y)
circularConvolutionResult=ifft(fft(hZeroPad).*fft(xZeroPad));
circularConvolutionResult=real(circularConvolutionResult)
circularConvolutionResultLength=length(circularConvolutionResult)
```

In this listing, line 2 changes the display format to suppress trailing zeros while lines 6 and 7 pad $h[n]$ and $x[n]$ by appending 4 zeros onto the original sequences. The MATLAB command window results are similar to:

```
y = 1   5   7   11   6   5   -3   -2   -3
yLength = 9
circularConvolutionResult = 1   5   7   11   6   5   -3   -2   -3
circularConvolutionResultLength = 9.
```

Notice that the results, and therefore the lengths of both techniques, are the same. Did you also notice that in line 13 of the listing the MATLAB command `real` was added to remove unwanted imaginary terms from the answer? Since we know that the convolution of two real-valued sequences, namely $h[n]$ and $x[n]$, is another real valued sequence, these imaginary terms are the result of numerical "noise" that occurs due to limited numerical precision during the transform process. Since the imaginary part of the filter's output should be zero, the resulting imaginary noise should be ignored.

Note that the padding may also be driven by the requirements of the FFT. For a radix-2 FFT, the input array length must be a power of two. In our example above, if we were using a radix-2 FFT we would need to pad M and N so that their padded length L is the next higher power of two such that $L \geq (N + M - 1)$. In the above example $N + M - 1 = 9$, so we would need to pad to $L = 16$. A time-domain convolution requires on the order of NM operations, but the fast convolution using an L-point radix-2 FFT requires on the order of $(8L \log_2(L) + 4L)$ operations [51].

8.1.7   Real-Time Fast Convolution

The techniques developed above work well for filtering short or medium length sequences, but what about filtering very long sequences? How about real-time systems, where nearly infinite length sequences (sequences where the input may persist for days or months) are common? We need to find a variation on the fast convolution described above to perform the filtering operation since we don't want to store all the input samples in memory before processing. Additionally, the wait or latency associated with gathering all the samples of a long sequence before we commence the actual filtering operation defeats the intent of real-time DSP.

To filter very long signal sequences, we must, therefore, partition the signal into shorter length sequences that we can filter individually and then recombine into a complete, filtered version of the original signal. A number of techniques have been developed to perform this operation, but we will limit our discussion to the two most common: overlap-add and overlap-save techniques. In our brief discussion of these two techniques we will use only short input sequences for both $h[n]$ and $x[n]$ in the hope that this will improve the understandability of the processes required, but the technique is intended for very long sequences of data.

Overlap-Add

In Figure 8.6, $x[n]$ is a 30-point sequence that is padded out to 36 points ($0 \leq n \leq 35$) to allow for a uniform x-axis labeling for all of the subplots. In this example, we wish to convolve $x[n]$ with the impulse response $h[n]$ of the sixth order lowpass filter shown in Figure 8.7. Obviously, the length of $h[n]$ is $M = 7$. The second ($x_0[n]$), third ($x_1[n]$), and fourth ($x_2[n]$) subplots in Figure 8.6 partition $x[n]$ into 3 non-overlapping segments, each segment being 10 samples in length ($N = 10$). This partition length ($N = 10$ samples) was chosen based on the desire to use a 16 point FFT. Remember that the output of a convolution operation is equal to $L = N + M - 1$, and so for this example, total length is $L = 16$ and filter length is $M = 7$, therefore our data length is required to be $N = 10$ samples. The fifth ($y_0[n]$), sixth ($y_1[n]$), and seventh ($y_2[n]$) subplots in Figure 8.6 show the results of the transform-based filtering operation (fast convolution using the FFT and IFFT) of $x_0[n]$,

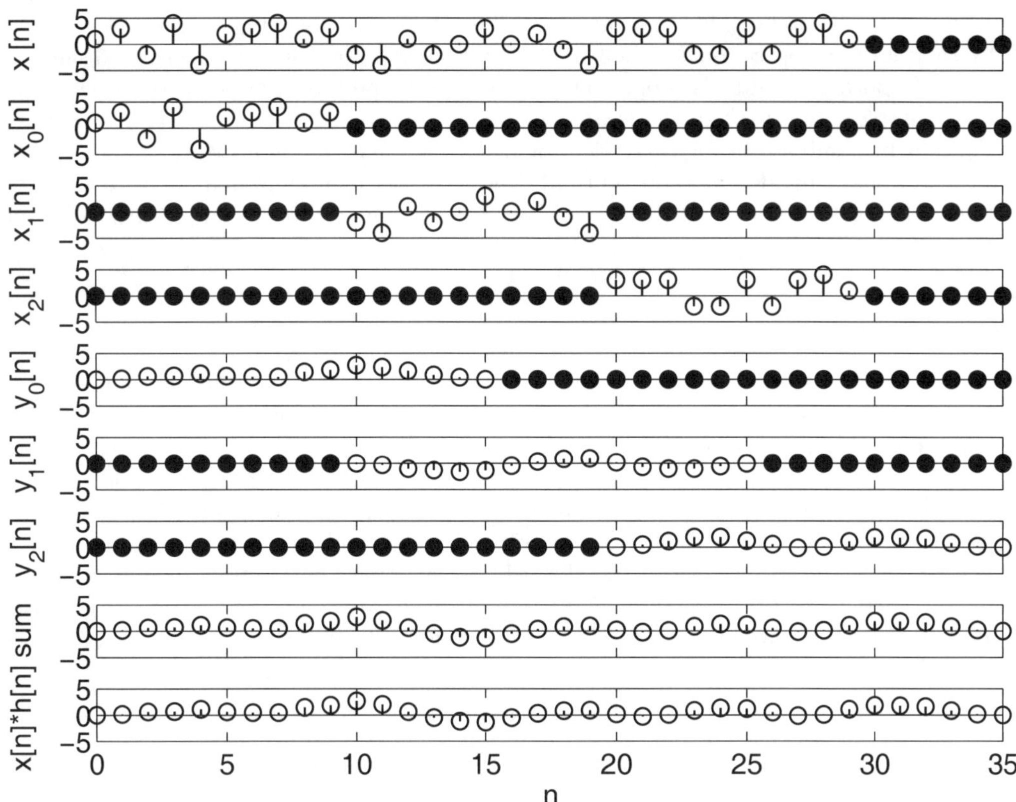

Figure 8.6: The overlap-add fast convolution process. Filled circles are zero padded values.

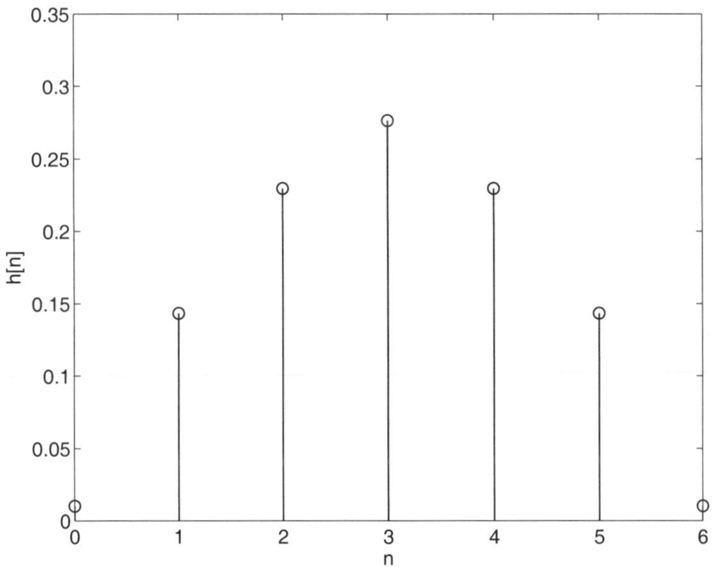

Figure 8.7: The impulse response associated with the lowpass filter used in the overlap-add fast convolution process.

$x_1[n]$, and $x_2[n]$ with $h[n]$. Notice that the filtering of $x_0[n]$, a 10 sample sequence, using $h[n]$, results in the 16 sample sequence $y_0[n]$. The last 6 samples of $y_0[n]$ overlap with the first 6 samples of $y_1[n]$. Similarly, the last 6 samples of $y_1[n]$ overlap with the first 6 samples of $y_2[n]$. To obtain the filter's proper output, the overlapping regions must be added *prior* to these samples being sent to the DSP system's output device (hence the name, "overlap-add"). The final subplot in Figure 8.6 is provided for comparison and is the system's output using traditional time-domain convolution. These last two subplots demonstrate that the transform-based technique and the traditional convolution sum technique return the same result.

Overlap-Save

In Figure 8.8, $x[n]$ is once again a 36 point input sequence. In this example, our desire is to convolve $x[n]$ with the impulse response of the same sixth order lowpass filter as before, shown in Figure 8.7; therefore the length of $h[n]$ is $M = 7$. The second ($x_0[n]$), third ($x_1[n]$), and fourth ($x_2[n]$) subplots in Figure 8.8 partition $x[n]$ into 3 *overlapping* segments, each segment being 16 samples in length. This partition length was chosen based on the desire to use a 16 point FFT. The fifth ($y_0[n]$), sixth ($y_1[n]$), and seventh ($y_2[n]$) subplots in Figure 8.8 show the results of the transform-based filtering operation of $x_0[n]$, $x_1[n]$, and $x_2[n]$ with $h[n]$. The first 6 samples ($M - 1 = 6$, which is the order of the LP filter) of $y_0[n]$, $y_1[n]$, and $y_0[n]$ are not accurate and are not used. To indicate this fact, these values

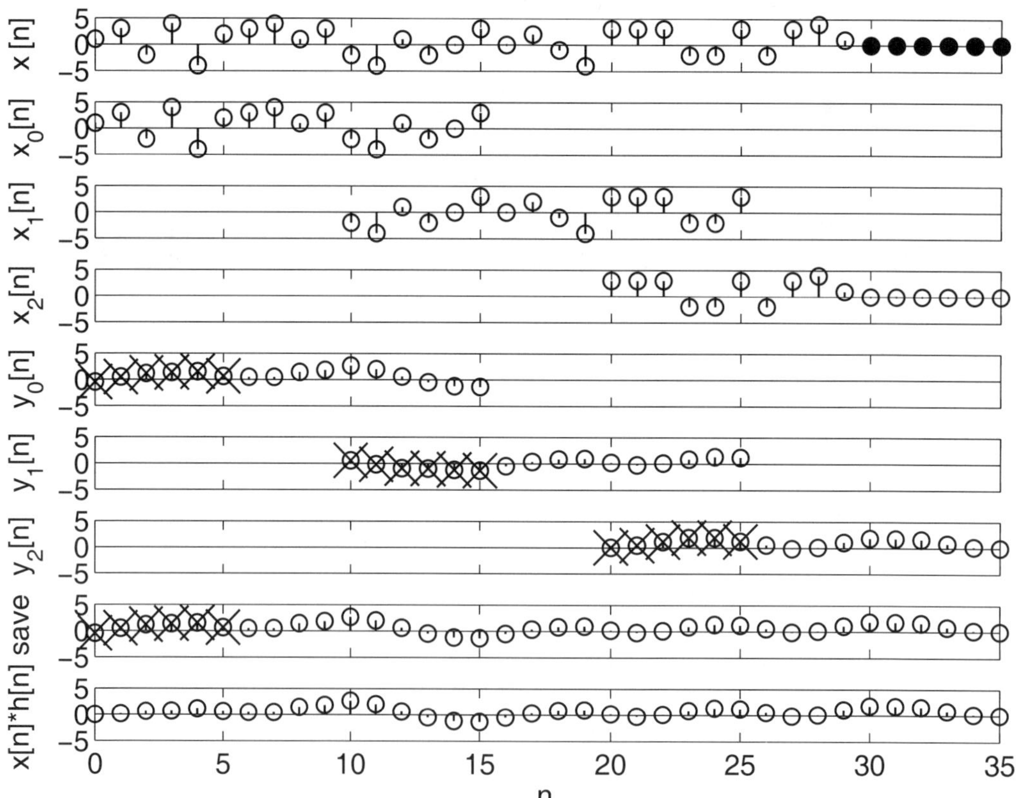

Figure 8.8: The overlap-save fast convolution process. Filled circles are zero padded values.

have large X's drawn through them. To obtain the filter's output, the portions of $y_0[n]$, $y_1[n]$, and $y_0[n]$ that do *not* have large X's drawn through them are concatenated. The final subplot in Figure 8.8 is provided for comparison and is the system's output using traditional time-domain convolution. These last two subplots demonstrate that, with the exception of the filter's initial transient, the transform-based technique and the traditional convolution sum technique return the same result.

In summary, the *overlap and add* technique has no overlap of the inputs but must add the overlapped portions of the output segments for a correct result. The *overlap and save* technique overlaps the input segments then throws away the resulting overlapped regions of the output segments and seamlessly "stitches together" the remaining portions. Either technique allows real-time filtering of long duration input sequences even if you're using high order filters.

8.2   winDSK6 Demonstration

There is no example of fast convolution in winDSK6.

8.3   MATLAB Implementation

There are many ways to demonstrate the FFT using MATLAB, and a basic example of fast convolution was shown on line 12 of Listing 8.2. The overlap techniques for real-time DSP can be simulated in MATLAB, but because this chapter is already rather lengthy, we prefer to move on to the C code.

8.4   DSK Implementation in C

We are dealing with several new ideas in this chapter, namely the FFT and its inverse, fast convolution in general, and the two overlap methods of real-time fast convolution. It would be unwise of us to attempt to cover them all in detail within a single chapter. We therefore choose to concentrate here on helping you become familiar with just the FFT. The other concepts are left for follow-on exercises, and if you understand the FFT they are not hard to implement on your own.

To ensure we understand and have implemented the FFT algorithm correctly, we can test it in non-real-time using the DSK as the CPU and compare the output to known correct values. The C program to implement this can be found in the `ccs\fft_example` directory of Chapter 8. It is by no means a fully optimized FFT routine, but is useful for understanding the concepts.

The code in `fft_example` is fairly straightforward C programming. We incorporate a common "trick" in the code: since the FFT must be able to handle complex numbers, and C does not support complex numbers directly, we use a structure.

Listing 8.3: A structure for implementing complex numbers in C

```
typedef struct {
    float real, imag;
} COMPLEX;
```

The `float` datatype that we use above is typical for DSP CPUs to save memory and gain speed if we don't really need the full precision of the `double` datatype; if we were implementing this on a PC or other general purpose processor we would probably have used `double` for the structure.

The calculation of the twiddle factors is accomplished by the function `init_W()`. Recall the twiddle factors are $W_N^n = \exp\left(-j\frac{2\pi n}{N}\right)$. By using the very helpful Euler's formula from complex analysis,

$$e^{jx} = \cos x + j\sin x,$$

our function `init_W()` can use the trigonometric equivalent of the complex exponential to make it easy to separate the real and imaginary parts of the twiddle factors. This works very well with the structure we defined above for complex numbers. Function `init_W()` is shown below; it runs only once and stores all the twiddle factors that will be needed for the specified length FFT.

Listing 8.4: A function for calculating the complex twiddle factors.

```
1  void init_W(int N, COMPLEX *W)
   {
3      int n;
       float a = 2.0*PI/N;
5
       for(n = 0 ; n < N ; n++) {
7          W[n].real = (float) cos(-n*a);
           W[n].imag = (float) sin(-n*a);
9      }
   }
```

In the listing above, `N` is the length of the FFT, `PI` was defined earlier in the program and `W` is a global array of complex numbers. It should be clear to you that function `init_W()` creates all the complex numbers needed for W_N^n.

The actual N-point butterfly for the FFT is performed by the code shown below excerpted from the `fft_c()` function.

Listing 8.5: The C code for performing the FFT butterfly operation.

```
   // perform fft butterfly
2  Windex = 1;
   for(len = n/2 ; len > 0 ; len /= 2) {
4      Wptr = W;
       for (j = 0 ; j < len ; j++) {
6          u = *Wptr;
           for (i = j ; i < n ; i = i + 2*len) {
8              temp.real = x[i].real + x[i+len].real;
               temp.imag = x[i].imag + x[i+len].imag;
10             tm.real = x[i].real - x[i+len].real;
               tm.imag = x[i].imag - x[i+len].imag;
12             x[i+len].real = tm.real*u.real - tm.imag*u.imag;
               x[i+len].imag = tm.real*u.imag + tm.imag*u.real;
14             x[i] = temp;
           }
16         Wptr = Wptr + Windex;
       }
18     Windex = 2*Windex;
   }
```

The input data is in array `x`, which is made up of complex datatype elements as defined by the structure discussed earlier, and the twiddle factors are accessed via pointer `u`. The variable `len` is used to successively split the data sets in half (see line 3 of the listing). The remaining lines simply perform the additions and multiplications required by the butterfly operations. You should work your way through this code with a small number of data elements (such as $N = 8$), while looking back frequently at the appropriate butterfly figure (such as Figure 8.5).

Finally, since the data coming out of the butterfly is in bit-reversed addressing order, we need to re-order the data back to "normal" order by "unscrambling" the elements of array `x`, which by now contains the FFT result, not the input data. This is accomplished by the code shown below.

Listing 8.6: A routine for "unscrambling" the order from bit-reversed addressing to normal ordering.

```
// rearrange data by bit reversed addressing
// this step must occur after the fft butterfly
j = 0;
for (i = 1; i < (n-1); i++) {
    k = n/2;
    while(k <= j) {
        j -= k;
        k /= 2;
    }
    j += k;
    if (i < j) {
        temp = x[j];
        x[j] = x[i];
        x[i] = temp;
    }
}
```

Now that you understand the code...

Go ahead and copy all of the project files into a separate directory. Note that this is a very simple program, without any real-time requirements, so the project is quite small. When ready, open the project in CCS and "Rebuild All." Once the build is complete, "Load Program" into the DSK and click on "Run." The program runs on the DSK and the output can be seen on the StdIO window at the bottom of the Code Composer Studio interface.

To keep things as simple as possible, the input data is hard-coded into the program file as $x = \{0, 1, 2, 3, 4, 5, 6, 7, 0, 0, 0, 0, 0, 0, 0, 0\}$. This means you will be calculating a 16-point FFT. Based upon our underlying knowledge of the Fourier transform, we can easily predict a few things about the result of an FFT on this data: 1) since the average value of the input data is non-zero, the frequency domain DC (or zero Hertz) value will also be non-zero; 2) since we have 16 input values we should have 16 output values (although the output will consist of 16 complex numbers); and 3) since the input data is real, the output data will be symmetrical about the $F_s/2$ point. While you are free to change the input data or modify the code to accept input data as an argument passed to the function, we wrote it this way so you could easily verify the correctness of the algorithm. All you need to do is compare the output of the C code running on the DSK to the result of the following MATLAB commands.

Listing 8.7: MATLAB commands used to confirm the correctness of your FFT.

```
x=[0 1 2 3 4 5 6 7 0 0 0 0 0 0 0 0];
X=fft(x)'
```

We used the transpose operator ($'$) at the end of line 2 so the output would line up better as a column. The MATLAB command window results should be similar to:

```
X =

  28.0000
  -9.1371 +20.1094i
  -4.0000 -  9.6569i
   2.3801 +  5.9864i
  -4.0000 -  4.0000i
   3.2768 +  2.6727i
  -4.0000 -  1.6569i
   3.4802 +  0.7956i
  -4.0000
   3.4802 -  0.7956i
  -4.0000 +  1.6569i
   3.2768 -  2.6727i
  -4.0000 +  4.0000i
   2.3801 -  5.9864i
  -4.0000 +  9.6569i
  -9.1371 -20.1094i
```

These numbers should be essentially identical to the output of the FFT when run via C code on the DSK. Not only is our algorithm functioning correctly but our predictions about the FFT result are also verified: the DC value is non-zero (28.0000), there are 16 output values, and there is symmetry on either side of the $F_s/2$ value (the $F_s/2$ value is where $X[8] = -4.0000$). Feel free to try other input data sets and compare with the FFT from MATLAB.

There is much more to using the FFT and interpreting its results than we have room to discuss here. We will revisit this topic in the context of spectral analysis in Chapter 9.

8.5   Follow-On Challenges

Consider extending what you have learned.

1. Modify the non-real-time FFT code given in this chapter to run inside a `ProcessBuffer` function. Extend it to create and test a frame-based implementation of an overlap-add filter that uses real-time fast convolution.

2. Modify the non-real-time FFT code given in this chapter to run inside a `ProcessBuffer` function (if you haven't already as part of the previous challenge). Extend it to create and test a frame-based implementation of an overlap-save filter that uses real-time fast convolution.

3. Determine a specific filtering situation where overlap-add or overlap-save is faster than one of the filtering techniques discussed earlier in this text.

4. Implement a Hilbert transform filter using the fast convolution approach.

Chapter 9

Spectral Analysis and Windowing

9.1 Theory

AS discussed in Chapter 8, the Fast Fourier Transform (FFT) allows us to transform a discrete-time signal from the sample (i.e., time) domain to the frequency domain. It is useful and often necessary to view the frequency content of a signal, and test equipment manufacturers market spectrum analyzers costing many thousands of dollars to perform this task. Nearly any engineer or technician working, for example, with communications or audio systems such as satellite up/down links, cellular telephone networks, radio/television stations, home theaters, or any high-end sound system *must* be able to analyze the frequency domain representation of signals. Rather than use a dedicated spectrum analyzer, we will explore how we can apply some basic DSP algorithms to achieve the same purpose.

Spectral analysis and estimation is a very broad topic in signal processing, and we will only scratch the surface of it here. If you wish to learn more about it, there are many fine texts available that cover the topic at different levels of detail (see, for example, [1, 2, 15, 55–57]). Note that spectral estimation can be divided into nonparametric methods and parametric methods. By using the FFT we are using a nonparametric method, which is not optimal with certain special types of signals, but it is very easy to use and can be efficiently calculated. For this reason, FFT-based spectral analysis is by far the most common technique in use today. Digital oscilloscopes and spectrum analyzers typically implement a form of FFT-based spectral analysis. Parametric methods (such as those using ARMA, MUSIC, or ESPIRIT models) may be more sophisticated, but a discussion of them is well beyond the scope of this text.

9.1.1 Power Spectrum of a Signal

Our goal in this chapter is to obtain the distribution of power versus frequency for a given signal; this is called the *power spectrum*. Recall that the output values of an FFT are complex numbers; we will concentrate here only on the magnitude of the power spectrum, and not concern ourselves with the phase. While the phase response may be important for some applications, we don't pursue it here in order to keep the discussion to a reasonable length.

If a discrete-time signal $x[n]$ is provided as input to the FFT, the output is $X[k]$, i.e., $\text{FFT}\{x[n]\} = X[k]$. Recall that just as each increment of n in the time domain equates to a difference of $T_s = 1/F_s$ seconds in the signal, each increment of k in the frequency domain equates to a difference of $\Delta f = F_s/N$ where N is the length of the FFT. The value of Δf is called the *frequency resolution* of the spectrum. If $x[n]$ was originally a voltage versus time signal or a current versus time signal, then the resulting $X[k]$ would be voltage versus frequency or current versus frequency, respectively. Thus to get the normalized power spectrum (i.e., normalized to an impedance of 1 Ω), we just use the relationship

$$|X[k]|^2 = (X[k]_{\text{real}})^2 + (X[k]_{\text{imaginary}})^2$$

which will yield the squared magnitude in watts versus frequency.

For examples of how to interpret the result of the FFT-based power spectrum, let's recall a few things about the FFT. The FFT (just like the DFT) will output the same number of data points, in the form of complex numbers, as it is provided as input data points. If the input signal is real, the output magnitude will be symmetrical about the $F_s/2$ point, which is where $k = N/2$. From $k = 0$ up to $k = N/2$ (i.e., up to the $F_s/2$ point), we interpret that value of k as corresponding to $f = k\Delta f = kF_s/N$. Beyond the $k = N/2$ point, the k values correspond to negative frequencies, and we interpret them as $f = -[(N - k)\Delta f] = -[(N - k)F_s/N]$. Now let's proceed to the examples.

Assume a signal $x(t)$ is sampled at $F_s = 48$ kHz for 2 seconds; we would obtain 96,000 data points such that $x[n]$ exists for $0 \leq n \leq 95,999$. If we take a fairly large FFT of $x[n]$, for example FFT$\{x[n]\}$ of just the first $N = 65,536$ data points[1], then $X[k]$ would be created for $0 \leq k \leq 65,535$. In this case, $\Delta f \approx 0.732$ Hz. If the squared magnitude of this FFT showed a significant spike at $k = 100$ (and also at $k = N - 100 = 65,436$ for the negative frequency component), it would mean the signal had significant power at $f \approx 732$ Hz. If instead we took a more reasonably sized FFT, for example $N = 4096$, then $\Delta f \approx 11.7$ Hz. If the squared magnitude of *this* FFT showed a significant spike at $k = 100$ (and also at $k = N - 100 = 3,996$ for the negative frequency component), it would mean the signal had significant power at $f \approx 1170$ Hz, not 732 Hz as before. Thus changing only the length of the FFT changes the frequency resolution Δf, which changes how we interpret the FFT results.

For real-time spectral analysis, we need to keep the FFT size reasonably small for two reasons.

1. A large FFT may take too long to calculate to meet the real-time schedule.

2. A large FFT may take too long to present results (even if it can be calculated in real time), may exceed the desired response time, and may not respond well to signals containing quickly changing frequency content.

As seen above, large FFTs result in far greater frequency resolution (that is, where Δf is very small) than we are likely to need: a resolution of less than 7/10 of one hertz as in the first example above may be overkill. On the other hand, if we choose a value for N that is *too* small, then while the FFT is calculated quickly the frequency resolution Δf is too coarse to be useful. As with most engineering trade-offs, the choice for the "best" FFT size is not clear cut.

In general, for real-time spectral analysis you would write a frame-based program (because the FFT needs more than one sample at a time) that continuously calculates the FFT of each new frame of data and provides the power spectrum output for presentation on the PC display. However, it's not quite that simple.

It turns out that to get a more accurate estimate of a signal's power spectrum, it is sometimes better to add a few more steps to the computation. The Welch periodogram for

---

[1]We assume a radix-2 FFT, which requires N to be a power of two.

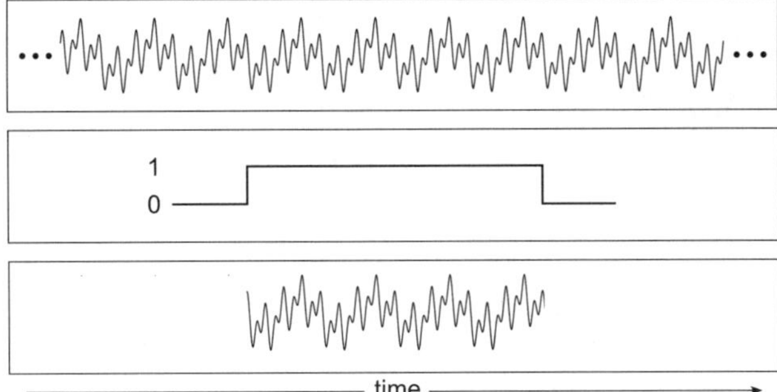

Figure 9.1: The time domain effect of applying a rectangular window. Top: a signal that lasts for an infinite time. Middle: a finite duration rectangular window. Bottom: The result of multiplying the infinite signal with the rectangular window is a finite duration signal.

example, a very popular method of calculating the power spectrum with the FFT, uses a smoothing window on each frame of data (as discussed below), averages the power spectrum of multiple frames of data, and overlaps the data by some percentage (often 50%) from one frame to the next. To discuss these finer points further would be beyond the scope of this text; see [15] for a very clear discussion.

9.1.2   The Need for Windowing

In many situations, we need to apply a smoothing window to each frame of data to get the best results from the FFT. If we don't apply a smoothing window to our data, then we have in effect applied a *rectangular* window, which has a constant value of 1 for its entire length.

Why is this true? As far as the FFT "knows," all data lasts an infinite amount of time, repeating for infinity with a period equal to the time duration of the given data. When we present only a finite length of data (as we must) to be transformed, this is the same as presenting an infinite length of data that has been multiplied with a finite length window having a constant value equal to 1. A visual example of this is shown in Figure 9.1. Why do we care? Recall that multiplication in the time domain is equivalent to convolution on the frequency domain. So the spectrum of the signal is effectively convolved with the spectrum of the window, and there's *always* a window of some sort—even if it's an unintentional rectangular window. What does the spectrum of a rectangular window look like?

A rectangular window is just like a single rectangular pulse (sometimes called a "rect function") that has a value of 1 for some finite region of time and a value of 0 everywhere else. You may recall from your signals and systems course that if we take the Fourier transform of a rectangular pulse, the resulting frequency spectrum is in the shape of a "sinc function," which is defined as $\mathrm{sinc}(x) = \sin(\pi x)/(\pi x)$. Another helpful recollection from Fourier theory is the reciprocal spreading property, which tells us in this case that as the rectangular pulse gets wider, the width of the sinc lobes get more narrow. A "wider" rectangular pulse in this sense is equivalent to a rectangular window with a larger value of N (i.e., a "longer" window with more data points). This "rect $\leftrightarrow$ sinc" relationship can be observed in Figure 9.2.

Suppose you take the FFT of a signal with no smoothing window applied to it first; then you have in effect used a rectangular window. Because the result is as if the true spectrum of the signal is convolved with the spectrum of the window, the observable signal spectrum gets unavoidably "blurred," as shown in Figure 9.3. This has two consequences of particular

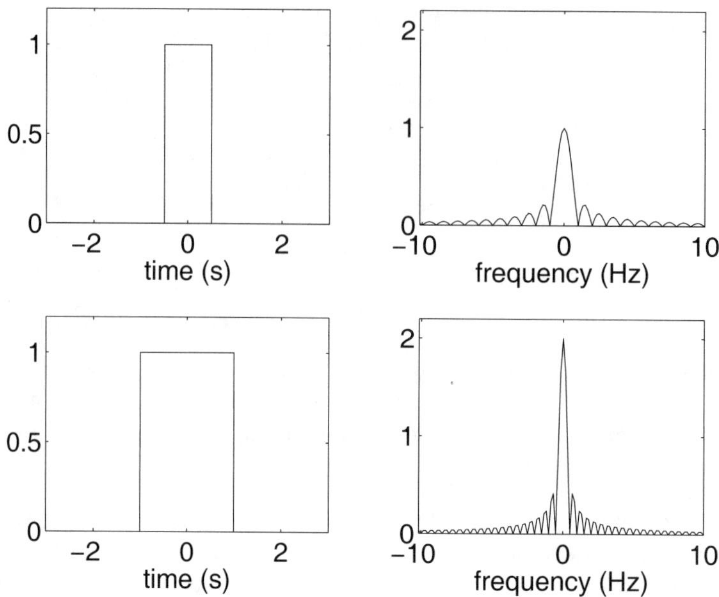

Figure 9.2: The "rect $\leftrightarrow$ sinc" Fourier transform pair. The left two plots are rectangular pulses in the time domain. The right two plots are the corresponding magnitude spectra of the Fourier transform; the phase plots are not shown. Note that a wider pulse in the time domain results in narrower lobes of the sinc in the frequency domain.

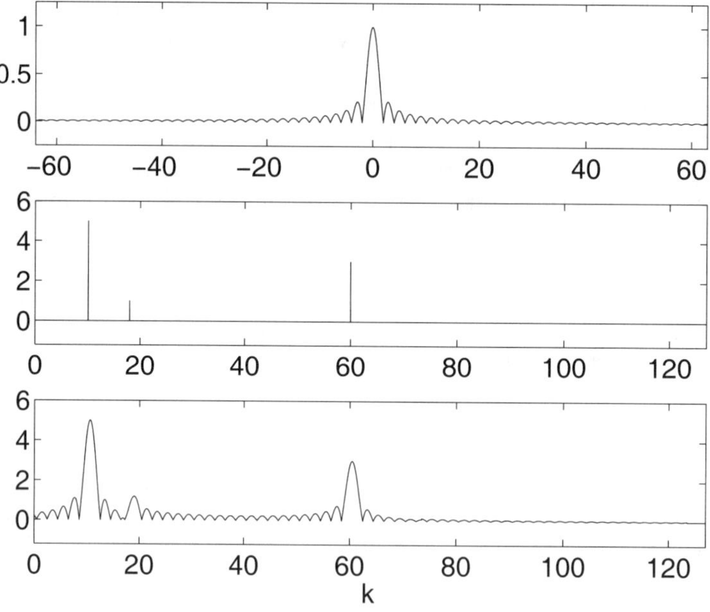

Figure 9.3: The frequency domain effect of applying a rectangular window. Top: the magnitude spectrum of a rectangular window is a sinc pulse. Middle: the theoretical spectrum (only positive frequencies shown) of some arbitrary infinite time duration signal that has three frequency components at $k = 10$, $k = 18$, and $k = 60$. Bottom: The result of convolving the window spectrum with the signal spectrum.

importance to spectral analysis that can be seen in the figure. First, the minimum width of a frequency component is limited by the width of the main lobe of the window's spectrum. Second, the ability to detect a weaker signal near a stronger signal is limited by the "height" of the sidelobes (called the sidelobe level).

The first effect is obvious in the bottom of Figure 9.3 where the frequency component "spikes" have been smeared to become wider lobes. If two frequency components are closer than half the main lobe width of the window, they will "blend into" each other and it will be impossible to distinguish them as separate components. The second effect can also be seen in the bottom of Figure 9.3 where the frequency component at $k = 18$ is almost obscured by the sidelobes from the component at $k = 10$. If the component at $k = 18$ had been slightly weaker in amplitude it would have been "covered up" by the nearby sidelobes. Remember that we *always* have a window applied to your data; if we didn't explicitly apply a smoothing window to our data then we have in effect used a rectangular window. From the discussion above, it should be obvious that we need to know two critical characteristics about any windows we may use: the main lobe width and the sidelobe level.

9.1.3 Window Characteristics

Many smoothing windows have been developed over the years, most taking the name of the person who first proposed them. In addition to the rectangular window, there is the Bartlett (a triangular window), the Hamming, the von Hann (also called the Hanning or the Hann), the Blackman (also called the Blackman-Harris), the Kaiser, and the Dolph (also called the Chebyshev or the Dolph-Chebyshev). Some of these are shown in the time domain in Figure 9.4.

The shape of the spectrum of each of these windows are somewhat similar to the sinc shape of the rectangular window's spectrum, in that there is a main lobe and some sidelobes. But the width of the main lobe and the sidelobe level differ from window to window; these are the two main criteria you use to select a window for a given spectral analysis application.

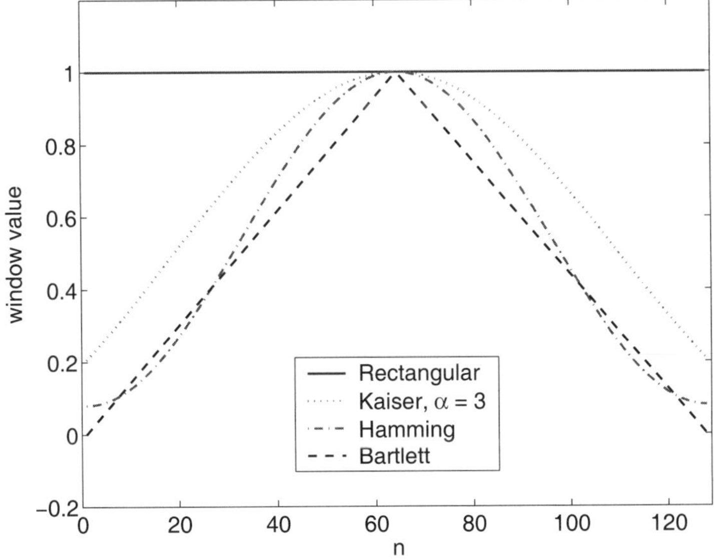

Figure 9.4: A few windows in the time domain. Note that all the windows smooth the data toward zero at the beginning and end of the data set, except for the rectangular window.

You can spend a great deal of time tracking down all the window characteristics in various DSP books, but we've collected the most important aspects of commonly used windows for you in Table 9.1.

In keeping with the practice of other DSP texts, the table shows main lobe width in normalized radian frequency (where π equates to $F_s/2$), with N being the length of the window, and the sidelobe level is shown in decibels. We don't show all the defining equations for the windows (which can be found in most theoretical DSP texts [2, 15]) because we'll discuss how you can easily create the windows you need using MATLAB.

For spectral analysis, only the "Main Lobe Width" and "Sidelobe Level" columns in Table 9.1 are important. The other columns are useful if you ever design FIR filters using the window method; we included the extra columns here to collect all the window information in one place. What we most desire in a window is: 1) a narrow main lobe width so that we can resolve closely spaced frequency components and 2) a low sidelobe level so that we can resolve a weak signal near a strong signal. Unfortunately, these are two conflicting requirements. Take a moment to notice in Table 9.1 that, in general, the main lobe width (for a given data length N) tends to get wider as the sidelobe level gets lower. This can also be seen in Figure 9.5. Furthermore, the main lobe width will get narrower as the window length (which must be equal to the data length) gets longer—but the sidelobe level is independent of window length. As with most engineering decisions, the "best" choice of a window will be a tradeoff of these characteristics based upon your specific needs.

The main lobe width of the rectangular window is the "best" that we can do, since half the main lobe width is $2\pi/N = F_s/N$ which is the best resolution that we can ever get from an N-point FFT. But the sidelobe level of the rectangular window is the "worst" that we can do. Windows other than the rectangular window are very often recommended for spectral analysis. There is another reason that we may want to avoid the rectangular window: a phenomenon called "bias" in which the center peaks of two closely spaced frequency components appear to be further away from each other than they really are (see [1] for a nice example). Applying one of the common smoothing windows instead of a rectangular window eliminates the bias problem.

Table 9.1: A summary of the characteristics of the most commonly used window functions.

Window[a] (length N)	Main Lobe Width	Sidelobe Level (dB)	Transition Bandwidth	Passband Ripple (dB)	Stopband Attenuation (dB)
rectangular	$4\pi/N$	-13.5	$1.8\pi/N$	0.75	21
Bartlett	$8\pi/N$	-27	$6.1\pi/N$	0.45	25
von Hann	$8\pi/N$	-32	$6.2\pi/N$	0.055	44
Hamming	$8\pi/N$	-43	$6.6\pi/N$	0.019	53
Blackman	$12\pi/N$	-57	$11\pi/N$	0.0017	74
Kaiser, $\alpha = 4$	$6.8\pi/N$	-30	$5.2\pi/N$	0.049	45
Kaiser, $\alpha = 8$	$10.8\pi/N$	-58	$10.2\pi/N$	0.00077	81
Kaiser, $\alpha = 12$	$16\pi/N$	-90	$15.4\pi/N$	0.000011	118
Dolph, $\alpha = -40$	$7.4\pi/N$	-40	NA	NA	NA
Dolph, $\alpha = -60$	$10.1\pi/N$	-60	NA	NA	NA
Dolph, $\alpha = -80$	$13.2\pi/N$	-80	NA	NA	NA

[a]Other window names: rectangular = boxcar, Bartlett = triangular, von Hann = Hann = Hanning, and Dolph = Chebyshev = Dolph-Chebyshev. In some books, the parameter α is called β instead. NA: not applicable, as the Dolph window is not often used for FIR filter design.

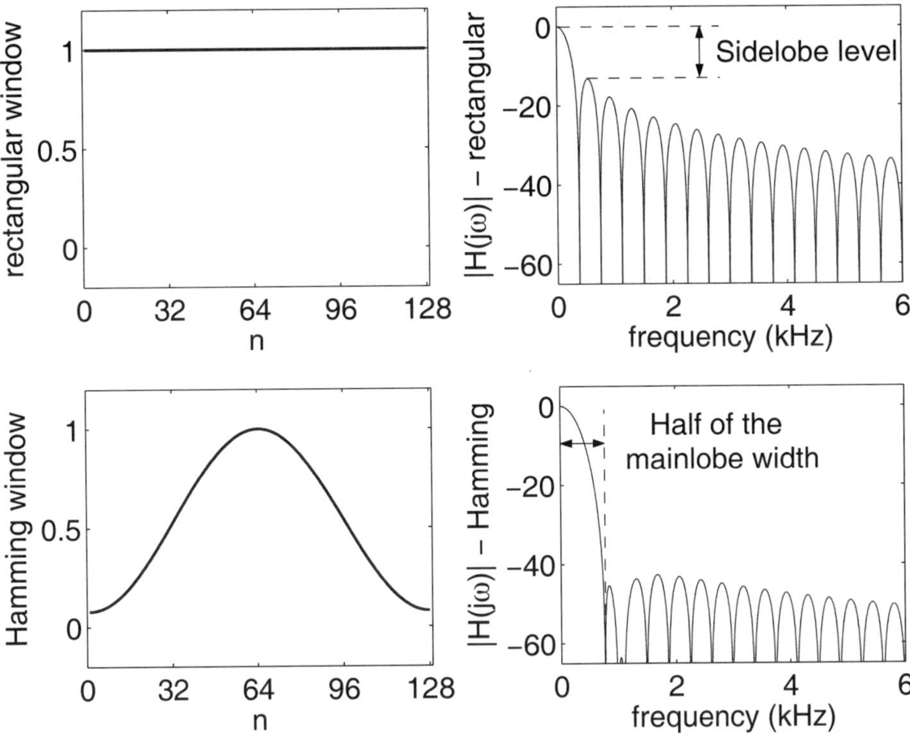

Figure 9.5: Two windows compared in the time and frequency domains. Note that a narrow main lobe also results in a high sidelobe level.

To bring these windowing ideas into focus, let's try a couple of simple examples. Suppose we want to perform spectral analysis on a signal, sampled at 48 kHz, for which we expect there to be a frequency component at 14.0 kHz and another frequency component of nearly equal strength at 14.1 kHz. Thus the frequency separation of the two components of interest is 100 Hz. Assume that for some other reason our data frame length is going to be fairly short at $N = 512$, so our window length must also be $N = 512$. We want to use a smoothing window to eliminate bias but don't want to smear these two components together with too wide a main lobe width. The sidelobe level is less important for this example because the magnitudes of the two signals are nearly equal. Can we use a Hamming window? From Table 9.1, half of the main lobe width of a Hamming window is $4\pi/N = 2F_s/N = 96000/512 = 187.5$ Hz, which is wider than the frequency separation of the two components of interest, so the answer is "No," we can't use a Hamming window—it will smear the two frequencies together—unless something changes. The only narrower window is the rectangular, but then we'll have a bias problem. The best alternative would be to increase the frame length to 1024 points, which would decrease the half main lobe width for the Hamming window to 93.75 Hz, less than the separation of the components of interest. In this case, the Hamming window would work.

Now suppose the second example is the same as before but that the amplitude ratio between the two components of interest is 100:1 (or 1:0.01) instead of being nearly equal. The frame size is still 1024 so the Hamming window won't smear the two components into each other. But what about the sidelobes? When amplitudes will be very different, we need to check the sidelobe level. In decibels, a 1:0.01 amplitude ratio is $20\log(0.01) = -40$ dB. From Table 9.1, we see that this eliminates the rectangular, Bartlett, and von

Hann windows but that the Hamming window will work, although we're cutting it a bit close. The Blackman window would be better in terms of the sidelobe level, but the main lobe width would be too wide.

Given these examples, perhaps you can appreciate the thought process that goes into choosing an appropriate smoothing window. If resolving closely spaced frequencies of similar magnitude is more important, we will tend to choose a window with a narrow main lobe width. If resolving frequency components of very different magnitudes is more important, we will tend to choose a window with a low sidelobe level. In off line applications, it is common to examine a signal with at least two windows, one with a narrow main lobe, the other with a low sidelobe level. However, we don't have that luxury if we are performing real-time spectral analysis, so in that situation we have to make an educated compromise when selecting the window. If we don't really know what frequencies and amplitudes may be lurking in the signal, at least we should know what could be hidden from us due to the window we choose!

9.2   winDSK6 Demonstration

The winDSK6 application has the ability to perform real-time spectral analysis, using a frame size of 512 samples. Plug an input signal into the DSK, start up winDSK6, and click the Oscilloscope button. Two screens similar to Figure 9.6 appear; be sure the Spectrum Analyzer function is selected.

Note that the magnitude of only the positive frequencies are displayed by the Spectrum Analyzer function. The frequency range for the x-axis is automatically adjusted to be 0 Hz to $F_s/2$ Hz, so to read the frequency scale we must know what sample frequency is being used. Under the Display category, you can also select a logarithmic y-axis (most common for displaying spectra) and choose to average the spectrum of a specified number of frames. Under the Data Window category you can choose which window type to apply; the available choices span the windows most commonly used for spectral analysis.

A logarithmic display of the spectrum of a 7 kHz sinusoidal signal sampled at 48 kHz is shown in the figure. The small spike close to DC is a 60 Hz power line artifact. You can experiment with a variety of input signals to the DSK and observe the associated spectral analyzer display.

9.3   MATLAB Implementation

Spectral analysis of stored signals using MATLAB is very easy and flexible, but real-time spectral analysis with MATLAB is more difficult. Let's discuss topics related to non-real-time analysis first.

If you want to see how to generate, analyze, or use various types of data windows, explore MATLAB functions such as `window`, `wintool`, `wvtool`, `sptool`, and `fdatool`. Functions which create only a single type of window include `rectwin`, `bartlett`, `hamming`, `hann`, `kaiser`, and `chebwin` (which creates a Dolph window). Use MATLAB's `help` command with any of these functions to get more detail.

For non-real-time spectral analysis in MATLAB, we'll assume you already have a discrete-time signal $x[n]$ stored on your computer in the MATLAB workspace as x. If you use the command `pwelch(x)` you'll quickly obtain a plot showing the power spectrum of signal $x[n]$ using the Welch periodogram method. For example, Figure 9.7 shows the output obtained from the `pwelch` command when x contains 512 samples of a 7 kHz sinusoid sampled at 48 kHz. Type `help pwelch` to get information on this command and all of its many options.

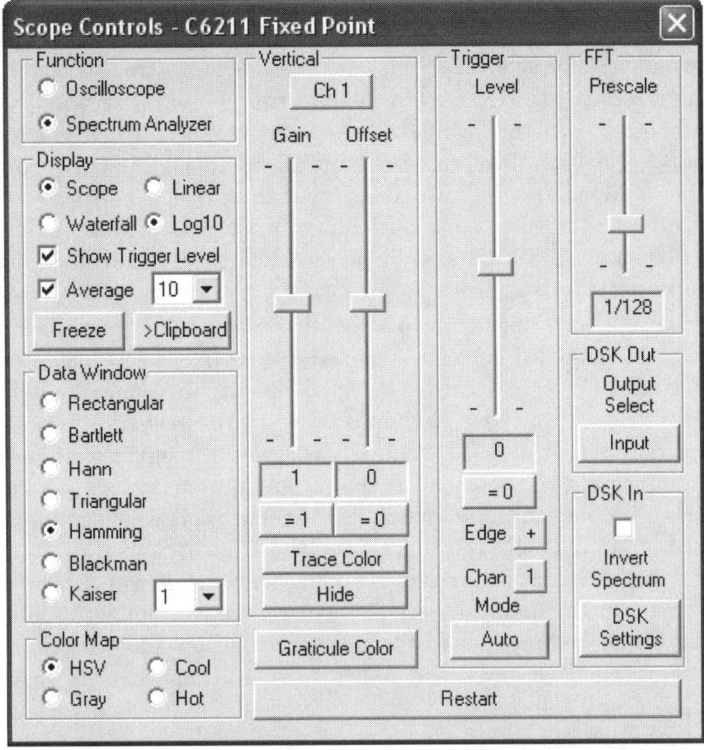

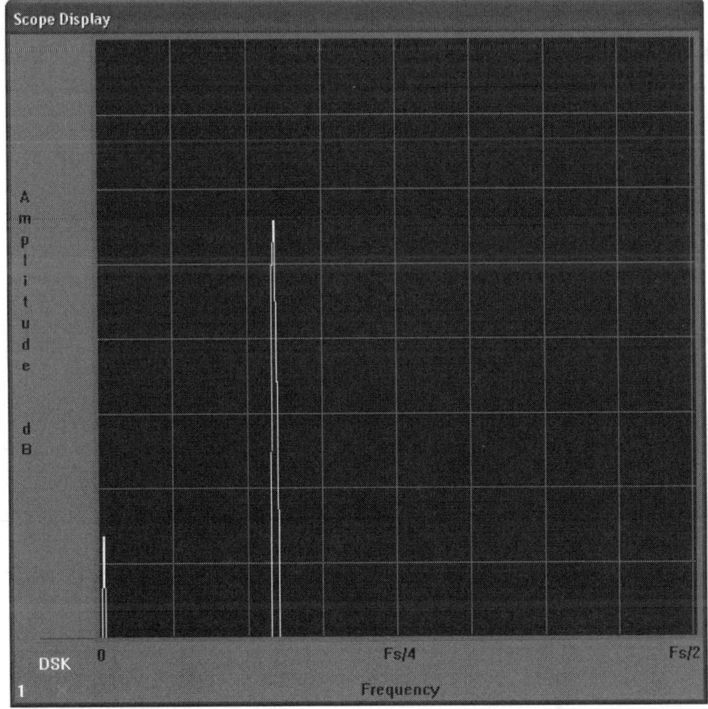

Figure 9.6: Spectrum analyzer windows for winDSK6. Note that you can select averaging, window type, log or linear y-axis, etc.

You may also want to explore functions such as `periodogram`, `pburg`, and `pmusic` for other methods of spectral analysis.

Obviously, what we have shown so far in MATLAB is not real-time. If you have the Data Acquisition Toolbox for MATLAB (available from The MathWorks), you can bring a signal in via your sound card and calculate the spectrum in real time. For those of you who do have the Data Acquisition Toolbox, we've included on the CD-ROM (in the Chapter 9 `matlab` directory) a MATLAB program called `specAn.m` which is a very simple real-time spectrum analyzer. It uses the sound card input of the PC to obtain the signal, with a default sample frequency of 8 kHz, and displays a real-time spectrum in a figure window very similar in appearance to Figure 9.7.

To use the program, copy `specAn.m` to some directory on your hard disk, and make that directory visible to MATLAB. Connect a signal source such as a CD player or a microphone to the input of your PC's sound card. In the MATLAB workspace, enter the command `specAn`, and press "Enter" on your keyboard to start the real-time process. To end the real-time process, click either mouse button anywhere on the figure window. If you want to use some other sample frequency F_s, specify that as an input argument (e.g., `specAn(44100)` for $F_s = 44.1$ kHz), keeping in mind that both MATLAB and your sound card will impose certain practical limits on F_s. If you don't have the Data Acquisition Toolbox, there is a MATLAB function called `audiorecorder` (see Chapter 2) that can bring samples into the MATLAB workspace via the sound card. However, we have found this function to be so much less capable than the functions in the Data Acquisition Toolbox (especially for a real time demonstration), that we haven't included a program using it. If possible, get access to the Data Acquisition Toolbox.

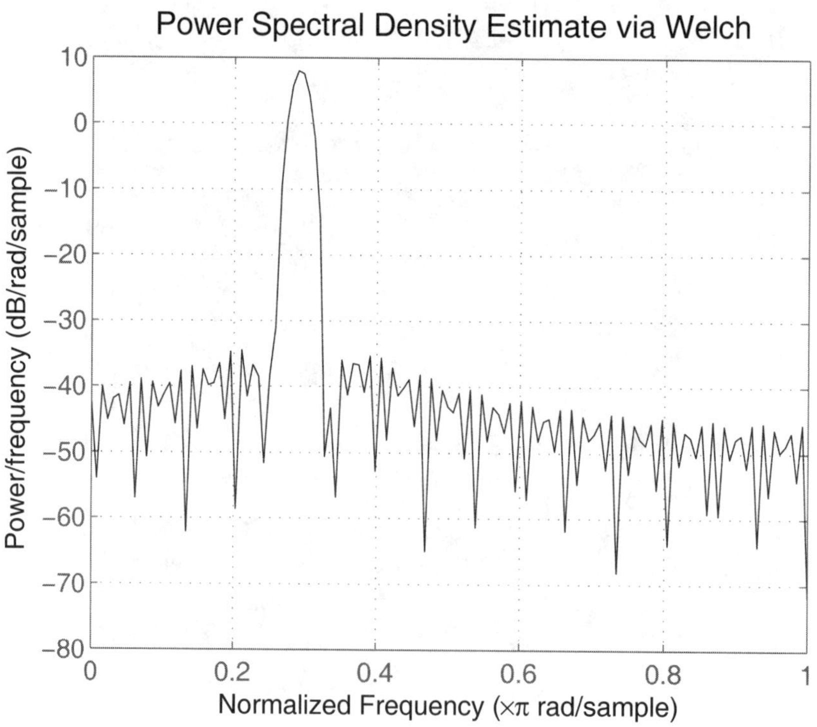

Figure 9.7: Spectrum plot from the `pwelch` command in MATLAB. The signal is a 7 kHz sinusoid, and the sample frequency is $F_s = 48$ kHz.

9.4 DSK Implementation in C

To achieve higher performance real-time spectrum analysis, we need to transition to a frame-based C program running on the DSK. However, we ideally want to view a live plot of the spectrum on the monitor of the host PC, which introduces an additional programming challenge. We have a general idea of how to program an FFT from Chapter 8, but how does the spectrum information get passed from the DSK back to the PC, and plotted on the monitor?

We could use some of the built-in capabilities of Code Composer Studio to do this, such as inserting a probe point at the appropriate place in the DSK program and using the rudimentary plot window available in Code Composer Studio. However, this approach will only work in DEBUG mode, and it halts the DSK's CPU for each plot update on the PC. While this technique might be useful in many situations (particularly for debugging purposes), it is not real-time operation.

Because this problem is more a problem of how to pass information back from the DSK to the PC and plot it than it is a problem of how to do spectral analysis on the DSK, we need to put off this discussion for now. In Appendix E you'll find a discussion of how to pass information back from the DSK and plot it on the PC monitor. This is explained in the appendix with a very basic real-time oscilloscope application first, followed by a short discussion and example of how to make the transition to real-time spectral analysis.

9.5 Conclusion

This concludes our coverage of what we call the *enduring fundamentals* of DSP, presented in the context of real-time operation. The next section of the book presents a series of projects that we hope you will find as fun and exciting as our students have. Don't stop now, for the fun is just beginning...

9.6 Follow-On Challenges

Consider extending what you have learned.

1. Use the winDSK6 spectrum analyzer to view a sinusoidal test signal. Increase the amplitude of this input signal until you exceed the limits of the ADC and clipping occurs. Does the spectrum change as you expect when the signal is clipped?

2. Use the winDSK6 spectrum analyzer to view a square wave test signal. Does the spectrum match what you expect from Fourier theory (i.e., do you see just odd harmonics that decrease in amplitude as they increase in frequency)?

3. Using the winDSK6 spectrum analyzer, select different windows and different numbers of frames for averaging. How does the spectral result change?

4. Explore all the options available for the `pwelch` function in MATLAB. Do the same for `periodogram`, `spectrogram`, `pburg`, and `pmusic`. Use various known input signals both with and without random noise added, and compare these different methods of spectral estimation.

5. Add features to the real-time spectrum analyzer example shown in Appendix E, such as those that are available in winDSK6.

Projects

Chapter 10

Project 1: Guitar Special Effects

10.1 Introduction to Projects

SINCE this is the first chapter in the Projects section, let's explain how this part of the book differs from the previous section. All the the previous chapters cover topics we have called the "enduring fundamentals" of DSP. The topics were relatively broad, and the examples shown were meant only to illustrate that particular topic.

In the Projects section, we change our focus somewhat; we will now provide in each chapter at least one fully functional version of an interesting DSP application we have found to be popular with students over the years. In particular, the two general areas students seem to be drawn to when first learning DSP are audio (such as the special effects of this chapter and the graphics equalizer of the next chapter) and communications (such as the various receiver and transmitter related projects of later chapters). While we provide fully functional code to get you started, we intentionally leave it to you to implement and improve the applications further. That's why they're called *projects*!

We remind the reader that most of the code shown for the projects in this and the following chapters is intentionally *not* fully optimized, because maximally efficient code is often very hard to understand—and our goal is for you to *understand* the code! You may want to explore ways in which you can improve the efficiency of these projects on your own. This first project chapter involves concepts that are fairly easy to understand; later project chapters progress to increasingly more involved examples.

10.2 Theory

10.2.1 Background

Special effects for electric guitars (and microphones) are a fun application of DSP. You can create all sorts of interesting sound variations quite easily with fairly simple DSP algorithms. Some of the more familiar effects include echo, chorus, flanger, phasing, reverb, tremelo, frequency translation, subharmonic generation, ring modulation, fuzz, compression/expansion, equalization, noise gating, and others. Whether you know them by these common names or not, you probably know what each of these effects "sounds" like if you've listened to even a moderate amount of popular music over the last forty years or so. We will discuss a representative group of these effects in this chapter.

Special effects for electric guitars started to become more popular in the 1960s, not long after electric guitars themselves became widely available. Of course in the early days these special effects were typically created with hardwired analog circuit designs, and thus a single "effects box" (as they were called) could only produce a single type of effect. As a result, an electric guitar player who needed multiple effects used to be surrounded by a gaggle of effects boxes and pedals on the floor, each requiring its own foot switch, power, and signal cable. Later, some analog effects boxes were redesigned to include multiple related effects such as echo and reverb, or chorus and flanger, but the "rat's nest" of cables was still a problem. Another serious problem was the noise (particularly hum) which was added to the analog guitar signal, due in large part to using so many cables and connections between the guitar and the amplifier.

When DSP was first being taught in engineering colleges, algorithms were quickly derived that would produce the special effects being used by so many musicians. Unfortunately, the cost of the required hardware remained prohibitive for many years. But in the 1990s, this all changed and digital implementation of guitar special effects quickly replaced analog designs. A single DSP-based effects box is able to produce many variations of special effects, with improved signal-to-noise ratio and a single cable. As more people became familiar with DSP, new effects never produced with analog boxes were invented; this continues to this day.

10.2.2   How the Effects Work

A brief introduction to a few of these types of special effects was first given in Chapter 3 in the context of FIR filters. In this chapter we'll discuss these and other special effects in a more general way.

The simplest special effect is an echo. A block diagram of how this can be implemented with a simple delay is shown in Figure 10.1. Notice the similarity of Figure 10.1 to Figure 3.10. The two diagrams in Figure 10.1 are equivalent, where the bottom diagram uses a more common representation for delay and gain. To set the amount of delay time, the value R specifies the number of sample periods of delay. For example, if the sample frequency is $F_s = 48$ kHz, then one sample period is $t_s = 1/F_s$ or 20.83 μs. Recall that in most DSP implementations, each sample period of delay requires a memory location; a one second delay in this case would thus require $R = 48,000$ and need a memory array 48,000 locations in length. This has some practical ramifications that will be discussed later. Note that

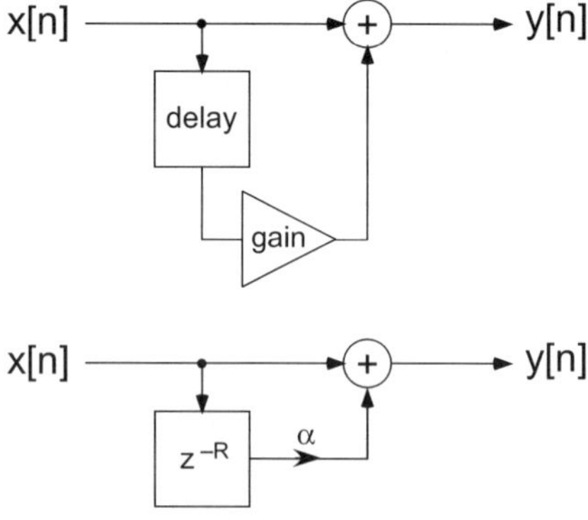

Figure 10.1: Block diagrams of a simple echo (or delay) effect using an FIR filter. The delayed version of the sound (delayed by R sample times), with amplification specified by gain value α, is added back to the non-delayed sound.

Figure 10.1 shows what is essentially an FIR filter, since only feedforward signal paths are used. This filter will result in a single echo, which may or may not be what you want. The filters in Figure 10.1 are of a type that are sometimes called "comb filters" for reasons that will become apparent.

To achieve multiple echos, we need to use a feedback signal path (sometimes called "regeneration" by musicians), which means we are now talking about an IIR filter. A block diagram of two simple ways this can be implemented is shown in Figure 10.2. To be consistent with most musical special effects references, the sign of the feedback coefficient (α) is the opposite of what was used in Chapter 4, but the transfer function equations presented later in this chapter will take this sign change into account. In IIR implementations, the "echoes" repeat forever, but the volume of the delayed sound decreases each sample time due to the stability condition of $|\alpha| < 1.0$. The repeated sound will thus "fade away" and be inaudible after some finite number of sample times. The filters in Figure 10.2, just like the one in Figure 10.1, are also called "comb filters." Both FIR and IIR comb filters are very common building blocks for the special effects used by musicians.

In addition to comb filters, another common filter used for special effects is the "allpass filter," which uses a combination of feedforward and feedback (with complementary gain values). Generic block diagrams of an allpass filter are shown in Figure 10.3. The version of the filter shown at the top of Figure 10.3 is the most basic implementation; the version

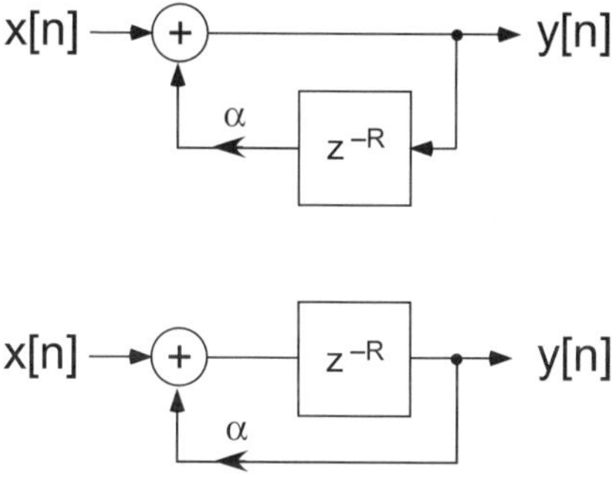

Figure 10.2: Block diagrams of a multiple echo effect using IIR filters. Ensure $|\alpha| < 1.0$ for stability.

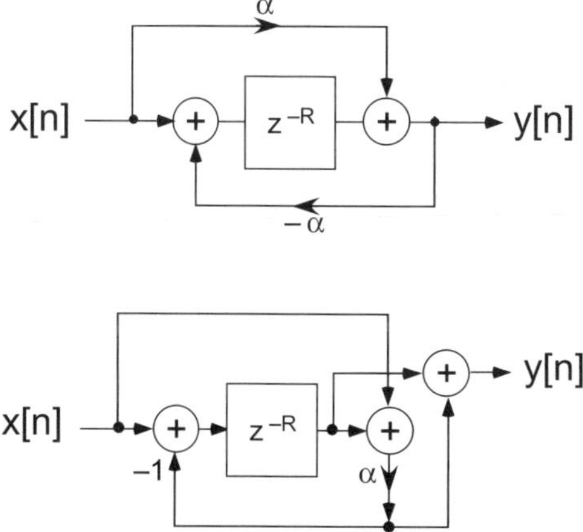

Figure 10.3: Block diagrams for allpass filters. Ensure $|\alpha| < 1.0$ for stability.

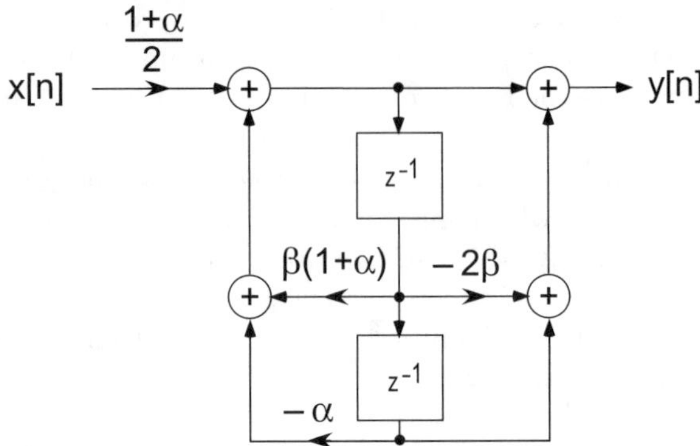

Figure 10.4: Block diagram for a second order IIR notch filter, implemented as Direct Form II. The notch frequency is determined by β and the width of the notch is determined by α.

shown at the bottom of the figure implements an identical filter (see [4]) but only requires a single multiplication operation.

The final type of filter we'll discuss is the "notch filter." While it is easy to create an FIR notch filter, an IIR implementation allows far more flexibility and is more commonly used. A block diagram for a versatile second order IIR notch filter is shown in Figure 10.4.

FIR Comb Filters

Since comb filters, both FIR and IIR, are used so often in creating special effects, we will examine them in more detail. We begin with the FIR version shown earlier in Figure 10.1. The transfer function of the filter in Figure 10.1 is

$$H(z) = 1 + \alpha z^{-R}$$

and from this equation the frequency response can be calculated. For example, using the `freqz` command in MATLAB where $R = 10$ and $\alpha = 1$ yields the frequency response seen in Figure 10.5.

The magnitude of the frequency response (plotted here on a logarithmic (dB) scale) shows multiple evenly spaced passbands and stopbands of the filter. The magnitude of the frequency response resembles the teeth of a comb, which is where the name originated. Being a symmetric FIR filter, the phase response is linear in each of the passbands.[1] For our purposes, this means the delay time for all frequencies in the passband will be equal (i.e., constant group delay). The stopbands arise from the fact that the delay or phase shift of certain frequencies approaches 180 degrees, so when this is added back to the original signal these frequencies tend to cancel each other out. These stopbands, or nulls, occur at the frequencies along the unit circle of the z-plane where the zeros of the transfer function occur. To see this on a z-plane plot in MATLAB where α is represented by `alpha`, we can use the command `zplane([1 zeros(1, R-1) alpha], 1)`. Whether you prefer to think of it as due to phase cancellation or as due to zeros, the result is that frequencies in the stopbands are attenuated, so this filter provides both a delay *and* an associated change in the "tone" of the sound. The number of passbands and stopbands is directly related to the value of R, which can be easily seen in Figure 10.6.

---

[1] If you change the value of α to something other than 1.0, then the filter is no longer truly symmetric and the phase will thus no longer be truly linear.

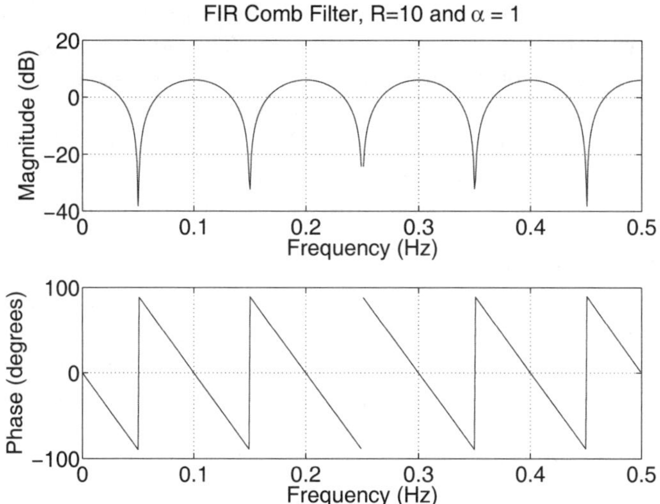

Figure 10.5: FIR comb filter response, with a normalized sample frequency of $F_s = 1$ Hz.

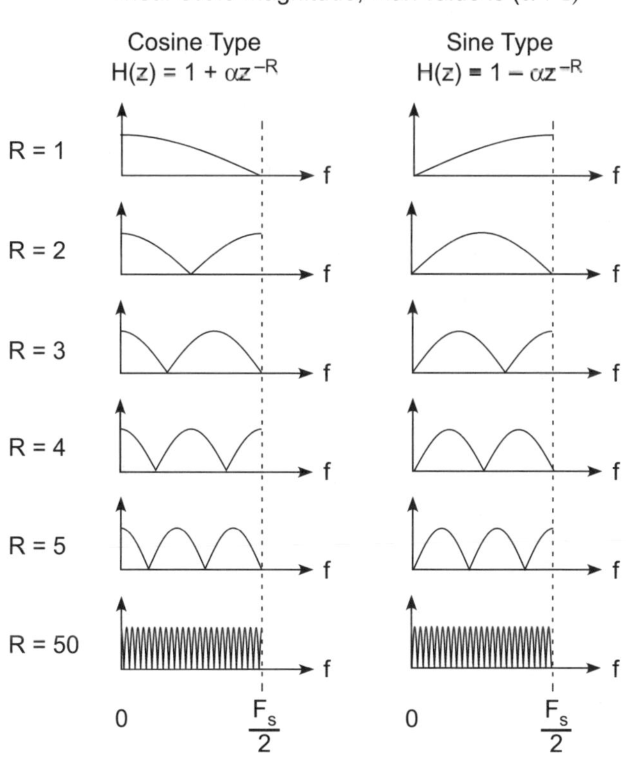

Figure 10.6: Effect of changing delay value R for an FIR comb filter. The linear scale magnitude of the frequency response is shown for two types of FIR comb filters.

We can infer many things from Figure 10.6. First, the number of stopbands is equal to $R/2$; for example when $R = 5$ we have two and "one half" stopbands in the magnitude response of the filter. For "cosine type" comb filters the stopbands are located at $0.5(F_s/R), 1.5(F_s/R), 2.5(F_s/R), \ldots$ and so on, however many fit between 0 and $F_s/2$. For "sine type" comb filters the stopbands are located at $0, (F_s/R), 2(F_s/R), 3(F_s/R), \ldots$ and so on, however many fit between 0 and $F_s/2$. Second, α effects the maximum value of the magnitude; for example, for the common setting of $\alpha = 1.0$ the maximum magnitude is 2.0 (on a logarithmic scale this would be +6 dB). When using real-world DSP hardware, we have to be careful in setting *alpha* not to exceed the dynamic range of the DAC; an overall scale factor of less than one may be needed. Third, if we *subtract* the delayed sample from the non-delayed sample instead of *adding* it, we get a different response, as seen on the right hand side of Figure 10.6. Fourth, as we use longer delay times such as is shown for $R = 50$, the width of the passbands and stopbands is so small that the frequency response is essentially flat, which is similar to an allpass filter. An allpass filter passes all frequencies equally well (so the "tone" of the sound is unaffected), but due to its phase response a pure delay is provided. The difference in this respect between Figure 10.1 and Figure 10.3 is that Figure 10.3 provides an allpass response at *all* values of R, which will be very useful for some types of special effects.

IIR Comb Filters

The transfer function of the filter at the top of Figure 10.2 is

$$H(z) = \frac{1}{1 - \alpha z^{-R}} \qquad |\alpha| < 1$$

which you should recognize as being IIR in nature. Again using the `freqz` command in MATLAB where $R = 10$ but $\alpha = 0.8$ (for stability), the frequency response can be seen in Figure 10.7. The magnitude of the frequency response shows multiple evenly spaced passbands and stopbands of the filter, similar to the teeth of a comb. As an IIR filter, the disadvantages are that the phase response cannot be truly linear in each of the passbands

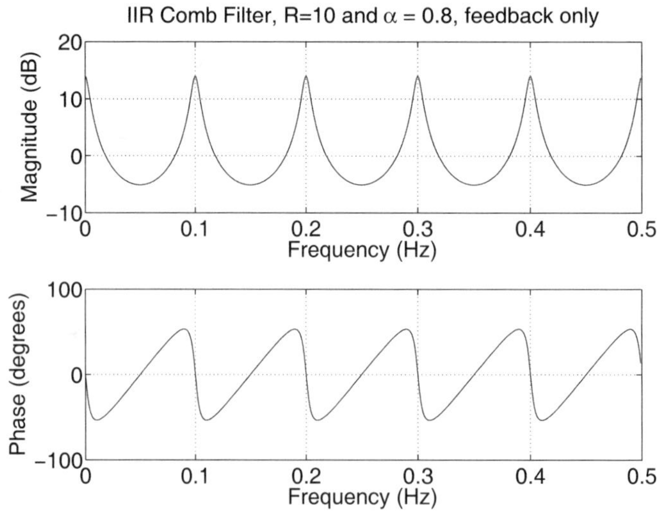

Figure 10.7: Response for the IIR comb filter shown at the top of Figure 10.2. The sample frequency has been normalized to $F_s = 1$ Hz.

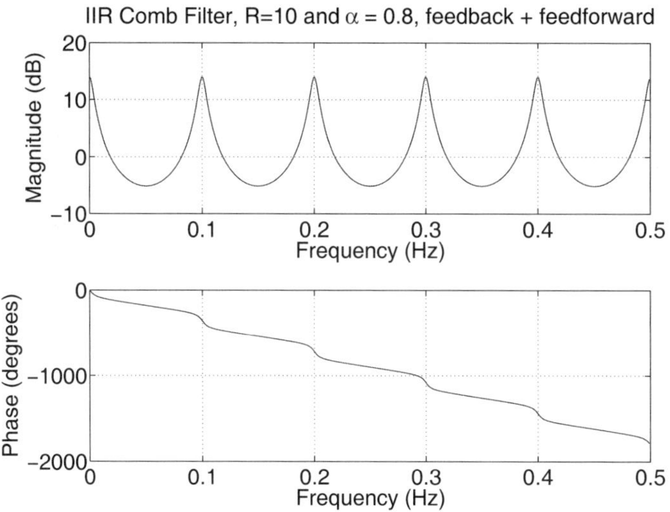

Figure 10.8: Response for the IIR comb filter shown at the bottom of Figure 10.2. The sample frequency has been normalized to $F_s = 1$ Hz.

and the filter can be unstable (if $\alpha \geq 1.0$ through design error or coefficient quantization). However, an advantage of the IIR version is that the passbands can be much sharper than is possible with the FIR version for similar memory size and/or computational requirements.

How is the the filter at the bottom of Figure 10.2 different from the filter at the top of Figure 10.2? The transfer function of the filter at the bottom of Figure 10.2 is

$$H(z) = \frac{z^{-R}}{1 - \alpha z^{-R}} \qquad |\alpha| < 1$$

which results in the frequency response shown in Figure 10.8. Note that while the magnitude of the frequency response shows essentially identical passbands and stopbands as was seen in Figure 10.7, the phase response of this filter is much closer to linear. This "almost linear" phase response provides a more uniform delay across the frequency range. For this reason, when an IIR comb filter is needed for certain audio special effects (particularly reverb), the version shown at the bottom of Figure 10.2 is often used. However, in this version the present time input $x[n]$ is delayed and not passed through directly to the output (i.e., there is no $x[n]$ term in the filter's difference equation, only an $x[n - R]$ term). Thus there is no "present time" output from this filter, which for some other types of effects would be undesirable (e.g., if you plucked a note on your guitar, nothing would be heard until after the delay time has passed). Both versions shown in Figure 10.2 have their uses.

Allpass Filters

The transfer function of both of the filters shown in Figure 10.3 is

$$H(z) = \frac{\alpha + z^{-R}}{1 + \alpha z^{-R}} \qquad |\alpha| < 1$$

and from this the frequency response can be easily shown. The **freqz** command in MATLAB, using $R = 10$ and $\alpha = 0.8$, provides the frequency response shown in Figure 10.9.

Note that the magnitude of the frequency response shows a "flat" gain of 1 (i.e., 0 dB) for all frequencies from DC to $F_s/2$, from which the "allpass" name is derived. However,

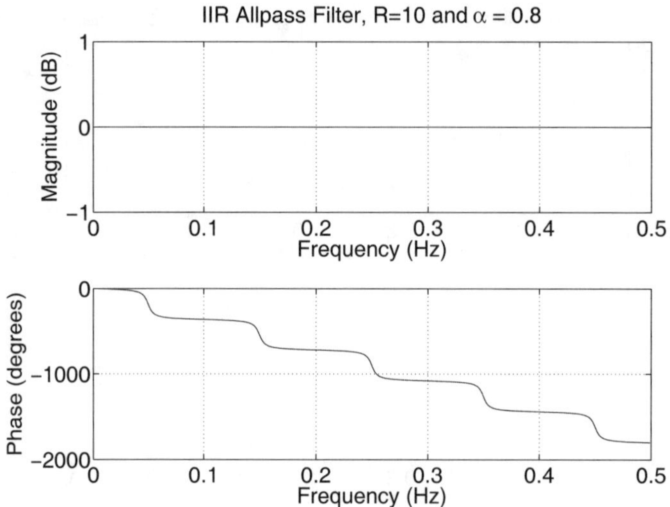

Figure 10.9: Response for the IIR allpass filters shown in Figure 10.3. The sample frequency has been normalized to $F_s = 1$ Hz.

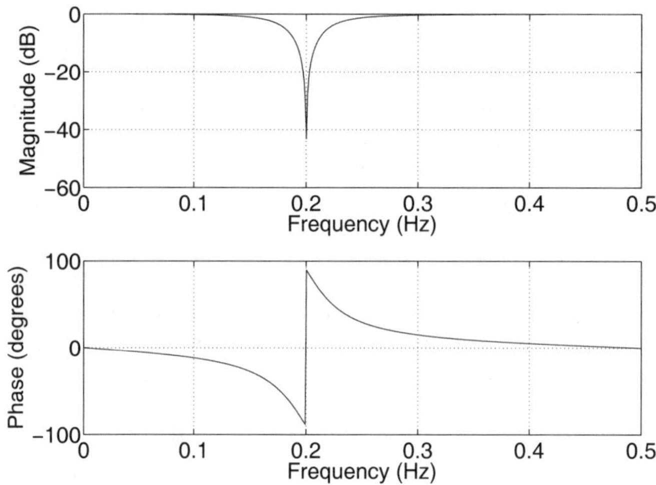

Figure 10.10: Response for the IIR notch filter of Figure 10.4. The sample frequency has been normalized to $F_s = 1$ Hz.

from the phase response we can see that this filter provides delay (note that the delay will be a different value in the narrow regions where the stopbands would exist for a comb filter using the same value of R). Thus this filter is useful for situations such as reverb effects where you need to delay the sound without unduly "coloring" the tone of the sound.

Notch Filters

A notch filter is similar to a comb filter, but rather than having multiple evenly spaced stopbands, it has only a single stopband. An example frequency response of a notch filter is shown in Figure 10.10. Ideally, both the "sharpness" of the stopband and the location of

Table 10.1: Typical methods for creating various special effects with basic filters. Delay times listed are only suggestions.

Effect	Filter	Type	Delay (ms)	Type of delay
single echo	comb	FIR	> 100	constant
multiple echos	comb	IIR	> 100	constant
doubling	comb	FIR or IIR	50–100	constant
chorus	comb	FIR	20–30	slowly varying
flanger	comb	FIR	1–10	slowly varying
flanger (metallic)	comb	IIR	1–10	slowly varying
phasing	allpass or notch	IIR	< 20	slowly varying
reverb	comb and allpass	FIR and IIR	multiple	constant

the stopband on the frequency axis would be adjustable. An efficient and versatile design that achieves this is defined by the notch filter shown in Figure 10.4, which has the transfer function of

$$H(z) = \frac{1+\alpha}{2} \frac{1 - 2\beta z^{-1} + z^{-2}}{1 - \beta(1+\alpha)z^{-1} + \alpha z^{-2}} \qquad 0 < \alpha < 1, \ -1 \le \beta \le 1$$

where the notch frequency is determined by β, the width of the notch is determined by α, and the first term of $(1 + \alpha)/2$ is a scale factor used to normalize the overall filter gain. To set a particular notch frequency f_0 in the allowed range of 0 to $F_s/2$, simply set $\beta = \cos(2\pi f_0/F_s)$. Regarding the notch width, the closer α approaches 1.0, the more narrow will be the notch (but if $\alpha \ge 1.0$ the filter will be unstable). Experiment with different values.

Putting It All Together

Now that we have the building blocks (comb filters, allpass filters, notch filters) used for many of the most common special effects, let's put it all together. While up until now we have assumed a constant delay value of R (or a constant notch frequency determined by β), we will see that for many of the special effects we will need to vary the delay or notch frequency slowly over time. Table 10.1 lists the ways in which certain special effects can be created using the basic filters discussed above. In may cases, a very different sounding effect can be created with an identical filter form by simply changing the delay time or notch frequency (or by changing the range over which it varies). Some effects typically need just a single filter stage, while others (such as reverb) often need multiple filters to achieve the desired sound.

For example, a block diagram of the flanging effect is shown in Figure 10.11, where only a single comb filter is used. As before, α is the gain or scale factor, but instead of showing

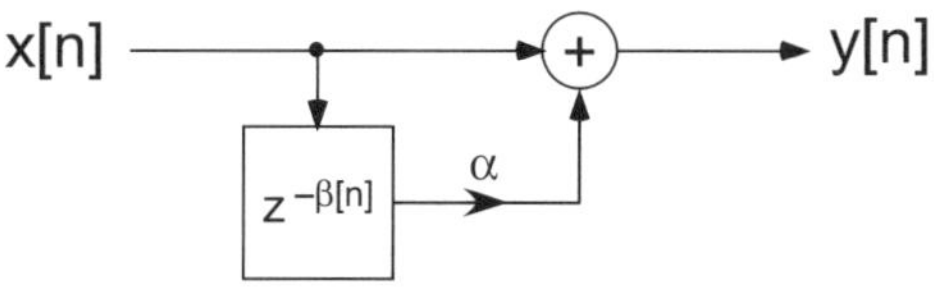

Figure 10.11: A block diagram of the flanging effect using a single comb filter.

a constant R for delay, we now use $\beta[n]$, which represents a periodically varying delay. A method to vary the delay sinusoidally is described by

$$\beta[n] = \frac{R}{2}\left[1 - \cos\left(2\pi\frac{f_0}{F_s}n\right)\right].$$

In the equation above, R is the maximum number of sample delays, f_0 is a relatively low frequency (often less than 1 Hz), and of course F_s is the sample frequency. This equation results in a slowly changing sinusoidally varying delay that ranges from 0 to R samples. Note that as an alternative to sinusoidally varying the delay, musicians sometimes choose to use a triangle, sawtooth, or even an exponential function for $\beta[n]$ in order to obtain a somewhat different sound.

A block diagram of the chorus effect is shown in Figure 10.12. To generate a chorus effect that makes one musician sound similar to four musicians playing the same notes, three separate chorus signals (identical to flanging except having longer delay times) are summed with the original signal. For the best sound, each of the β's and α's should be independent.

The phasing effect can be achieved in various ways. One method uses the output of an allpass filter with a slowly varying delay that is added back to the original signal; due to the phase shift, some frequencies will tend to cancel (creating notches). This effect is very similar to the comb filter used in the flanger. Another method, which is easier to fine tune to get exactly the sound you desire, uses the output of a notch filter with a slowly varying notch frequency that is added back to the original signal. Because it is easy to independently control the notch frequency and the notch width, this second method has many advocates.

Two block diagrams that can implement the phasing effect are shown in Figure 10.13. To generate a rich phasing sound, more than one separately phased signal can be summed with the original signal. For the best sound, it is recommended that the notch frequencies *not* be evenly spaced or harmonically related; experiment with all the parameters to obtain the sound you wish.

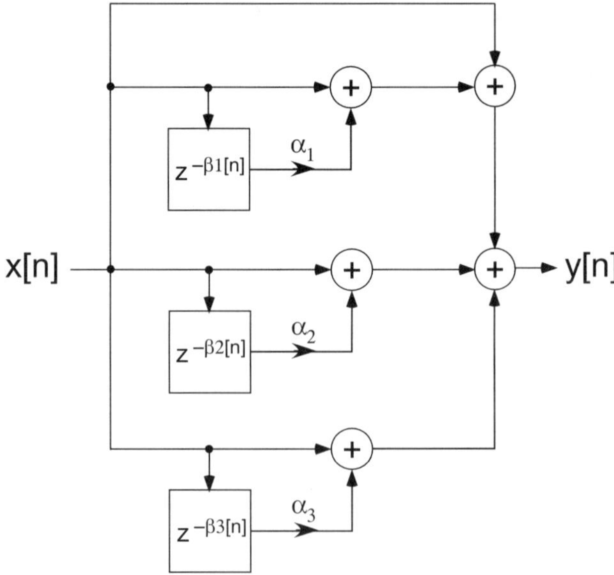

Figure 10.12: A block diagram of the chorus effect using three comb filters.

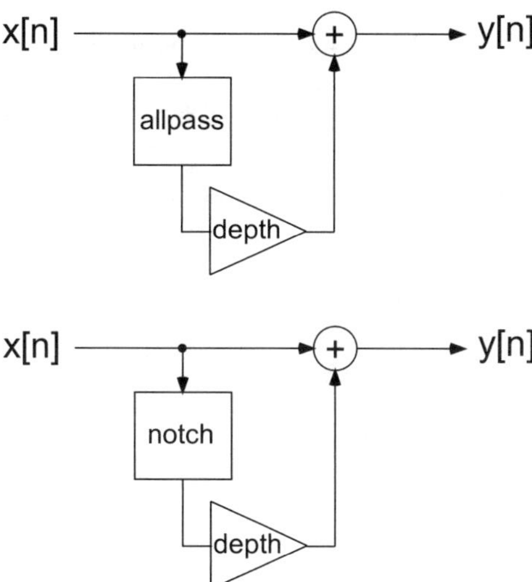

Figure 10.13: Block diagrams of the phasing effect using an allpass filter or a notch filter.

Reverb is an effect that we hear every day and in most cases take for granted. One proposed design [58] for a realistic-sounding reverb effect is shown in Figure 10.14. Modern music studios add some amount of reverb to almost every recording to compensate for consumers listening to the playback in relatively small rooms; without any added reverb the recording would have a "dead" sound to it. Reverb is caused by a multitude of sound

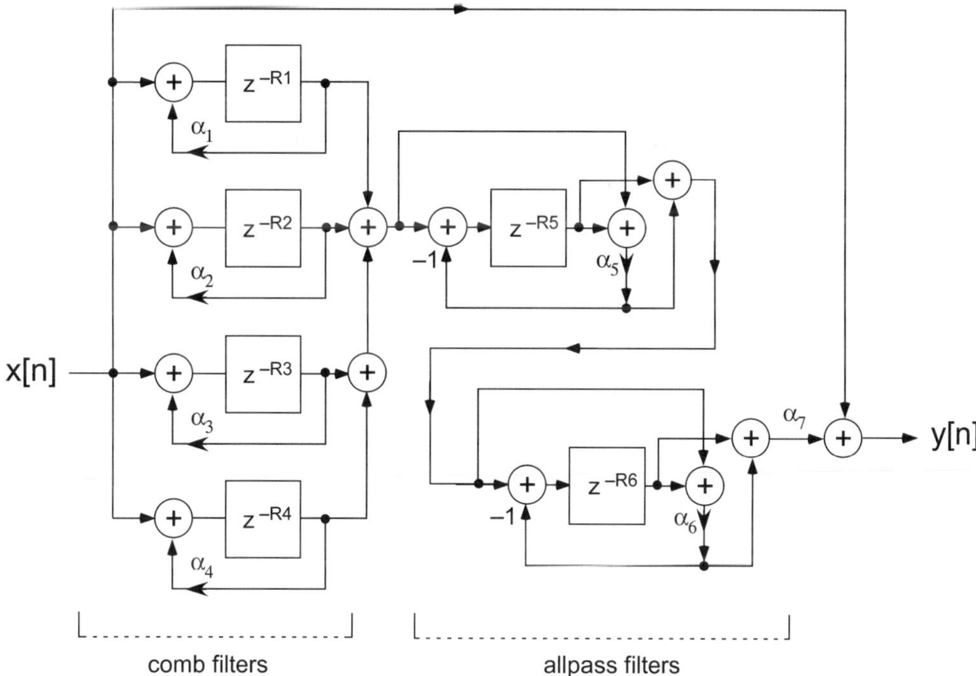

Figure 10.14: A proposed block diagram for the reverb effect using a combination of comb and allpass filters [58].

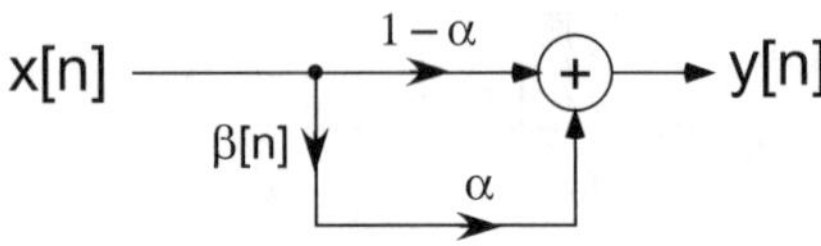

Figure 10.15: Block diagrams of the tremelo effect. The bottom block diagram, while less simple than the top, allows control of the amount or "depth" of the tremelo effect compared to the original signal.

reflections coming from the walls of the room or concert hall in which it is played; in a small room the delay times are too short to notice. In a larger room we hear these reflections arrive at many different times, so for a realistic effect we need to use multiple delay times. While the block diagram of Figure 10.14 may seem complex, it is actually much simpler than some of the modern studio-quality reverb algorithms. Discussion of these studio-quality algorithms would be beyond the scope of this chapter.

We now see that echo, chorus, flanger, phasing, and reverb are all created with some combination of delays that effect the phase of the signal, which is easy to do in DSP. Other effects such as tremelo, fuzz, compression/expansion, and noise gating are created by altering the amplitude (rather than the phase) of the signal.

In general, intentional variations in the amplitude of a signal is called "amplitude modulation" or AM. This is used in radio communications systems, but also in special effects. The two most common special effects that use AM are tremelo and ring modulation.

Tremelo[2] (also spelled "tremolo") is simply a repetitive up/down variation in the volume of the signal.[3] Listen to the original version of the classic song "Crimson and Clover" by Tommy James and the Shondells for a great example of tremelo. Block diagrams that show how simple it is to implement tremelo are shown in Figure 10.15. The rate of the variation in volume is controlled by the time changing nature of β, and the amount or "depth" of the tremelo effect compared to the original signal is controlled by α, where $0 \le \alpha \le 1$. Tremelo usually varies the volume at a constant sinusoidal rate, at a frequency below 20 Hz (some claim 7 Hz to be "ideal"), which can be expressed as

$$\beta[n] = \frac{1}{2}\left[1 - \cos\left(2\pi\frac{f_0}{F_s}n\right)\right].$$

In this equation, f_0 is the frequency of variation and F_s is the sample frequency. A communications engineer would view tremelo as form of amplitude modulation called "double sideband large carrier" (DSB-LC) because the "carrier" frequency (in tremelo this is just the original signal) will always show up in the output.

Ring modulation is a special effect whereby the guitar signal is multiplied by some other signal, usually an internally generated constant frequency sinusoidal signal such as $\beta[n] = \cos(2\pi(f_0/F_s)n)$. A block diagram that shows how simple it is to implement ring modulation is shown in Figure 10.16. Pure multiplication of any two frequencies f_1 and f_2 results in the sum $(f_1 + f_2)$ and the difference $(f_1 - f_2)$ frequencies. In communications

---

[2]This is not to be confused with the "tremelo technique" in classical guitar and flamenco guitar, which is a particular right hand effect for the bass and treble strings.

[3]Some electric guitar and amplifier manufacturers confuse "tremelo" with "vibrato" but the correct definition of vibrato is an up/down variation in *pitch* not volume.

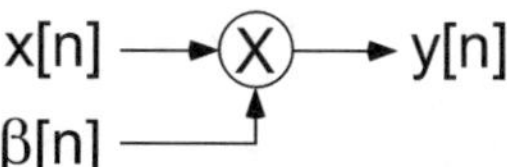

Figure 10.16: Block diagrams of the ring modulation effect. Signal $\beta[n]$ is usually an internally generated sinusoid, such as $\beta[n] = \cos(2\pi(f_0/F_s)n)$.

theory this technique is a form of amplitude modulation called "double sideband suppressed carrier" (DSB-SC). The "suppressed carrier" modifier refers to the fact that the "carrier" (in Figure 10.16 this would be $\beta[n]$) does not show up in the output (nor does the original guitar signal $x[n]$, for that matter). In fact, a ring modulator can create tremelo by setting $\beta[n] = [1 + \alpha\cos(2\pi(f_0/F_s)n)]$, where, as before, α controls depth for $0 \le \alpha \le 1$. In ring modulation, the frequency of $\beta[n]$ is usually higher than it is for tremelo; selecting a frequency somewhere in the 500 Hz to 1 kHz range is fairly common. Be aware that aliasing will occur if the ring modulator's sum frequencies exceed $F_s/2$ (a result which you may or may not want, depending upon the sound you seek).

Fuzz is an intentionally introduced distortion in the signal, typically caused by "clipping" or limiting the amplitude variations of the signal (see Figure 10.17 for a simple example). This effect was first discovered accidentally when the dynamic range of a tube-based amplifier stage was exceeded, which caused clipping, which in turn resulted in higher frequency harmonics (harmonic distortion) being added to the original signal. Heated debates continue today about the "tube" sound of fuzz versus the "solid-state" or "transistor" sound of fuzz. Keep in mind that Figure 10.17 is a very simplistic example of clipping. The signal can be clipped so that the positive portion is limited to a different magnitude than the negative portion, or the clipping can be gradual rather than "flat;" these variations (and others) will all produce a different sound of fuzz effect. Furthermore, the clipping is often followed by a frequency selective filter (lowpass or bandpass are the most common) to adjust the "harshness" or "color" of the fuzz sound. More sophisticated fuzz effects also have the option to engage a frequency selective filter before the clipping stage, in order to clip just a certain band of frequencies, then add this back to the unmodified or fully clipped signal. The possibilities are endless; experiment for the sound you like!

This has been a very brief discussion of the theory underlying the special effects used for electric guitars or microphones. Note that when combining multiple effects, the order of the effects in the signal chain will result in different sounds at the output. There are many articles and even entire books written on the subject of special effects, both at an engineering level and at the hobbyist level. See [4, Chapter 11] for an excellent introduction, and [58–60] for other treatments. A web search on the Internet will also yield a plethora of information;

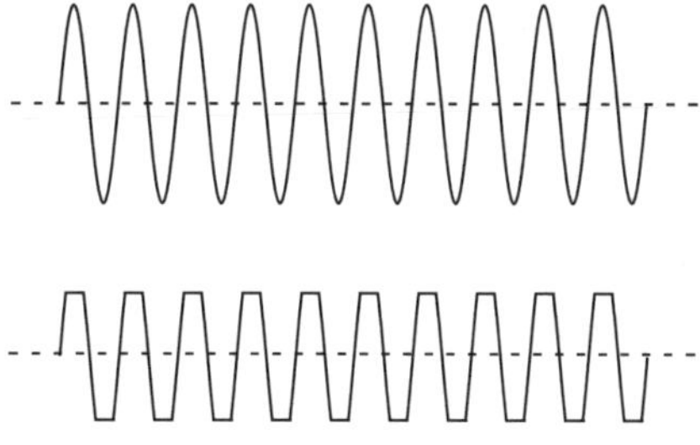

Figure 10.17: Clipping a signal produces the fuzz effect. Top: original signal. Bottom: symmetrically clipped signal. Asymmetrical clipping can be used for a somewhat different fuzz sound.

at the time of this writing one of the best sources was `http://www.harmony-central.com/Effects/`, particularly the "Effects Explained" link. A more theoretical treatment can be found at the site `http://www.sfu.ca/sonic-studio/handbook/index.html`. Both of these web sites have many audio files that demonstrate the various effects.

10.3   winDSK6 Demonstration

See Chapter 3 for a discussion of the Audio Effects that can be produced by winDSK6. At the click of a mouse, you can create effects such as echo, chorus, flanger, tremelo, frequency translation, subharmonic generation, and ring modulation (by selecting the DSB-SC option for tremelo). Equalization is available in the Graphic Equalizer application provided with winDSK6. Use these functions of winDSK6 to quickly compare with your own special effects so you'll know if you're getting the kind of sound associated with a particular effect.

10.4   MATLAB Implementation

The next step in the path to real-time DSP is to explore these filters in MATLAB to be sure we understand them. At first, we are free to take advantage of the "vectorized" optimizations available in MATLAB along with various built-in functions and toolbox commands. As we've mentioned in previous chapters, this is handy when you want to quickly check aspects of a DSP algorithm such as the output signal, output spectrum, pole-zero plot, and so on. But before moving on to a C program running on real-time DSP hardware, we must "de-vectorize" the MATLAB code and stop using built-in functions and toolbox commands so that our m-file is as close as possible to a C implementation, while making sure the program still works as expected. Then, and *only* then, will we be ready to move on to the next step of creating a real-time program in C. This is the pattern that has consistently resulted in student success time after time, and is the pattern we'll continue to follow for the projects.

In this section, we'll examine a subset of the filters described earlier in the chapter. None of these filters running under MATLAB operate in real time; in a later section we'll show example C code for real-time operation. In this section we'll show two versions of an FIR comb filter, three versions of an IIR comb filter, two versions of an IIR notch filter, and one version of a flanger. Given this foundation, the reader should be able to create in MATLAB any of the special effects filters described in this chapter.

10.4.1   FIR Comb Filter

The program listed below, `fir_comb1.m`, takes advantage of the built-in function called `filter`.

Listing 10.1: A MATLAB FIR comb filter example.

```
% Method using the filter command

R=round(R);   % ensure R is an integer before proceeding
A=1; % the "A" vector is a scalar equal to 1 for FIR filters
B=zeros(1,R+1);   % correct length of vector b
B(1)=1; B(R+1)=alpha;
% B vector is now ready to use filter command
y=filter(B,A,x);
```

This program can be found in the `matlab` directory of Chapter 10, along with all the other MATLAB code for this chapter; only the key parts of the programs are shown in this section.

The program below implements the filter shown at the bottom of Figure 10.1, with the response shown in Figure 10.5. The input variables are x (input vector), R (number of sample time delays desired), and *alpha* (feedforward coefficient α).

The program `fir_comb1.m` processes all the samples in the entire x input vector, but of course not in real time. The program was easy to write, but it's not suitable for getting us to a C implementation in real time. Note that lines 5 and 6 ensure the B coefficient vector has the proper length and values.

The next program, `fir_comb2.m`, is an identical filter that while also not capable of real-time operation is much closer to C code and will help us transition to C later. This program has the same input variables as before.

Listing 10.2: A MATLAB FIR comb filter example closer to C.

```
% Method using a more "C-like" technique

R=round(R);  % ensure R is an integer before proceeding
N1=length(x);  % number of samples in input array
y=zeros(size(x));  % preallocate output array
% create array and index for "circular buffer"
buffer=zeros(1,R+1);
oldest=0;
% "for loop" simulates real-time samples arriving one by one
for  i=1:N1
     buffer(oldest+1)=x(i);  % read input into circular buffer
     oldest=oldest + 1;  % increment buffer index
     oldest=mod(oldest,R+1);  % wrap index around
     y(i)=x(i) + alpha*buffer(oldest+1);
end
```

Instead of processing one sample at a time as we demonstrated for IIR filters in Chapter 4, this program uses a "for loop" to simulate samples arriving one by one, thus processing all the samples in the entire x input vector. This "for loop" would not be used in the real-time C program, but the code inside the "for loop" would be used in an interrupt service routine (ISR) for sample-by-sample processing. The use of the modulus operator in line 13 causes the index value `oldest` to "wrap around" creating a circular buffer as was discussed in Chapter 3 (see Figure 3.17). The use of "oldest + 1" in lines 11 and 14 is due to MATLAB's rule of the array index values starting at 1 instead of zero. Incrementing the index value in line 12 causes the index to point at the *oldest* sample in the buffer. If you aren't sure about this, draw a short circular buffer on paper and run through the whole circle a few times, placing samples in the buffer one by one. Now are you convinced?

You may want to use the handy program `demo_fir_comb2.m` to read in a WAV audio file, run the comb filter, and play the result. Experiment with different values, especially the delay value, referring to Table 10.1.

10.4.2   IIR Comb Filter

The program listed below, `iir_comb1.m`, again takes advantage of the built-in function called `filter`. It implements the IIR comb filter shown at the bottom of Figure 10.2, with the response shown in Figure 10.8. The input variables are the same as before (x is the input vector, R is the number of sample time delays desired), and *alpha* is now the feed**back** coefficient.

Listing 10.3: A MATLAB IIR comb filter example.

```matlab
% Method using the filter command

R=round(R);   % ensure R is an integer before proceeding
A=zeros(1,R+1);   % correct length of vector A
A(1)=1; A(R+1)=-alpha;
% A vector is now ready to use filter command
B=zeros(1,R+1);   % correct length of vector B
B(R+1)=1;  % use "B=1;" for other IIR version
% B vector is now ready to use filter command
y=filter(B,A,x);
```

Lines 4 and 5 ensure the A coefficient vector has the proper length and values (take note of the sign needed for **alpha**). Lines 7 and 8 set up the B vector properly. If you would rather implement the IIR comb filter shown at the top of Figure 10.2, comment out line 7 and change line 8 to read "B=1;" as noted in the comment for line 9.

The next program, **iir_comb2.m**, is an identical filter that is much closer to C code and has the same input variables as before. It implements the filter in "Direct Form I" (see Figure 4.14).

Listing 10.4: A MATLAB IIR comb filter example closer to C, in Direct Form I.

```matlab
% Method using a more "C-like" technique

R=round(R);   % ensure R is an integer before proceeding
N1=length(x); % number of samples in input array
y=zeros(size(x));   % preallocate output array
% create array and index for "circular buffer"
bufferx=zeros(1,R+1); % to hold the delayed x values
buffery=zeros(1,R+1); % to hold the delayed y values
oldest=0; newest=0;
% "for loop" simulates real-time samples arriving one by one
for i=1:N1
    bufferx(oldest+1)=x(i); % read input into circular buffer
    newest=oldest; % save value of index before incrementing
    oldest=oldest + 1; % increment buffer index
    oldest=mod(oldest,R+1); % wrap index around
    y(i)=bufferx(oldest+1) + alpha*buffery(oldest+1);
    buffery(newest+1)=y(i);
end
```

We again use a "for loop" to simulate samples arriving one by one, thus processing all the samples in the entire x input vector. Since this filter is derived from the block diagram (not the transfer function), we don't want a negative sign this time on the value of **alpha** for feedback (see line 16). The code shown implements the IIR comb filter shown at the bottom of Figure 10.2. To implement the IIR comb filter shown at the top of Figure 10.2, just make a change in line 16 such that the index used for **bufferx** is **newest** rather than **oldest**.

While a Direct Form I filter implementation is easy derive from the difference equation or transfer function, its key disadvantage is the need for two buffers (circular buffers as before): one to hold the delayed x values and one to hold the delayed y values. We can eliminate this inefficiency easily by using a Direct Form II implementation similar to Figure 4.16. This is shown below in **iir_comb3.m**.

Listing 10.5: A MATLAB IIR comb filter example closer to C, in Direct Form II.

```matlab
% Method using a more "C-like" technique

R=round(R);  % ensure R is an integer before proceeding
N1=length(x); % number of samples in input array
y=zeros(size(x));  % preallocate output array
% create array and index for "circular buffer"
buffer=zeros(1,R+1); % to hold the delayed values
oldest=0; newest=0;
% "for loop" simulates real-time samples arriving one by one
for i=1:N1
    newest=oldest; % save value of index before incrementing
    oldest=oldest + 1; % increment buffer index
    oldest=mod(oldest,R+1); % wrap index around
    buffer(newest+1)=x(i) + alpha*buffer(oldest+1);
    y(i)=buffer(oldest+1);
end
```

While this implements exactly the same filter as before, only one circular buffer is needed. You may want to use the program `demo_iir_comb3.m` to read in a WAV audio file (perhaps from the `test_signals` directory), run the comb filter, and play the result. Experiment with different values, especially the delay value, referring to Table 10.1. Once again, the code shown in the listing implements the IIR comb filter shown at the bottom of Figure 10.2. To implement the IIR comb filter shown at the top of Figure 10.2, just make a change in line 15 such that the index used for `buffer` is `newest` rather than `oldest`.

Notice that all the filter programs above that use circular buffers and "for loops" run *much* faster than those using the built-in MATLAB `filter` command. The `filter` command is very general and is therefore not optimized to our specific use of it.

10.4.3 Notch Filter

We provide a notch filter program `notch1.m` that uses the built-in MATLAB `filter` command, but we don't bother to show its code here. A notch filter program `notch2.m` that implements a Direct Form II version in a very "C-like" manner is shown below. Its input variables are x (the input vector), *Beta* (to set the notch frequency), and *alpha* (to set the width of the notch).

Listing 10.6: A MATLAB IIR notch filter example, in Direct Form II.

```matlab
% Method using a more "C-like" technique

N1=length(x); % number of samples in input array
y=zeros(size(x));  % preallocate output array
% create array and index for "circular buffer"
% This second order filter only needs a buffer 3 elements long
buf=zeros(1,3); % to hold the delayed values
oldest=0; nextoldest=0; newest=0;
x=x*(1+alpha)/2; % scale input values for unity gain
% Set coefficients so calculation isn't done inside "for" loop
% If sweeping Beta over time, do this inside "for" loop
B0=1; B1=-2*Beta; B2=1;
A0=1; A1=Beta*(1+alpha); A2=-alpha;
```

```matlab
14  % "for loop" simulates real-time samples arriving one by one
    for i=1:N1
16      newest=oldest; % save value of index before incrementing
        oldest=oldest + 1; % increment buffer index
18      oldest=mod(oldest,3); % wrap index around
        nextoldest=oldest + 1; % increment buffer index again
20      nextoldest=mod(nextoldest,3); % wrap index around
        buf(newest+1)=x(i)+A1*buf(nextoldest+1)+A2*buf(oldest+1);
22      y(i)=B0*buf(newest+1)+B1*buf(nextoldest+1)+B2*buf(oldest+1);
    end
```

This program uses similar techniques to the previous examples, but you should verify by
comparing the code above to Figure 10.4 that the program really does result in the proper
filter being implemented. Note that because this type of notch filter is always second order,
we only need a circular buffer three elements long, and we access the current (**newest**) value,
the **nextoldest** value which has been delayed by one sample time, and the **oldest** value
which has been delayed by two sample times.

10.4.4   Flanger

Below is the MATLAB code for a flanger; it implements the filter shown in Figure 10.11. The
input variables are: x (input vector), t (maximum delay in seconds), $alpha$ (feedforward
coefficient), $f0$ (variation frequency delay time), and Fs (sample frequency). This program
is similar in many ways to Listing 10.2, except that the delay time is varied sinusoidally.

Listing 10.7: A MATLAB flanger example.

```matlab
1   % Method using a more "C-like" technique

3   Ts=1/Fs;  % time between samples
    R=round(t/Ts); % determine integer number of samples needed
5
    N1=length(x); % number of samples in input array
7   Bn=zeros(1,N1); % preallocate array for B[n]
    arg=0:N1-1; arg=2*pi*(f0/Fs)*arg;
9   Bn=(R/2)*(1-cos(arg)); % sinusoidally varying delays from 0-R
    Bn=round(Bn); % make the delays integer values
11
    y=zeros(size(x));  % preallocate output array
13  % create array and index for "circular buffer"
    buffer=zeros(1,R+1);
15  oldest=0;
    % "for loop" simulates real-time samples arriving one by one
17  for i=1:N1
        offset=R-Bn(i); % adjustment for varying delay
19      buffer(oldest+1)=x(i); % input sample into circular buffer
        oldest=oldest + 1; % increment buffer index
21      oldest=mod(oldest,R+1); % wrap index around
        offset=oldest+offset; % if delay=R this equates to fir_comb2
23      offset=mod(offset,R+1);
        y(i)=x(i) + alpha*buffer(offset+1);
25  end
```

Notice that lines 7 to 10 create $\beta[n]$, an array of sinusoidally varying integers ranging from 0 to R to be used as delay values. There are many ways to use $\beta[n]$ to make the index for the circular buffer to point to the value with the correct amount of delay. In the program above, we use a simple technique that clearly shows how this filter operation differs from the comb filter of Listing 10.2. Line 18 determines by how many locations the circular buffer index will need to be adjusted (compared to the comb filter of Listing 10.2) to account for the variable delay. For example, if $\beta[n] = R$ for a particular n then the offset calculated in line 18 will be $R - R = 0$ and the index value calculated in line 22 will be the same as that used in Listing 10.2, which results in a delay of R. At the other extreme, if $\beta[n] = 0$ then the offset calculated in line 18 will be $R - 0 = R$ and the index value calculated in line 22 will be such that it will "wrap around" the circular buffer back to the current sample, which results in a delay of 0. Thus the filter uses a sinusoidally varying delay, as needed for the flanger.

In Listing 10.7 above, the length of the array for $\beta[n]$ is the same as the length of the input vector, to keep things simple. When we transition to real-time C code, this would be impractical, as we can't predict how many input samples we'll be processing and probably wouldn't want to use an array that long anyway. So how do we overcome this? We just implement $\beta[n]$ as a separate circular buffer (with its own index variable) filled with sinusoidally varying values between 0 and R. How long should this buffer be? You don't need to store values in $\beta[n]$ that exceed one period of the sinusoid. A bit of thought should convince you that one period of the sinusoid must have a length of $Fs/f0$ elements. For example, if the sample frequency is 48 kHz and the frequency of the delay variation is 0.5 Hz (remember $f0$ is typically a very low frequency), then the array for $\beta[n]$ would need to be 96,000 elements long. There are techniques to cut this size to a half or a fourth of $Fs/f0$, but we leave that up to your imagination.

10.4.5　Tremelo

The concept of implementing $\beta[n]$ as a circular buffer is demonstrated below in the program tremelo.m, which implements the tremelo effect from the bottom of Figure 10.15.

Listing 10.8: A MATLAB tremelo example.

```
1  % Method using a "C-like" technique

3  N1=length(x); % number of samples in input array
   N2=Fs/f0;  % length for one period of a sinusoid
5  Bn=zeros(1,N2); % preallocate array for B[n]
   arg=0:N2-1; arg=2*pi*(f0/Fs)*arg;
7  Bn=(0.5)*(1-cos(arg)); % sinusoidally varying numbers from 0-1
   scale=1-alpha; % to scale the non-modulated component

9
   y=zeros(size(x));  % preallocate output array
11 % create index for "circular buffer" of Bn
   Bindex=0;
13 % "for loop" simulates real-time samples arriving one by one
   for i=1:N1
15     y(i)=scale*x(i) + Bn(Bindex+1)*alpha*x(i);
       Bindex=Bindex + 1; % increment Bn index
17     Bindex=mod(Bindex,N2); % wrap index around
   end
```

The input variables are: x (input vector), *alpha* (the gain of the amplitude modulated part of the signal), $f0$ (frequency of the modulation), and Fs (the sample frequency). This example also shows how simple multiplication allows you to adjust the amplitude of the signal. The fuzz effect, which clips amplitude, would be even easier to implement than tremelo.

The MATLAB code shown above (from the `matlab` directory of Chapter 10) provides the basic building blocks for you to create any of the special effects filters described under the Theory section of this chapter. We now turn our attention to the transition from MATLAB to real-time C code that will run on a DSK.

10.5   DSK Implementation in C

In this section, we get you started with making the transition from MATLAB code that is not real time to C code that will run in real time on the DSK. Once you see how to convert to C for a few types of filters, you will be able to create all the effects discussed in this chapter. We stay with sample-by-sample processing to keep the code simple, but there is no reason why you can't implement these same ideas using frame-based code.

10.5.1   Real-Time Comb Filters

We provide three versions of C code that implement comb filters in real time. All three versions are written in such a way that you can easily change between an FIR or an IIR filter by choosing which lines of code to uncomment. If you choose FIR, the C code closely follows the MATLAB example of Listing 10.2; if you choose IIR, the C code closely follows the MATLAB example of Listing 10.5 (i.e., Direct Form II so only one circular buffer is required).

The files necessary to run this application are in the `ccs\Echo` directory of Chapter 10. The primary files of interest are `ISRs_A.c`, `ISRs_B.c`, and `ISRs_C.c`. **Important:** you must only have *one* of these ISR files loaded as part of your project at any given time. The ISR files contain the necessary variable declarations and perform the actual filtering operation.

One aspect that needs to be discussed is how some of the variables are declared. Just as with the "C-like" MATLAB examples, an array implemented as a circular buffer will provide the delays needed for the filter.

Listing 10.9: Excerpt of variable declarations for the `ISRs_A.c` comb filter.

```
  unsigned int oldest = 0; // index for buffer value
2 #define BUFFER_LENGTH 96000 // buffer length in samples
  #pragma DATA_SECTION (buffer,"CEO"); // buffer in external SDRAM
4 volatile float buffer[2][BUFFER_LENGTH]; // for left and right
  volatile float gain = 0.75; // set gain value for echoed sample
```

The index value for the array is declared in line 1. The length of the buffer is defined in line 2, and the gain is defined in line 5, which is equivalent to R and α respectively in previous examples. Assuming a 48 kHz sample frequency, the value 96000 for R will provide a delay of two seconds. Lines 3 and 4 allocate memory needed for the buffer. Since we must have room for both left and right channel samples, the array actually requires $2 \times 96000 = 192000$ elements. Note that line 3 is critical, and its omission is a common error. The linker, which takes the outputs of the C compiler, will try to fit everything in the internal RAM for speed purposes, but an array of 192000 floats will not fit in the internal memory. Without line 3, the linker will generate an error message and the array will not be created. Unfortunately this error message, which shows up in the bottom Code Composer

Studio window, is often missed because it usually scrolls up and out of sight. But when you load and run the program without line 3 the filter output will be zero (silent). Line 3 directs the compiler to place the array that will be called `buffer` in the memory region we call `CE0`, which is external SDRAM. Thus the buffer will have enough room to be created, the linker will not complain, and the program will run correctly. Another common error is not using the **volatile** keyword in lines 4 and 5 (see Appendix F for more about these issues).

The part of `ISRs_A.c` that performs the actual filtering operation is part of the transmit ISR called `McBSP_Tx_ISR()`. To allow for the use of a stereo codec (e.g., the native C6713 codec, the PCM3006-based daughtercard codec for the C6711, etc.), the program implements independent Left and Right channel filters. However, for clarity, only the Left channel will be discussed following the code listing.

Listing 10.10: Real-time comb filter from <code>ISRs_A.c</code>.

```
1  xLeft=CodecData.Channel[LEFT]; // current LEFT input to float
   xRight=CodecData.Channel[RIGHT]; // current RIGHT input to float
3
   buffer[LEFT][oldest]=xLeft;
5  buffer[RIGHT][oldest]=xRight;
   newest=oldest; // save index value before incrementing
7  oldest=(++oldest)%BUFFER_LENGTH; // modulo for circular buffer

9  // use either FIR or IIR lines below

11 // for FIR comb filter effect, uncomment next two lines
   yLeft=xLeft + (gain * buffer[LEFT][oldest]);
13 yRight=xRight + (gain * buffer[RIGHT][oldest]);

15 // for IIR comb filter effect, uncomment four lines below
   //buffer[LEFT][newest]=xLeft + (gain * buffer[LEFT][oldest]);
17 //buffer[RIGHT][newest]=xRight + (gain * buffer[RIGHT][oldest]);
   //yLeft=buffer[LEFT][oldest];  // or use newest
19 //yRight=buffer[RIGHT][oldest];  // or use newest

21 CodecData.Channel[LEFT]=yLeft;   // setup the LEFT value
   CodecData.Channel[RIGHT]=yRight; // setup the RIGHT value
```

The real-time steps involved in comb filtering

An explanation of Listing 10.10 follows.

1. (Line 1): The transmit ISR first converts the current sample (obtained from the codec as a 16-bit integer by the receive ISR) to a floating point value and assigns it as current input element, equivalent to `x[0]`.

2. (Line 3): The current (i.e., the newest) sample is written into the circular buffer, overwriting the oldest sample.

3. (Line 6): The index value pointing to the newest value is saved before the next line of code causes the index to be incremented. This is used only for the IIR implementation. In fact, when running the FIR version it's likely that you'll get a compiler warning about the variable "newest" being set but not used. You can safely ignore this warning.

4. (Line 7): This is the line that causes the buffer to be "circular" in that the modulus operator (the % character in C and C++) causes the index to "wrap around" in the same way that `mod()` did in several of the earlier MATLAB examples. This line also includes the prefix `++`, which increments the index value `oldest` *before* the modulus is applied. Thus this line of code ensures the value of the index points to what is now the oldest sample in the circular buffer.

5. (Line 12): This line performs the FIR filter operation. Comment these lines out if you want the IIR comb filter.

6. (Lines 16 and 18): These lines together perform the IIR filter operation in a Direct Form II manner. Comment these lines out if you want the FIR comb filter. As shown, the code implements the IIR comb filter shown at the bottom of Figure 10.2. To implement the IIR comb filter shown at the top of Figure 10.2, simply change the index variable `oldest` in Line 18 to be `newest`.

7. (Line 21): This line of code transfers the result of the filtering operation, $y[0]$, to the `CodecDataIn.Channel[LEFT]` variable for transfer to the DAC side of the codec via the remaining code of the transmit ISR.

Now that you understand the code...

Go ahead and copy all of the files into a separate directory. Open the project in CCS and "Rebuild All." Once the build is complete, "Load Program" into the DSK and click on "Run." Your comb filter is now running on the DSK. Use a guitar or even a microphone as input and listen to the output. You should hear an echo (if FIR) or echoes (if IIR) spaced two seconds apart, due to the buffer length of each channel being 96000 long (assuming a sample frequency of 48 kHz). Feel free to change the definition of `BUFFER_LENGTH` to some other value to use a different value of R. After making the change, save the ISR file, rebuild the project, reload the program, and run it. In a similar fashion, you can experiment with changing the value of the `gain` variable to use a different value of α.

A small improvement to ISRs_A

While the modulus operator used on Line 7 above makes it easy to implement a circular buffer, it is not the recommended method for real-time processing. Modulus is the remainder after division, so using modulus in a line of code forces a division operation, which is costly in terms of CPU cycles since the DSP has no hardware support for division. A *far* more efficient way to implement a circular buffer is used in `ISRs_B.c`, where Line 7 is replaces with the two lines shown below.

Listing 10.11: Efficient circular buffer in the `ISRs_B.c` comb filter.

```
if (++oldest >= BUFFER_LENGTH) // implement circular buffer
        oldest = 0;
```

This is the only change between `ISRs_A` and `ISRs_B`. Even more efficient code could be realized by adopting the common practice of sizing your circular buffers to be a power of two (i.e., 2^n) in length, since the index wrap-around can then be accomplished very quickly by "AND-ing" it with $2^n - 1$. But we leave it to you to implement that change if desired.

To switch from using `ISRs_A` to `ISRs_B`, right click `ISRs_A.c` in the left project window and select "Remove from project." At the top of the Code Composer Studio window click "Project," "Add Files to Project" and select `ISRs_B.c`. Then click "Rebuild All." Once the build is complete, "Load Program" (or "Reload Program") into the DSK and click on "Run." Your improved comb filter is now running on the DSK.

ISRs_C allows interactive control

The `ISRs_C.c` file has some minor changes compared to `ISRs_B.c` that allow the use of what is called a GEL file. GEL, or General Extension Language, is explained again in the Chapter 11; we only briefly introduce it here.

We want to be able to interactively change the delay time R and the gain value α, so we'll create two slider controls that can be used for this. While `gain` is already declared as a variable and can thus be changed during program execution, the definition of `BUFFER_LENGTH` (which sets the delay time) is hard-coded at compile time. Thus we add the following line to allow the delay time to be changed during program execution.

```
int MyDelaySamples = BUFFER_LENGTH;   // manipulated by GEL file
```

The GEL file itself is rather simple, and is shown below.

Listing 10.12: GEL file `Echo.gel` to be used with the `ISRs_C.c` comb filter.

```
1  menuitem "Echo Controls";

3  // assumes 48 kHz Fs
   // delay from 0 to 2 s in 50 ms increments
5  slider DelayControl(0, 96000, 2400, 1, mydelayfactor)
   {
7    MyDelaySamples = mydelayfactor;
   }
9
   // converts to a float from 0 to 1 in 0.1 increments
11 slider GainControl(0, 10, 1, 1, mygainfactor)
   {
13   gain = (float)mygainfactor * 0.1F;
   }
```

An explanation of the GEL file code follows.

1. (Line 1): Adds the item "Echo Controls" to Code Composer Studio's GEL pulldown menu.

2. (Line 5): Creates a slider that will be labeled "DelayControl," with a bottom value of 0, top value of 96000, a single scroll increment of 2400, a page scroll of 1, and this information is assigned to the local GEL integer variable `mydelayfactor`.

3. (Line 7): The value of `mydelayfactor` is assigned to the C variable `MyDelaySamples` from `ISRs_C`. Assuming a sample frequency of 48 kHz, this results in a delay adjustment that varies from 0.0 to 2.0 seconds in increments of 50 milliseconds.

4. (Line 11): Creates a slider that will be labeled "GainControl," with a bottom value of 0, top value of 10, a single scroll increment of 1, a page scroll of 1, and this information is assigned to the local GEL integer variable `mygainfactor`.

5. (Line 13): The value of `mygainfactor` needs to be divided by 10 before being assigned to the C variable `gain` from `ISRs_C`. This will result in a gain adjustment that varies from 0.0 to 1.0 in increments of 0.1. But as before we wish to avoid a costly division operation, so we multiply by 1/10 instead. Note that the capital "F" in 0.1F forces the compiler to treat this number as a float, not a double, which is more efficient and yields enough precision for our needs. The compiler would then automatically promote the integer `mygainfactor` to a float, making the cast to float in line 13 actually redundant. But the cast may improve the readability of the code.

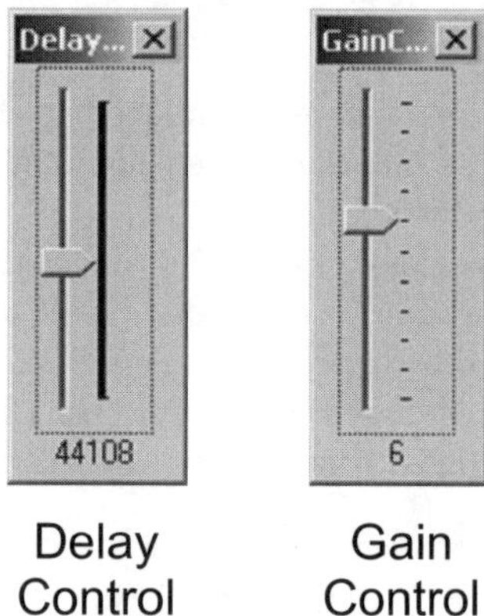

Figure 10.18: GEL slider controls available with ISRs_C.

To use the GEL file, first change the Project to remove ISRs_B and include ISRs_C in the same manner that you changed from ISRs_A to ISRs_B before. The GEL file must also be added to the Project. You can use "File," "Load GEL" or just right click "GEL files" in the left project window tree, then click "Load GEL." Select "Echo.gel" to be loaded. Rebuild, reload, and run the program as before. To activate the GEL controls, click Code Composer Studio's GEL pulldown menu, select item "Echo Controls," and select "DelayControl." Do the same for "GainControl." The sliders may end up on top of each other; if so, drag one a bit to the side to uncover the other slider underneath it. When you activate the two GEL sliders, you should see something similar to Figure 10.18 on your screen. The sliders display the integer GEL variable value at the bottom of the control, not the C variable value. In the figure, the delay slider is set to 44108 and the gain slider is set to 6. This equates to a delay of $44108/48000 \approx 919$ ms and a gain of 0.6.

As demonstrated, a GEL file allows you to interactively change variable values in your code without stopping and recompiling the program. GEL files operate only in the DEBUG version of a compiled program so be sure to load your program from the DEBUG subdirectory. Note that using the DEBUG link for GEL file communication momentarily halts the processor while a variable is being changed by a GEL file control. This results in a momentary loss of your output signal, and will sound like "clicks" or "popping" sounds while you're moving one of the sliders. If you're willing to invest just a bit more time and effort, you can avoid the disadvantages of the GEL files but still gain interactive control by using the functions called Windows Control Applications that accompany this book. They allow you to run a fully optimized RELEASE version of your compiled program and avoid any processor halts while controlling your program. Similar techniques were used to create the winDSK6 application. See Appendix E for more information.

10.5.2 Other Real-Time Special Effects

The demonstrations above of how to convert MATLAB examples of comb filters to real-time C code that will run on the DSK should allow you to create any of the other special effects described in this chapter. A notch filter, tremelo, fuzz, and so on can all be converted to C and run in real time. For creating sinusoidally varying delay times for effects such as a flanger or chorus, consider creating the sinusoidal values for $\beta[n]$ in the StartUp.c module. You *don't* want to put something like that inside an ISR which executes over and over again at the rate of the sample clock!

When you start to concatenate multiple special effects, you may exceed the limits of the real-time schedule because you're trying to do too much in one sample time. This situation calls for converting your code to frame-based processing, using EDMA to transfer data without CPU overhead, and other techniques to extract the maximum performance from the DSP.

10.6 Follow-On Challenges

Consider extending what you have learned.

1. Add a time varying (such as sinusoidal) delay time to the real-time comb filter, and choose an appropriate range of delay times to create the flanger effect.

2. Create three different comb filters similar to the flanger, but with longer delay times, and use them to create a chorus effect.

3. Whether you set it manually in the code or by using the GEL slider for gain on an IIR comb filter, what do you expect to happen if you set $\alpha = 1.0$? Try it and find out. How would you modify the code to avoid this problem?

4. Replace the GEL sliders with Windows Control Applications.

5. Implement a notch filter and use it for a phasing (i.e., phaser) effect.

6. Implement a tremelo effect.

7. Implement a ring modulator effect.

8. Implement a fuzz effect. Experiment with different ways of clipping the signal. Experiment with following and/or preceding the clipping operation with a frequency selective filters.

9. Try combining more than one effect, such as a flanger followed by reverb.

10. Convert the real-time comb filter to frame-based operation.

Chapter 11

Project 2: Graphic Equalizer

11.1 Theory

THE parallel implementation of a 5-band FIR-based equalizer was first discussed in Chapter 3. A block diagram of such an equalizer and the generalized extension of this parallel implementation are shown in Figure 11.1. While this extension to M bands may not seem like a major change, increasing the computational complexity of the DSP algorithm by adding additional parallel filters will eventually result in being unable to meet the real-time schedule. At this stage in the equalizer's development, we must either settle for the current level of system performance or rethink our approach to implementing the algorithm. This is very similar to the approach taken in Chapter 3 where we progressed from an easily understood brute force implementation of the filter's dot-product, to a much more efficient, but more complicated to understand, implementation using a circular buffer.

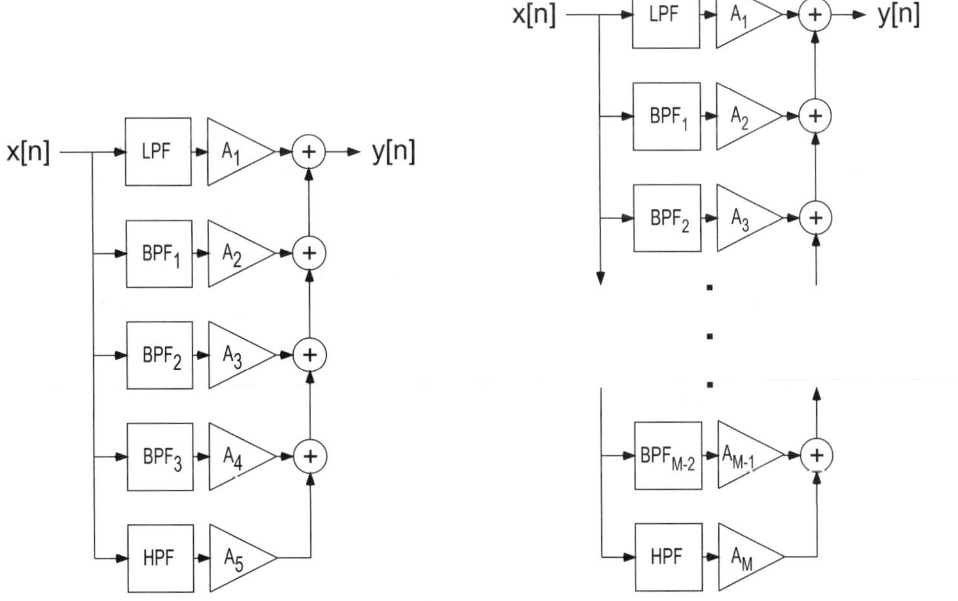

Figure 11.1: Left: Block diagram associated with winDSK6's 5-band Graphic Equalizer application. Right: generalized block diagram of an M-band graphic equalizer.

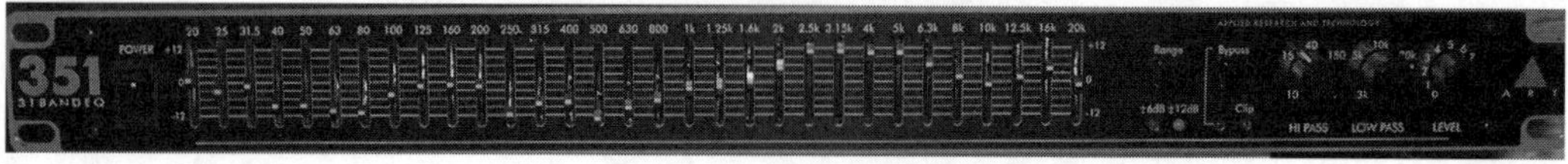

Figure 11.2: Photograph of an Applied Research and Technology, Inc. 31-band, 1/3 octave, ISO spaced, monaural graphic equalizer.

One of the first results of "rethinking" our approach to implementing the equalizer is achieved when we realize that the gains associated with each of the parallel filters are not changed very often. If we assume that this is the case, why can't we calculate an equivalent filter, and implement this single filter instead of summing the outputs of M parallel filters?

Consider a 31-band audio equalizer similar to the commercially available model shown in Figure 11.2. If we are confident that the gain controls of the individual filters are only occasionally adjusted, then we can reduce the computational complexity of our DSP implementation by a factor of nearly 31, by *first* calculating an equivalent filter. A reduction factor of 31 is achieved by implementing a single filter instead of 31 filters, but this ignores the addition of the outputs of the 31 filters. For large order filters, these extra additions become negligible. This equivalent filter technique will allow us to implement an almost unlimited number of parallel filters.

The 31-band monaural audio equalizer shown in Figure 11.2 has center frequencies of 20, 25, 31.5, 40, 50, 63, 80, 100, 125, 160, 200, 250, 315, 400, 500, 630, 800, 1000, 1250, 1600, 2000, 2500, 3150, 4000, 5000, 6300, 8000, 10000, 12500, 16000, and 20000 Hz, which constitute frequency bands spaced 1/3 octave apart. An equivalent stereo equalizer would have identical but independent frequency bands for both the left and right channels.

11.2   winDSK6 Demonstration

If you double click on the winDSK6 icon, the winDSK6 application will launch, and a window similar to Figure 11.3 will appear. Before proceeding, be sure that the appropriate selections appear for the DSK and Host Configuration options.

11.2.1   Graphic Equalizer Application

Clicking on the winDSK6 Graphic Equalizer button will load that program into the attached DSK, and a window similar to Figure 11.4 will appear. The Graphic Equalizer application implements a five-band audio equalizer, similar to the block diagram shown on the left side of Figure 11.1. If you're using a stereo codec on your DSK (and you have selected that codec from the winDSK6 main window), independently adjustable equalizers are active on both the left and the right channels.

The equalizer uses five FIR filters (a lowpass (LP) filter, 3 bandpass (BP) filters, and a highpass (HP) filter) operating in parallel. The gain sliders (A_1 to A_5) in the dialog box operate on memory locations used to control the gains of each filter and the overall system gain. The 5 FIR filters are designed as high-order ($N = 128$) filters; the resulting steep roll-off of these filters can be seen in Figure 11.5.

11.2.2   Effect of the Graphic Equalizer

There are a number of ways you can experience the effect of the graphic equalizer filtering. For example, you could connect the output of a CD player to the signal input of the DSK, and connect the DSK signal output to a powered speaker. Play some familiar music while

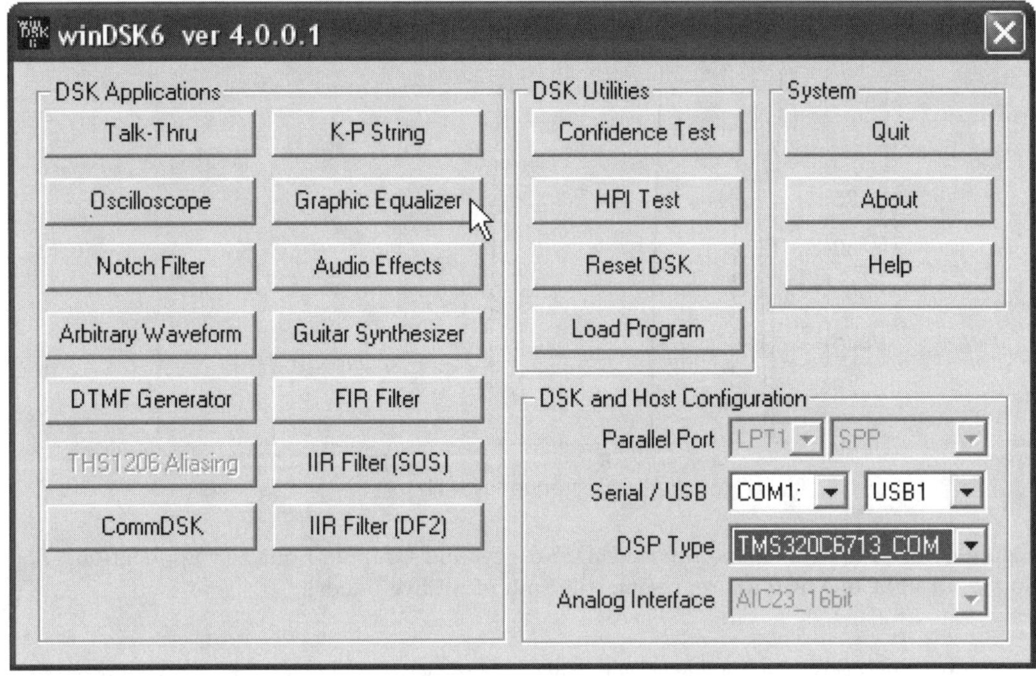

Figure 11.3: winDSK6 ready to load the Graphic Equalizer application.

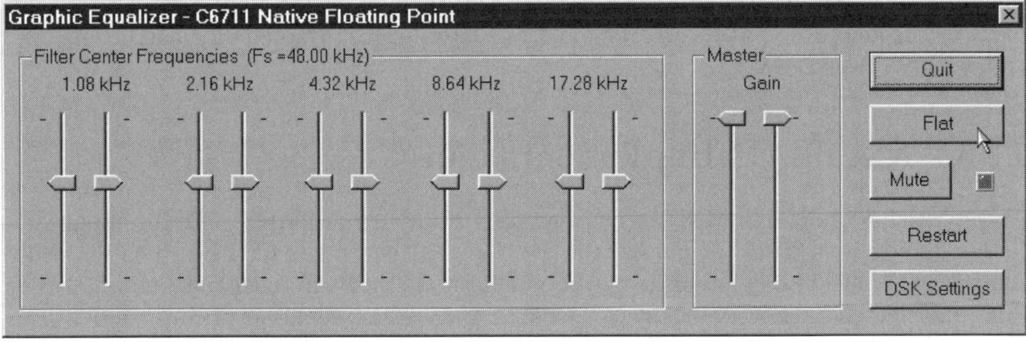

Figure 11.4: winDSK6 running the Graphic Equalizer application.

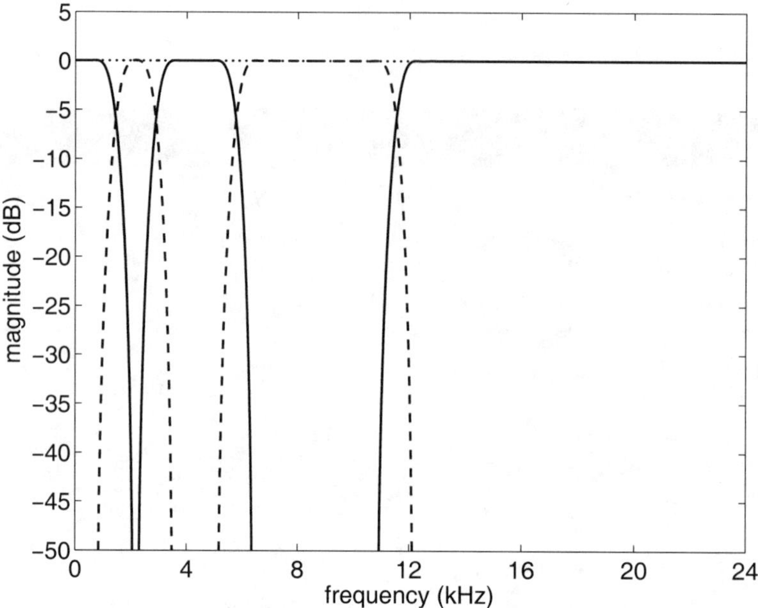

Figure 11.5: Frequency response of winDSK6's 5-band Graphic Equalizer application. The dotted straight line at 0 dB represents the sum of all five bands.

you adjust the graphic equalizer slider controls and listen to the result. A more objective experiment would be to play the track of additive white Gaussian noise (AWGN) on the CD-ROM that accompanies this book (in directory `test_signals` play the file `awgn.wav`), which theoretically contains all frequencies. If the DSK signal output is then connected to a spectrum analyzer, you could observe which band of frequencies is affected, and how much it is affected, as you adjust the slider controls. If you don't have a spectrum analyzer available, a second DSK running winDSK6 can be used in its place (select the "Oscilloscope" button from the main screen, select "Spectrum Analyzer" from the next screen, and select "Log10" to display the result in decibels). Alternatively, you can use your computer's sound card to gather a portion of the DSK's output. This can be accomplished using the Windows sound recorder, MATLAB's data acquisition (DAQ) toolbox, or the audio recorder that was recently introduced into MATLAB (version 6.1 or later). This recorded data can be analyzed and displayed using MATLAB.

11.3   MATLAB Implementation

As we stated in Chapter 3, MATLAB has a number of ways of performing filtering operations. In this chapter we will only emphasize the creation of an equivalent filter based on the scaled sum of the parallel filters that make up the equalizer. As shown in the listing below, creating this equivalent filter is very straightforward as long as the equalizer is constructed using filters of equal order. "Zero padding" will be required to sum filters of differing length.

Listing 11.1: Calculating an equivalent impulse response.

```
  % Simulation inputs
2 load('equalizer.mat')
  A = [1.0 1.0 1.0 1.0 1.0]; % graphic equalizer scale factors
```

```
4
% Calculated terms
6 equivalentFilter = A(1)*filt1.tf.num + A(2)*filt2.tf.num + ...
    A(3)*filt3.tf.num + A(4)*filt4.tf.num + A(5)*filt5.tf.num;
```

A few items need to be discussed concerning this code listing.

1. The stored filter coefficients need to be loaded into the MATLAB workspace (line 2).

2. The filter scale factors, A_1, A_2, ..., A_5 are specified (line 3). In this example, all of the scale factors are set equal to 1. This results in a flat response.

3. The scale factors are multiplied by each of the filters impulse response and then summed together (lines 6–7). The five filters used in this example were previously designed using the MATLAB function `sptool`. This function is capable of exporting structure-based variables to the MATLAB workspace. The variable `filt1.tf.num` contains the numerator coefficients `num` associated with the transfer function `tf` of `filt1`.

The first five subplots of Figure 11.6 show the impulse responses of each of the five FIR filters that make up the equalizer. The final subplot is the sum of all five impulse responses. The frequency response associated with these filters was shown in Figure 11.5.

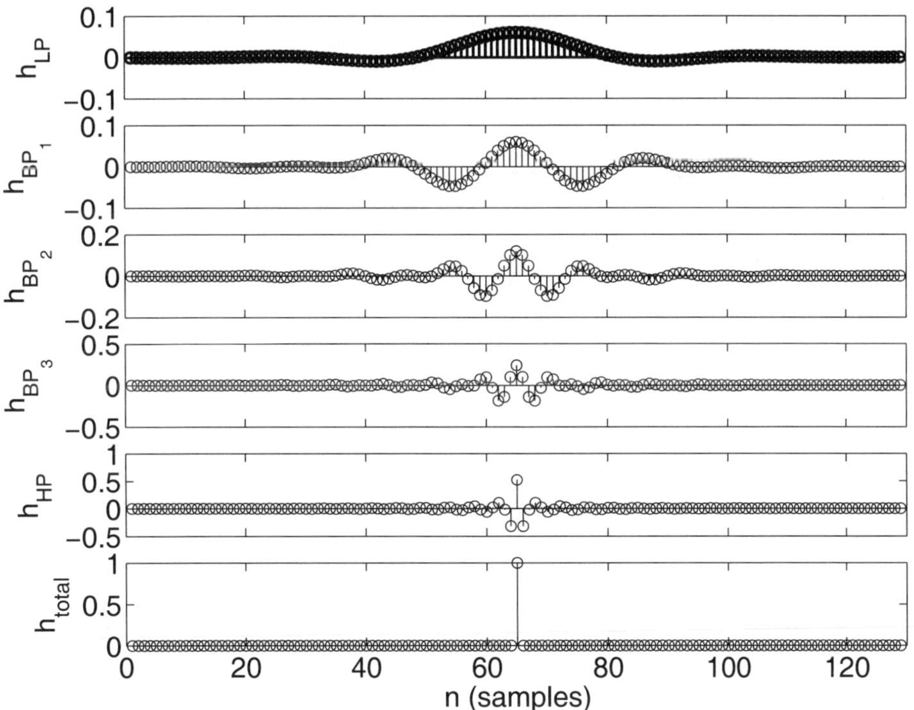

Figure 11.6: Impulse response of the five FIR filters and the sum (at bottom) of these impulse responses for unity gain at all bands.

With all of filter the gains set to 1.0 as they are in this example, the system should have a flat frequency response. Thus it should come as no surprise that equivalent impulse response of the sum of all the equalizer filter impulse responses is a single delta function.

That is, the frequency response and the impulse response are a Fourier transform pair, and the Fourier transform of a single delta function is a flat magnitude spectrum (equal power at all frequencies).

To change the frequency response of the equalizer all we need to do is adjust the individual filter gains. The single line modification to the previous code listing is shown below.

Listing 11.2: Calculating a new equivalent impulse response.

```
1  A = [0.1  0.5  1.0  0.25  0.1];  % new graphic equalizer scale factors
```

The individual impulse responses and their sum are shown in Figure 11.7. The resulting equalizer's frequency response is shown in Figure 11.8. Finally, Figure 11.9 shows the equivalent filter's impulse and frequency response magnitude together.

11.4   DSK Implementation in C

11.4.1   Applying Gain to Filter Bands

When you understand the MATLAB code, the conceptual translation into C is fairly straightforward. The equalizer is actually implementing a single equivalent FIR filter, and any of the techniques discussed in Chapter 3 could be used. The new portion of this project is applying the gains to each filter band's coefficients and calculating the equivalent filter. This is accomplished in the `main.c` file, which is shown in Listing 11.3.

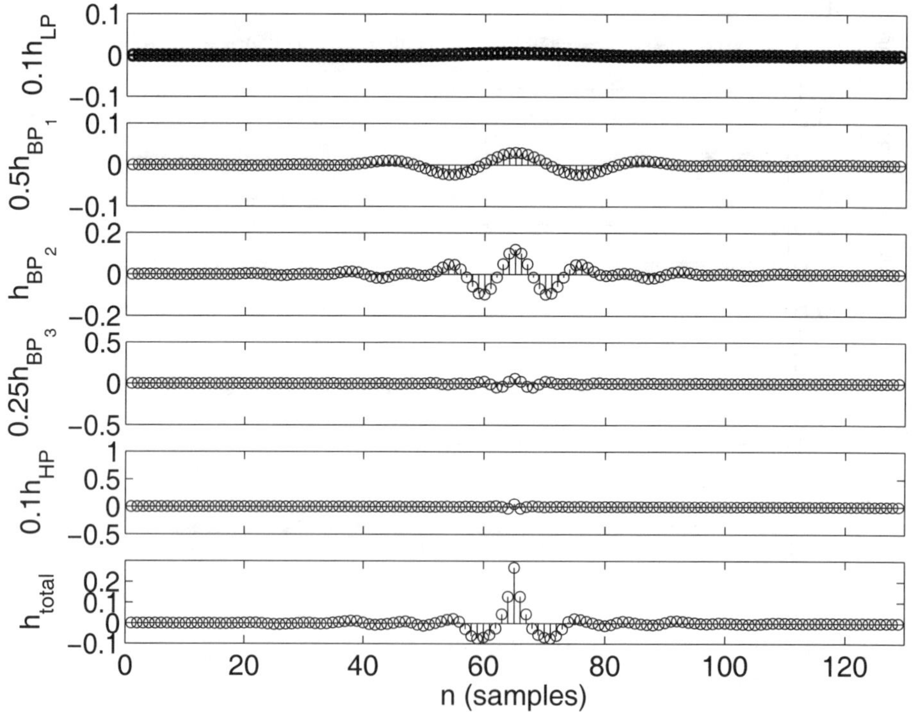

Figure 11.7: Impulse response of the five FIR filters and the sum of these impulse responses with unequal gain in the bands.

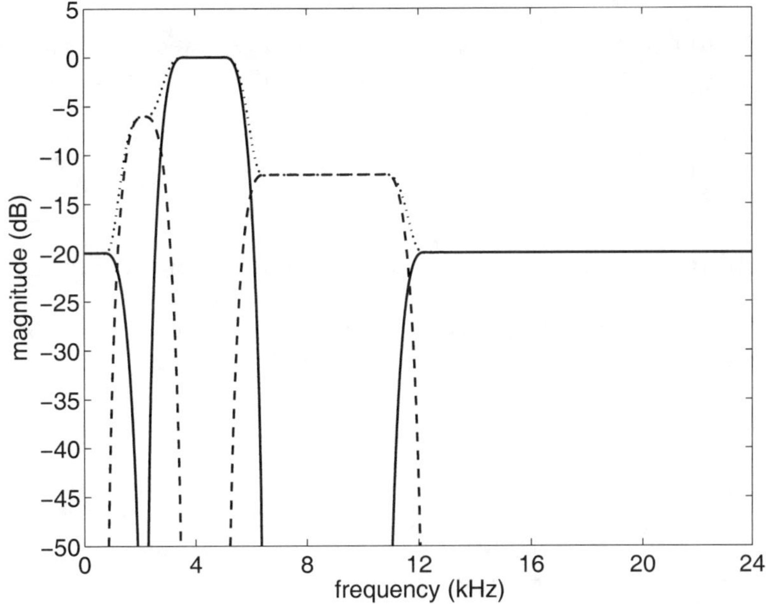

Figure 11.8: Frequency response of the five FIR filters and the equivalent filter.

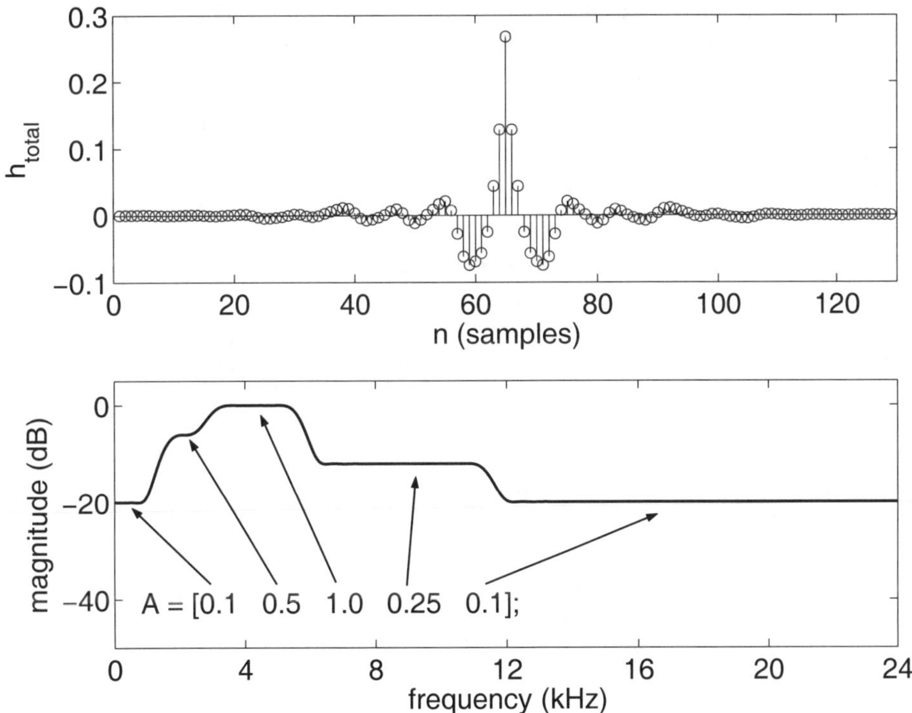

Figure 11.9: Impulse and frequency response of the equivalent filter.

Listing 11.3: Graphic equalizer project main.c code.

```c
#include "coeff.h"              // coefficients used by FIR filter
#include "coeff_lp.h"           // coefficients for equalizer
#include "coeff_bp1.h"
#include "coeff_bp2.h"
#include "coeff_bp3.h"
#include "coeff_hp.h"

volatile float new_gain_lp=1, new_gain_bp1=1, new_gain_bp2=1;
volatile float new_gain_bp3=1, new_gain_hp=1;
volatile float old_gain_lp=0, old_gain_bp1=0, old_gain_bp2=0;
volatile float old_gain_bp3=0, old_gain_hp=0;

void UpdateCoefficients()
{
    int i;

    old_gain_lp  = new_gain_lp; // save new gain values
    old_gain_bp1 = new_gain_bp1;
    old_gain_bp2 = new_gain_bp2;
    old_gain_bp3 = new_gain_bp3;
    old_gain_hp  = new_gain_hp;

    for(i = 0; i <= N; i++) { // calculate new coefficients
        B[i]=(B_LP[i]*old_gain_lp) + (B_BP1[i]*old_gain_bp1)
            + (B_BP2[i]*old_gain_bp2) + (B_BP3[i]*old_gain_bp3)
            + (B_HP[i]*old_gain_hp);
    }
}

int main()
{
    // initialize DSK for selected codec
    DSK_Init(CodecType, TimerDivider);

    UpdateCoefficients(); // update FIR filter coefficients

    // main stalls here, interrupts drive operation
    while(1) {
        // check if any gains have changed
        if((new_gain_lp != old_gain_lp)
            || (new_gain_bp1 != old_gain_bp1)
            || (new_gain_bp2 != old_gain_bp2)
            || (new_gain_bp3 != old_gain_bp3)
            || (new_gain_hp != old_gain_hp)) {
            UpdateCoefficients();
        }
    }
}
```

An explanation of Listing 11.3 follows.

1. (Lines 2–7): Include the header files associated with the filter coefficients.

2. (Lines 9–12): Declare the filter gains. There are `old_gain` values, which are the gains that are in use, and there are `new_gain` values, which are the gains that have been updated.

3. (Line 14): Beginning of the `UpdateCoefficients` function.

4. (Lines 18–22): Copy the `new_gain` values to the `old_gain` values.

5. (Lines 24–28): Calculate the new equivalent filter coefficients, `B[i]`.

6. (Line 36): Function prototype for the `UpdateCoefficients` function.

7. (Line 39): Stalled, waiting for interrupts.

8. (Lines 41–46): If any of the equalizer gains have changed, call the `UpdateCoefficients` function.

11.4.2 GEL File Slider Control

Code Composer Studio supports a general extension language (GEL) that allows for sliders, menu boxes, and other interfaces to be rapidly created. The GEL file system uses the DEBUG portion of the CCS/DSK communication link to update variable values. Using the DEBUG link halts the processor while the update is taking place. This will result in a momentary loss of your output signal. While this loss of signal is undesirable, the speed at which GEL file interfaces can be made to a CCS project still makes them a valuable tool. The GEL file that creates the 5-band filter sliders is shown in Listing 11.4.

Listing 11.4: GEL slider code (`graphicequ.gel`).

```
1  menuitem "Graphic Equalizer Controls";

3  slider LP(0, 100, 1, 1, lp_slider){
5    new_gain_lp = lp_slider * 0.01;
   }

7  slider BP1(0, 100, 1, 1, bp1_slider){
9    new_gain_bp1 = bp1_slider * 0.01;
   }

11 slider BP2(0, 100, 1, 1, bp2_slider){
13   new_gain_bp2 = bp2_slider * 0.01;
   }

15 slider BP3(0, 100, 1, 1, bp3_slider){
17   new_gain_bp3 = bp3_slider * 0.01;
   }

19 slider HP(0, 100, 1, 1, hp_slider){
21   new_gain_hp = hp_slider * 0.01;
   }
```

An explanation of Listing 11.4 follows.

1. (Line 2): Creates the pulldown menu item in Code Composer Studio as shown in Figure 11.10.

2. (Lines 4–6): Creates a slider that will be labeled "LP," have a bottom value of 0, top value of 100, a single scroll increment of 1, a page scroll of 1, and this information is assigned to the local variable `lp_slider`. The value of `lp_slider` is multiplied by 0.01 (same as being divided by 100 but avoids the costly division operation) and assigned to the variable `new_gain_lp`. This results in a LP filter gain adjustment that varies from 0.0 to 1.0.

3. (remaining lines): Accomplishes the same task as above for the remaining four filters.

Pulling down "GEL," "Graphic Equalizer Controls," and then clicking on each of the filters, "LP," "BP1," "BP2," "BP3," and "HP," will then bring up sliders that will look similar to Figure 11.11. You may have to drag the individual sliders with the mouse so they aren't on top of one another. After the sliders are positioned, load and run the program in the usual way to start the equalizer project. Moving the sliders will now vary the filter gains interactively.

Using GEL-based sliders is easy, but the fact that the DSK's CPU is halted each time you adjust a slider keeps this from being a fully real-time program. To avoid the CPU halts, Windows-based sliders would need to be created. See Appendix E for more information.

Figure 11.10: Screen capture of Code Composer Studio's GEL pulldown menu system.

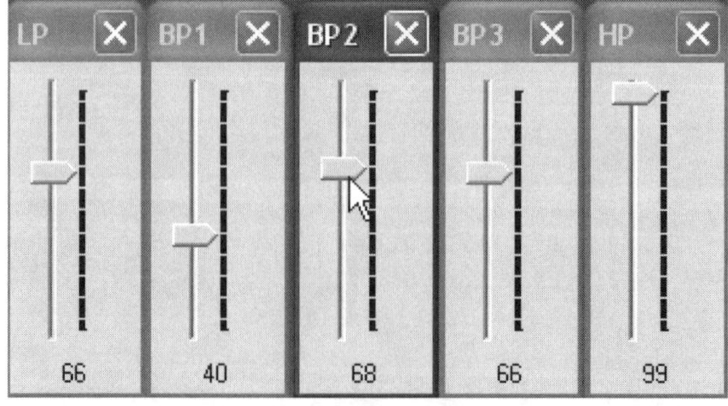

Figure 11.11: GEL sliders for the graphic equalizer.

11.5 Follow-On Challenges

Consider extending what you have learned.

1. Design and implement your own monaural graphic equalizer. Select the frequency bands you desire and create an equivalent filter that will run in real time, using GEL-based sliders to control the gain of each frequency band.

2. Design and implement your own stereo graphic equalizer in a fashion similar to the monaural version.

3. Implement Windows-based slider controls for your graphic equalizer.

Chapter 12

Project 3: Peak Program Meter

12.1 Theory

THE basic block diagram of the generic DSP system, shown in Figure 12.1, was first introduced in Chapter 2. We also emphasized in Chapter 2 that the analog signal that is digitized by the analog to digital converter (ADC) should not exceed the maximum voltage range of the converter. To avoid unintended distortion, be careful to ensure that the ADC is not driven beyond the maximum input voltage range, in either the positive or negative directions.

Even if the input analog signal remains within the proper range of the ADC, it is still possible to distort the signal by exceeding the output range of the digital to analog converter (DAC). As an example, the possible range of values for a 16-bit converter using two's complement representation is $+32,767$ to $-32,768$. Any operation within the DSP algorithm that results in an output value being written to the DAC that falls outside this range will also distort the signal. The only way the signal can exceed the output range for the DAC is if the DSP algorithm has a gain that exceeds 1.0. That is to say that the algorithm results in an amplification of the signal. While gain greater than one is not strictly prohibited, saturating the DAC should be avoided, unless for some reason you actually *want* a distorted output.

Historically, volume unit (VU) meters were used in audio systems to monitor the signal's level. VU meters, however, have display accuracy problems that are largely due to the fact that the meter takes an average measurement that is severely restricted by the ballistics of the mechanical metering system. This can result in short, but very loud transients, being missed or improperly displayed.

More recently, audio equipment manufacturers developed the peak program meter (PPM) to overcome the VU meter's lackluster performance at displaying peak signal levels. The PPM improves on the VU meter's performance problems by integrating the signal for 5 ms. This integration process will then only detect peaks that are long enough to be heard by a typical human listener.

Figure 12.1: A generic DSP system.

12.2   winDSK6 Demonstration: commDSK

If you double click on the winDSK6 icon, the winDSK6 application will launch, and a window similar to Figure 12.2 will appear. Clicking on winDSK6's commDSK button will load that program into the attached DSK, and a window similar to Figure 12.3 will appear.

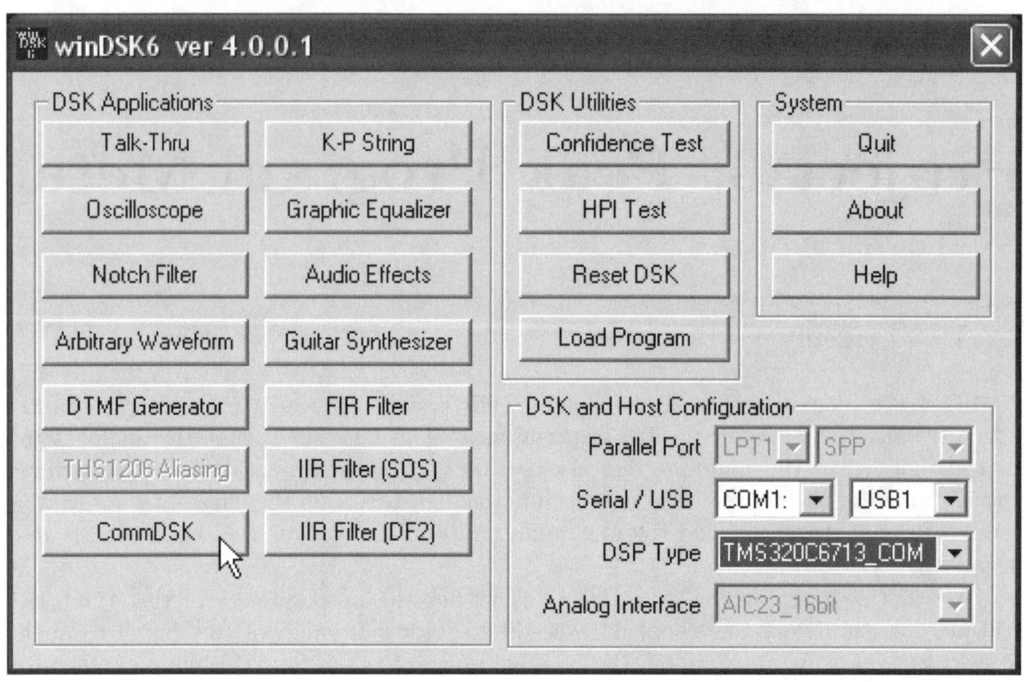

Figure 12.2: winDSK6 ready to load the commDSK application.

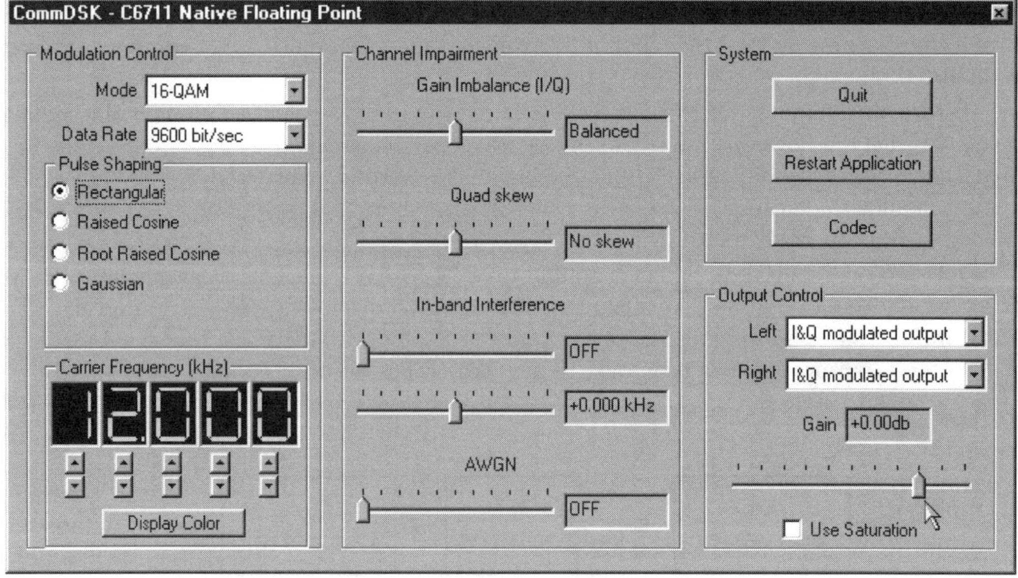

Figure 12.3: winDSK6 running the commDSK application.

In Figure 12.3 the "Output Control," "Right" box has been changed from "Symbol timing signal" to "I&Q modulated output."

The commDSK program is discussed in more detail in Chapter 16. Operation of a PPM can be seen by increasing the system gain. Increasing the system gain is accomplished by clicking on the gain slider located just below the box labeled "Gain," and then moving the slider to the right. This will result in positive gain numbers appearing in the "Gain" box. Additionally, as the "Gain" is increased, the three LEDs (on a C6711 DSK) or three of the four LEDs (on a C6713 DSK) will function as a PPM.

12.3 MATLAB Implementation

MATLAB does not have an equivalent function for turning on an LED, except through an indirect use of the Data Acquisition Toolbox. We therefore omit a MATLAB discussion for this chapter.

12.4 DSK Implementation in C

Like the VU meter and PPM, the primary function of this program is to detect and provide an LED indication/warning whenever an output value is approaching the range limit of the DAC. Since these conditions are checked every $T_s = 1/F_s$, a dwell time is also required to maintain the "ON" status of each of the LEDs. Without this dwell time, the LED would cycle on/off too rapidly to be visible. The output values above which an LED turns on were chosen to be $\pm 28,000$, $\pm 32,000$, and $\pm 32,767$. These turn-on levels are shown in Figure 12.4, where the sinusoidal signal is at maximum amplitude for the DAC.

12.4.1 Example PPM Code

The files necessary to run this application are in the ccs\PPM directory of Chapter 12. The primary files of interest are the PPM_ISRs.c, PPM_ISRs1.c, PPM_ISRs2.c, and PPM_ISRs3.c. **Important:** you must only have *one* of these ISR files loaded as part of your project at any given time. The ISR files contain the necessary variable declarations and perform the actual filtering operation.

We will discuss PPM_ISRs.c first. The declarations associated with the PPM code are shown in the listing that follows.

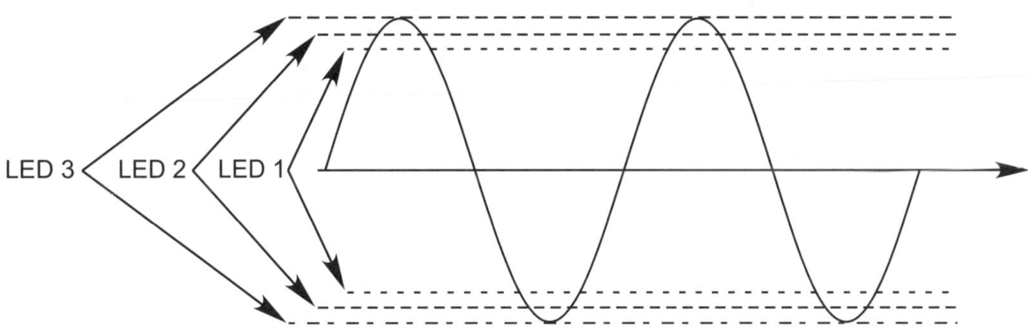

Figure 12.4: LED turn-on levels for the PPM.

Listing 12.1: Declarations associated with the PPM.

```
#define RESET 4800   // turns the LED off after 4800 samples

unsigned int LED_Mask = 0;   // all LEDs off
#define LED1_BIT 1
#define LED2_BIT 2
#define LED3_BIT 4
```

An explanation of Listing 12.1 follows.

1. (Line 1): Sets the minimum time that an LED will remain "ON." This time equals $(\text{RESET})(T_s)$, which in this case is $4800/48000 = 0.1$ seconds.

2. (Line 3): Defines a variable, LED_Mask, which will turn all of the LEDs "OFF."

3. (Lines 4–6): Define variables that represent the "ON" state for each of the 3 LEDs.

A portion of the receive ISR in PPM_ISRs.c is shown in the listing below.

Listing 12.2: Example PPM code from PPM_ISRs.c.

```
// LED 1 logic
if ((abs(outputLeft) > 28000)||(abs(outputRight) > 28000)) {
    LED_Mask |= LED1_BIT;   // LED1 on
    LED_1_counter = RESET;
}
else {
    if (LED_1_counter > 0)
        LED_1_counter -= 1;
    else
        LED_Mask &= ~LED1_BIT;   // LED1 off
}

// LED 2 logic
if ((abs(outputLeft) > 32000)||(abs(outputRight) > 32000)) {
    LED_Mask |= LED2_BIT;   // LED2 on
    LED_2_counter = RESET;
}
else {
    if (LED_2_counter > 0)
        LED_2_counter -= 1;
    else
        LED_Mask &= ~LED2_BIT;   // LED2 off
}

// LED 3 logic
if ((abs(outputLeft) > 32767)||(abs(outputRight) > 32767)) {
    LED_Mask |= LED3_BIT; // LED3 on
    LED_3_counter = RESET;
}
else {
    if (LED_3_counter > 0)
        LED_3_counter -= 1;
    else
```

```
34          LED_Mask &= ~LED3_BIT;   // LED3 off
   }
36
   UpdateLEDs(LED_Mask);   // update LED status
```

Explanation of the PPM code

An explanation of Listing 12.2 follows.

1. (Lines 2–5): If either or both of the left or right channel levels is greater than 28,000 in magnitude, then LED 1 is turned on. A counter is also set to 4800. This counter keeps the light on for 0.1 seconds.

2. (Lines 6–11): This block of code decrements LED 1's counter and whenever the counter reaches zero, the LED is turned "OFF."

3. (Lines 14–17): If either or both of the left or right channel levels is greater than 32,000 in magnitude, then LED 2 is turned on. A counter is also set to 4800. This counter keeps the light on for 0.1 seconds.

4. (Lines 18–23): This block of code decrements LED 2's counter and whenever the counter reaches zero, the LED is turned "OFF."

5. (Lines 26–29): If either or both of the left or right channel levels is greater than 32,767 in magnitude, then LED 3 is turned on. A counter is also set to 4800. This counter keeps the light on for 0.1 seconds.

6. (Lines 30–35): This block of code decrements LED 3's counter and whenever the counter reaches zero, the LED is turned "OFF."

7. (Lines 37): Calls the UpdateLEDs() subroutine, which as its name implies, updates the status of the system's LEDs.

12.4.2 DSK LED Control

The LEDs on the DSK are controlled by updating the bit pattern at IO_PORT (address 0x90008000) as shown in the table below. The table entries represent bit positions of the data word, so that D0 is the least significant bit.

DSK type	LED 1	LED 2	LED 3	LED 4	logic
C6711	D24	D25	D26	—	active low
C6713	D0	D1	D2	D3	active high

You may want to inspect the function UpdateLEDs() in the PPM_ISRs.c file to satisfy yourself that the bits get set in the proper manner.

12.4.3 Another PPM Code Version

As noted above, four different versions of the PPM ISR code are provided on the CD-ROM. We already discussed the pertinent parts of PPM_ISRs.c. We will now discuss PPM_ISRs3.c; explore the variations present in PPM_ISRs1.c and PPM_ISRs2.c on your own.

This implementation uses the same declarations as those shown earlier for PPM_ISRs.c, as well as the same function UpdateLEDs(). A portion of the receive ISR in PPM_ISRs3.c is shown in Listing 12.3.

Listing 12.3: Another approach to creating the PPM, excerpted from `PPM_ISRs3.c`.

```c
maxOutput = _fabsf(outputLeft);
if(maxOutput < _fabsf(outputRight))
    maxOutput = _fabsf(outputRight);

if (maxOutput > 32767) {
    LED_3_counter = RESET;
    LED_2_counter = RESET * 2;
    LED_1_counter = RESET * 3;
}
else if (maxOutput > 32000) {
    if(LED_2_counter < RESET)
        LED_2_counter = RESET;
    if(LED_1_counter < RESET * 2)
        LED_1_counter = RESET * 2;
}
else if (maxOutput > 28000) {
    if(LED_1_counter < RESET)
        LED_1_counter = RESET;
}

LED_Mask = 0; // all LEDs off
if (LED_3_counter) {
    LED_3_counter--;
    LED_Mask |= LED3_BIT;   // LED3 on
}

if (LED_2_counter) {
    LED_2_counter--;
    LED_Mask |= LED2_BIT;   // LED2 on
}

if (LED_1_counter) {
    LED_1_counter--;
    LED_Mask |= LED1_BIT;   // LED1 on
}

UpdateLEDs(LED_Mask);   // update LED status
```

An explanation of Listing 12.3 follows.

1. (Lines 1–3): Determines `maxOutput` (the maximum of the absolute value of both the left and right output channels).

2. (Lines 5–9): If `maxOutput` is greater than 32,767 in magnitude, then LEDs 1, 2, and 3 are turned on. Counters for LED 1, 2, and 3 are set at 14400, 9600, and 4800. These counter values keep LEDs 1, 2, and 3 "ON" for 0.3, 0.2, and 0.1 seconds, respectively.

3. (Lines 10–15): If `maxOutput` is greater than 32,000 in magnitude, then LEDs 1 and 2 are turned on. Counters for LED 1 and 2 are set at 9600 and 4800. These counter values keep LEDs 1 and 2 "ON" for 0.2 and 0.1 seconds, respectively.

4. (Lines 16–19): If `maxOutput` is greater than 28,000 in magnitude, then LED 1 is turned on. The counter for LED 1 is set at 4800. This counter value keeps the LED "ON" for 0.1 seconds.

5. (Lines 21–35): Updates the `LED_Mask` based on the status of the LED counters.

6. (Line 37): Calls the `UpdateLEDs()` subroutine, which as its name implies, updates the status of the system's LEDs.

12.5 Follow-On Challenges

Consider extending what you have learned. Remember, to fully test the PPM as part of a larger program, some part of the DSP algorithm (other than the PPM part) should have a gain greater than one. This can be accomplished easily by adding a multiplicative scale factor somewhere in the algorithm between the input and output.

1. Design and implement your own peak program meter.

2. If you have a C6713 DSK, implement a peak program meter that utilizes all 4 LEDs.

Chapter 13

Project 4: AM Transmitter

13.1   Theory

ONE of the simplest modulation schemes is amplitude modulation, which is normally just abbreviated as AM. The commercial AM broadcast radio stations in the United States use a version of AM called double sideband large carrier (DSB-LC), sometimes also called double sideband with carrier (DSB-WC). See [61, 62] for some general theoretical background on amplitude modulated signals, and [63, 64] for more DSP-specific background on AM communications.

For several decades now, commercial AM radio broadcasts can be received on almost any consumer radio sold in the United States. Most U.S. commercial AM radio stations, which occupy the 550 to 1600 kHz band, are used primarily for public service, news, talk radio, and sports reporting, but only for limited music broadcasting. The majority of music programming has shifted to the more noise immune (and higher fidelity) stereo frequency modulation (FM) systems in the 88 to 108 MHz band. This is not to imply in any way that broadcast AM is no longer important! In fact, AM systems are still used all around the world. Additionally, AM provides an easily understood modulation scheme that can be thought of as the starting point for many of today's more complicated modulation schemes.

There are several ways in which AM (DSB-LC) can be generated. One method that is particularly easy to explain uses two steps:

1. offset the message signal $m(t)$ by adding a D.C. bias signal B, then

2. multiply the offset message signal $[B + m(t)]$ by some higher frequency sinusoidal carrier signal.

This process can be seen in Figure 13.1. To express this process mathmatically, the AM

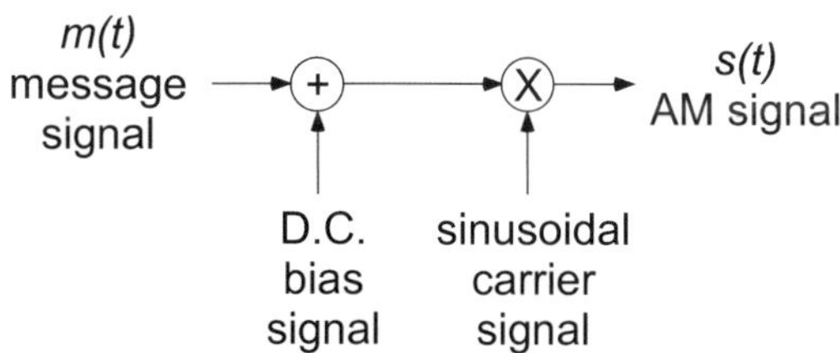

Figure 13.1: The block diagram for AM generation.

signal equation can be written as,

$$s(t) = A_c[B + m(t)]\cos(2\pi f_c t) \tag{13.1}$$

where A_c is the carrier amplitude, f_c is the carrier frequency, and t represents time.

For distortion-free message recovery using envelope detection techniques (see Chapter 14), the term $B + m(t)$ must be kept greater than zero. This can be accomplished by adjusting either the bias signal value or the amplitude of the message. Figure 13.2 shows 100 ms of a voice signal. Figure 13.3 shows the result of adding the voice signal from Figure 13.2 and 5 mV of bias. In Figure 13.3 it is clear that the voice signal plus 5 mV does not always remain positive. Either the message amplitude needs to be reduced or the bias value needs to be increased. Otherwise, this will result in distortion if an envelope detector is to be used for message recovery. Figure 13.4 shows the result of adding the voice signal from Figure 13.2 and 20 mV of bias; it is clear that the voice signal plus bias now remains positive for the time period shown.

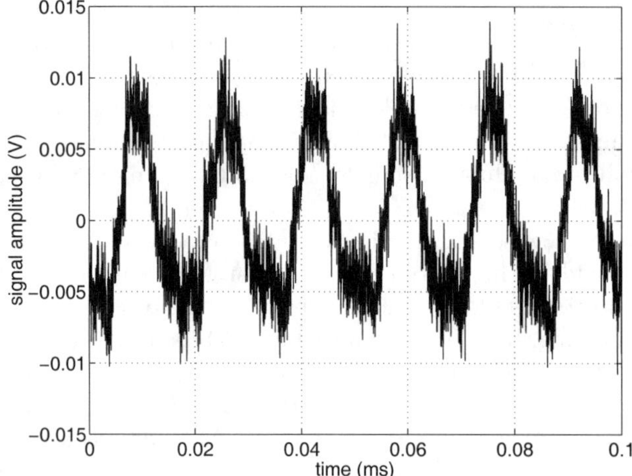

Figure 13.2: Plot of 100 ms of voice data.

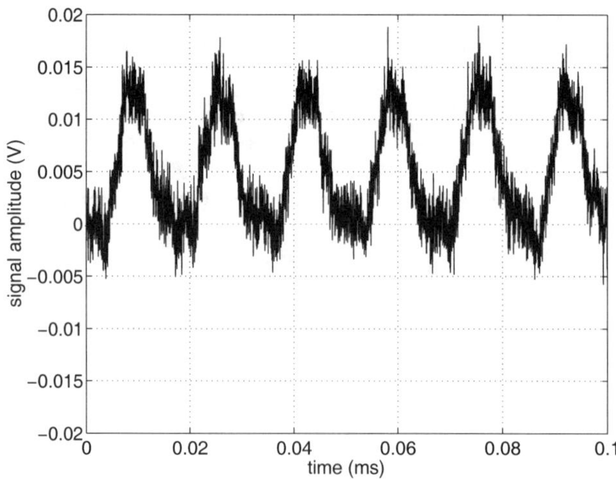

Figure 13.3: Plot of 100 ms of voice data with 5 mV of added bias.

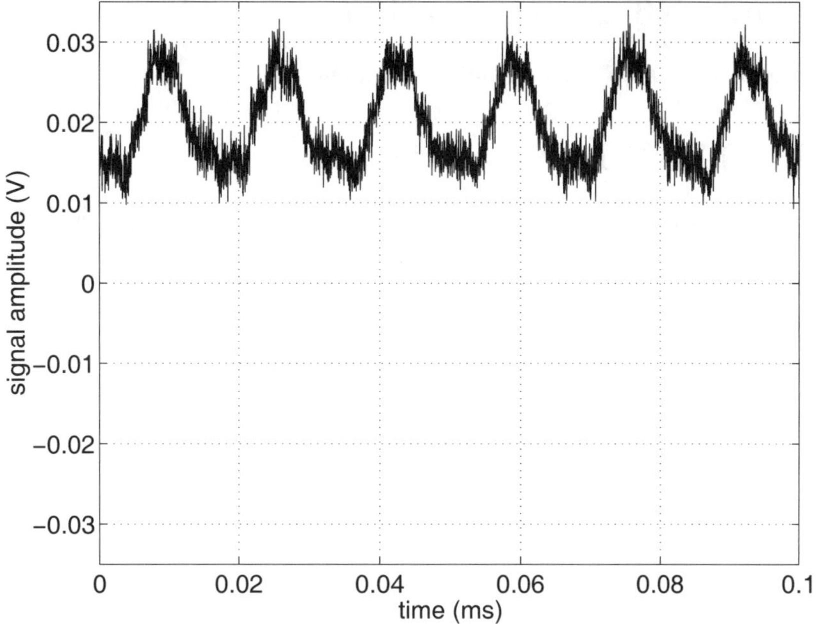

Figure 13.4: Plot of 100 ms of voice data with 20 mV of added bias.

The final step in the AM generation process is to multiply the properly biased message signal by a sinusoidal signal (called the *carrier*), as shown in Figure 13.5. The multiplication is a point-by-point operation, so the MATLAB operator ".*" must be used.

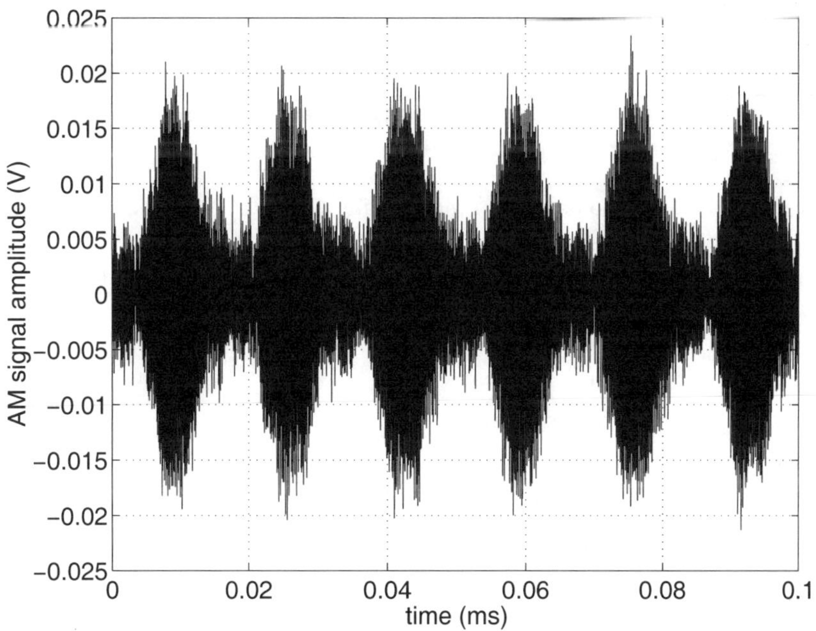

Figure 13.5: Voice signal modulating (DSB-LC) a 12 kHz carrier.

13.2   winDSK6 Demonstration

The winDSK6 program does not provide an equivalent function.

13.3   MATLAB Implementation

The output from a MATLAB simulation of the AM generation process is shown in Figure 13.6.

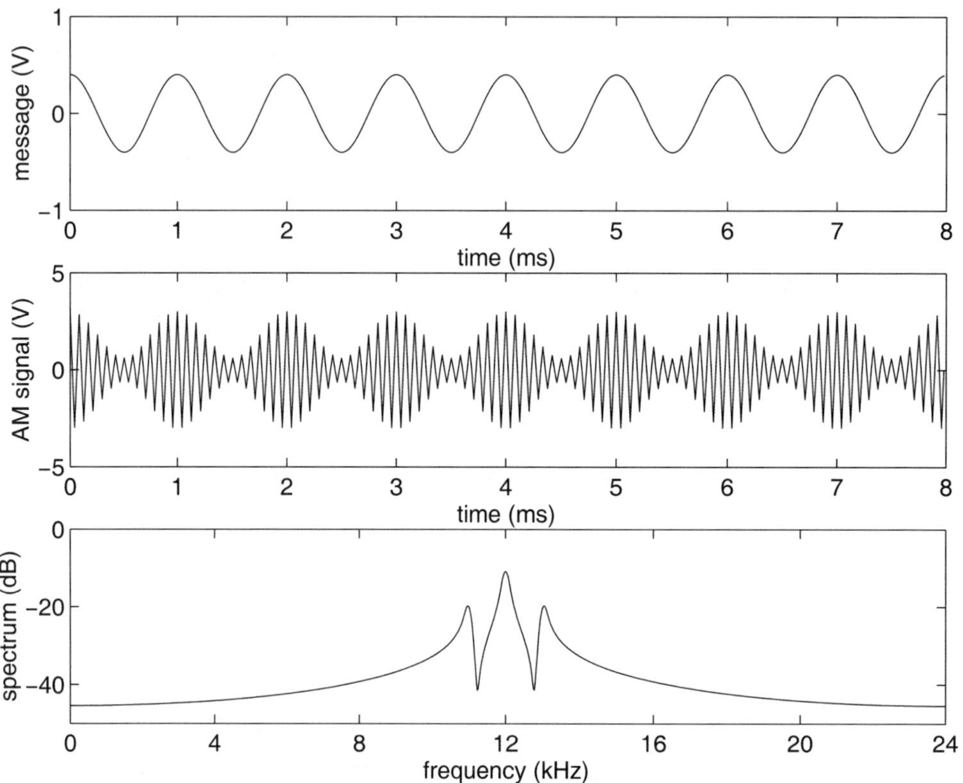

Figure 13.6: Sinusoidal signal modulating (DSB-LC) a 12 kHz carrier.

The simulation inputs, calculated terms, and simulation outputs are as follows.

- Simulation inputs

    - Simulation or sample frequency
    - Message frequency and amplitude
    - Bias level
    - Carrier frequency and amplitude
    - Simulation duration

- Calculated terms

    - Message value (we are calculating or simulating the message)
    - Carrier value

        – AM signal value

- Simulation outputs

        – Plots of the message signal and the AM modulated signal

        – Plot of the estimated power spectrum magnitude of the AM modulated signal

Figure 13.6 is divided into 3 subplots. Subplot 1 (top) plots the message signal (1000 Hz sinusoid with a 0.5 V amplitude) in the time domain. Subplot 2 (middle) plots the AM waveform in the time domain. Notice how the message waveform defines the AM signal's envelope. Also notice that the carrier appears to resemble a triangle wave. When a 12 kHz sinusoid is simulated using a 48 kHz sample frequency, only 4 samples per carrier cycle are present. For a unit amplitude cosine function, without any phase shift, the resulting carrier values are 1, 0, -1, and 0 per cycle. For a unit amplitude sine function, without any phase shift, the resulting carrier values are 0, 1, 0, and -1 per cycle. In either case, these values plot as a triangle wave. This is not a problem in a DSP system since the DAC incorporates a reconstruction (lowpass) filter. This filter will remove the high frequency components, only passing the fundamental frequency. This process turns the triangle wave back into a sinusoidal wave at 12 kHz.

Subplot 3 (bottom) plots the magnitude of the power spectral density (PSD) estimate of the AM signal. Additional details concerning spectral estimation can be found in Chapter 9. The m-file that created this plot is called **AM_SignalGeneratorAndPlotter.m** and it can be found in the **matlab** directory for Chapter 13. The code listing is given below.

Listing 13.1: MATLAB example of AM (DSB-LC) generation.

```
1  %
   %  Generates an AM modulation figure that has 3 subplots:
3  %
   %  1 - message signal (time domain)
5  %  2 - AM signal (time domain)
   %  3 - PSD estimate of the AM signal
7  %


9  % Simulation inputs
   Fs = 48000;            % sample frequency
11 Fmsg = 1000;           % message frequency
   Amsg = 0.4;            % message amplitude
13 bias = 0.6;            % bias (offset)
   Fc = 12000;            % carrier frequency
15 Ac = 3;                % carrier amplitude
   duration = 0.008;      % duration of the signal in seconds
17 Nfft = 2048;           % number of points used for PSD frames


19 myFontSize = 12;       % font size for the plot labels


21 % Calculated terms
   NumberOfPoints = round(duration*Fs);
23 t = (0:(NumberOfPoints - 1))/Fs;          % establish time vector
   message = Amsg*cos(2*pi*Fmsg*t);          % create message signal
25 carrier = cos(2*pi*Fc*t);                 % create carrier signal
   AM_msg = Ac*(bias + message).*carrier;    % create AM waveform

27
```

```matlab
   % Simulation outputs
29 subplot(3,1,1)
   set(gca, 'FontSize', myFontSize)
31 plot(t*1000, message)
   xlabel('time (ms)')
33 ylabel('message (V)')
   axis([0 8 -1 1])

35
   subplot(3,1,2)
37 set(gca, 'FontSize', myFontSize)
   plot(t*1000,AM_msg)
39 xlabel('time (ms)')
   ylabel('AM signal (V)')

41
   subplot(3,1,3)
43 set(gca, 'FontSize', myFontSize)
   [Pam, frequency] = psd(AM_msg, Nfft, Fs, blackmanharris(Nfft));
45 plot(frequency/1000, 10*log10(Pam))
   xlabel('frequency (kHz)')
47 ylabel('spectrum (dB)')
   set(gca,'XTick', [0 4 8 12 16 20 24])
49 set(gca,'XTickLabel', [0 4 8 12 16 20 24]')
   axis([0 24 -50 0])

51
   print -deps2 AM_SignalPlot          % save eps file of figure
```

13.4   DSK Implementation in C

When you understand the MATLAB code, the concept translation into C is fairly straight-forward. For the MATLAB simulation we listed the simulation inputs, calculated terms, and the simulation outputs. While these concepts are similar for the DSK's code, the names are modified. For the DSK, we will use the terms "declaration" and "algorithmic process." We will need few modifications to the MATLAB thought process, however.

- The DSP must process the data from the ADC in real-time; therefore, we cannot wait for all of the message samples to be received prior to beginning the algorithmic process.

- Real-time DSP is inherently an interrupt driven process and the input samples should only be processed using interrupt service routines (ISRs). Given this observation, it is incumbent upon the DSP programmer to ensure that the time requirements associated with periodic sampling are met. Also, remember the input and output ISRs are asynchronous. Nothing will go in and out of your DSP hardware unless you program the DSP with an appropriate receive and transmit ISR.

- The digital portion of both an ADC and a DAC are inherently *integer* in nature. No matter what the ADC's input range is, the analog input voltage is mapped to an integer value. For a 16-bit converter using two's complement representation, the possible values range from $+32,767$ to $-32,768$. Since $-32,768$ is the maximum negative value of the signal that can be received by the DSP, a bias level of no more than $+32,768$ will be necessary to prevent envelope distortion at the receiver. You can always make the bias value larger than $+32,768$ to explore the effects on AM signal generation.

Given these considerations, the program is broken into the following parts.

- Declaration

  - Bias level
  - Carrier frequency

- Algorithmic process

  - Read in a message sample from the ADC
  - Calculate the next carrier value
  - Calculate the AM signal value
  - Scale the AM signal value for the DAC
  - Write the AM signal value to the DAC

See the files associated with this project in the `ccs\AmTx` directory for Chapter 13. We provide two implementations of this project, a direct method (using the file `ISRs.c`) and a more efficient method (using the file `ISR_Table.c`). Select one (and only one) of these two files to include in your CCS project. If you're using `ISR_Table.c` then modify `StartUp.c` so that the function call to FillSineTable() is not commented out. The code is written for one of the higher-speed stereo codecs; if you're going to use the on-board codec of the C6711 you will have to lower the message and carrier frequencies.

If we directly implement the AM generation equation $s(t) = A_c[B + m(t)]\cos(2\pi f_c t)$ without considering the required scaling for the DAC, we will likely exceed the allowable range. We must assume that the full ADC range is possible for input data, so `CodecData.Channel[LEFT]` can range from $-32,768$ to $+32,767$. Thus `bias` must be at least $+32,768$ to prevent the combined term `bias+CodecData.Channel[LEFT]` from possibly becoming negative. Remember that this combined term must not be negative, or message distortion will occur at the output of the receiver when using an envelope detector. If `bias` is set to this minimum value of $+32,768$, then the maximum value of `bias+CodecData.Channel[LEFT]` will be $32,768+32,767 = 65,535$. The magnitude of the sine we will generate for the carrier is $|\sin(\cdot)| \le 1$, so that means that a scale factor of 0.5 is needed to prevent the AM value from exceeding the allowable range of the DAC.[1] The only flexibility we lose by using this method is that we are no longer being able to freely increase the "bias" value without also changing the 0.5 scale factor. The AM generation equation can therefore be implemented using the code statement show in the listing below. Watch for line wraps due to margins in this and the following listings.

Listing 13.2: C code for scaled implementation of DSB-LC AM.

```
CodecData.Channel[LEFT]=(float)0.5*(bias+CodecData.Channel[LEFT])
   [+]*sinf(phase);
```

After profiling this code, it becomes obvious that the transmitter ISR is still the major user of DSP computational resources. This is due to the function `sinf`, which although seemingly straightforward to calculate, is a call to a relatively "expensive" (in computational terms) routine prototyped in the header file `math.h`. As discussed in Chapter 5, numerous techniques exist to generate a sinusoid. The following code combines these concepts for

---

[1]Since division is more "expensive" computationally than multiplication, we multiply by 0.5 rather than divide by 2.

AM signal generation. The only change to StartUp.c is to uncomment the line that makes
the function call to FillSineTable(); this version of the project will use the interrupt service
routines contained in the file ISR_Table.c.

With the array SineTable filled, the line of code shown next extracts noninterpolated
sine function values from the lookup table.

Listing 13.3: C code to extract the sine function values from the lookup table.

```
sine=SineTable[(int)(index/GetSampleFreq()*NumTableEntries)];
```

The code associated with the final calculations of the AM waveform generation and scaling
follows.

Listing 13.4: C code for scaled implementation of DSB-LC AM with sine table lookup.

```
CodecData.Channel[LEFT]=(float)0.5*(bias+CodecData.Channel[LEFT])
    *sine;
```

Code profiling using the new carrier generation algorithm reveals approximately an 80%
reduction in computational resources used by the transmitter ISR.

13.5   Follow-On Challenges

Consider extending what you have learned.

1. Even though both output channels are used, the RIGHT channel data is just a copy
   of the LEFT channel data. Investigate how to use DSB-LC to transmit stereo information.

2. Since even momentary transmission with a modulation index greater than one results
   in distortion for a receiver using an envelope detector, how would you prevent this
   situation from occurring?

3. From a practical perspective, if the message signal that you are using to modulate the
   carrier has a baseband bandwidth that exceeds the carrier frequency, then aliasing
   will occur. Design and implement a system that digitally bandlimits (LP filters) the
   message signal *prior* to modulating the carrier signal.

4. What are the implications of setting the bias term to a value greater than $+32,768$?

5. How would you implement a single sideband (SSB) transmitter?

Chapter 14

Project 5: AM Receiver

14.1   Theory

$\mathbf{A}$MPLITUDE modulation (AM) is a very popular modulation scheme. As we discussed in Chapter 13, AM signals are carried in the envelope of the carrier signal. An AM signal with carrier frequency, $f_c = 550$ kHz, message frequency, $f_{msg} = 5$ kHz, and modulation index, $\mu = 0.8$, is shown in Figure 14.1. In this figure, the signal's envelope is clearly sinusoidal. Counting the envelope variations shows that the signal envelope experiences five periods in the displayed time duration of one millisecond. The message frequency can now be verified to be, $f_{msg} = 5/0.001 = 5$ kHz. Since the carrier is displayed as solid shading in this figure, it is impossible to precisely determine its exact frequency without rescaling the plot. In fact, the carrier frequency is at $f = 550$ kHz, which means that this AM signal has $550,000/5,000 = 110!$Hz/period. That is, 110 cycles of the carrier occur for every cycle of the message. This represents the worst-case ratio for US-based commercial AM detection

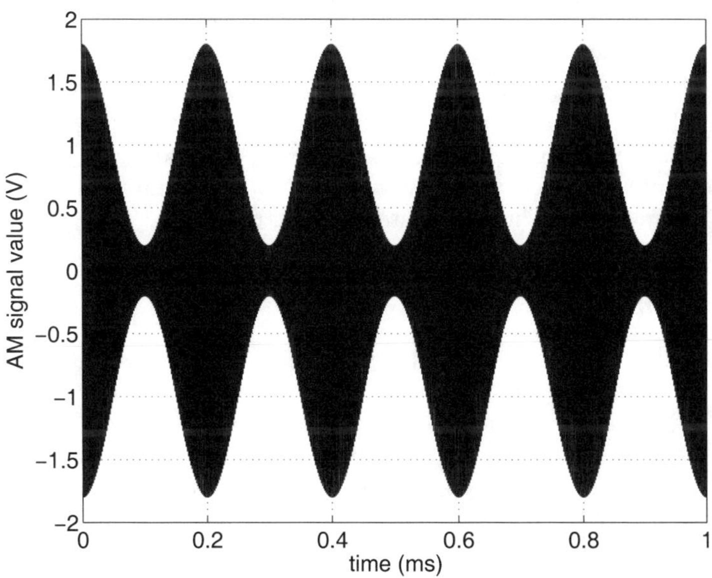

Figure 14.1: An AM signal in the time domain ($f_c = 550$ kHz, $f_{msg} = 5$ kHz, and $\mu = 0.8$).

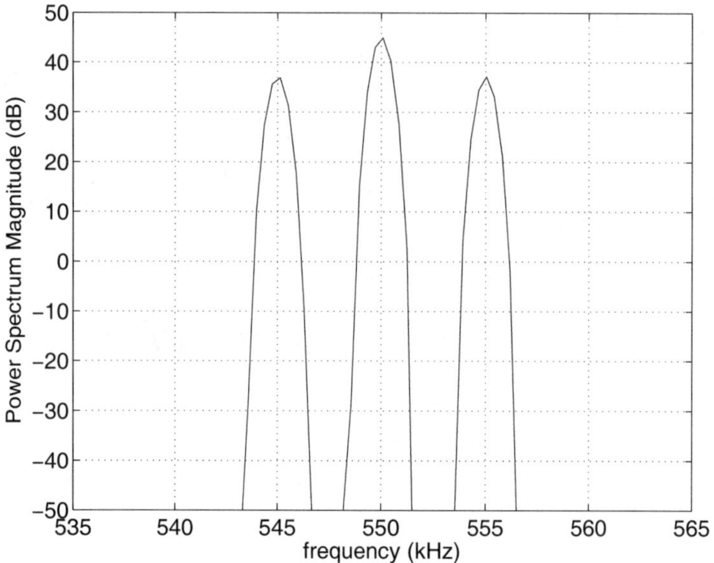

Figure 14.2: An AM signal shown in the frequency domain ($f_c = 550$ kHz, $f_{msg} = 5$ kHz, and $\mu = 0.8$). These are the only three frequency components in the signal.

since the minimum authorized carrier frequency is 550 kHz and the maximum allowed message frequency is 5 kHz. We are discussing messages that are sinusoidal in nature. We will use these sinusoidal tones as example messages to illustrate a number of different points. In actual radio systems the messages will be much more complex and usually will occur at more than one frequency. It is also very common when discussing AM systems to treat the maximum message frequency as the message's bandwidth.

The AM signal shown in the time domain in Figure 14.1 is also shown in the frequency domain as Figure 14.2. Shown from left-to-right in this figure is the lower sideband (LSB), carrier frequency, and the upper sideband (USB). The LSB occurs at $f_c - f_{msg} = 545$ kHz, the carrier frequency is at $f_c = 550$ kHz, and the USB occurs at $f_c + f_{msg} = 555$ kHz. Theoretically, these spectral components should occur as delta function (spectral lines with no width); however, this figure was created using the MATLAB `psd` function[1] using a Blackman-Harris windowing function. As discussed in Chapter 9, the spectrum of the windowing function in use is convolved with the *actual* spectrum. This is why the spectral components are displayed as "humps" instead of delta functions.

14.1.1   Envelope Detector

One of the most inexpensive AM demodulation techniques employs the envelope detector. Traditional circuit-based implementations of the envelope detector utilize a diode and an analog lowpass (LP) filter to demodulate the AM signal. The diode halfwave rectifies the incoming signal; that is, it passes either the positive half of the AM signal or the negative half of the AM signal, depending upon how the diode is connected in the circuit. The analog LP filter extracts the relatively low frequency message from the AM signal's envelope. The effect of halfwave rectification can be seen in Figure 14.3 (time domain) and Figure 14.4 (frequency domain), where an ideal diode is assumed to pass only the positive half of the

---

[1]At this writing, The MathWorks plans to replace the `psd` function with the `pwelch` function.

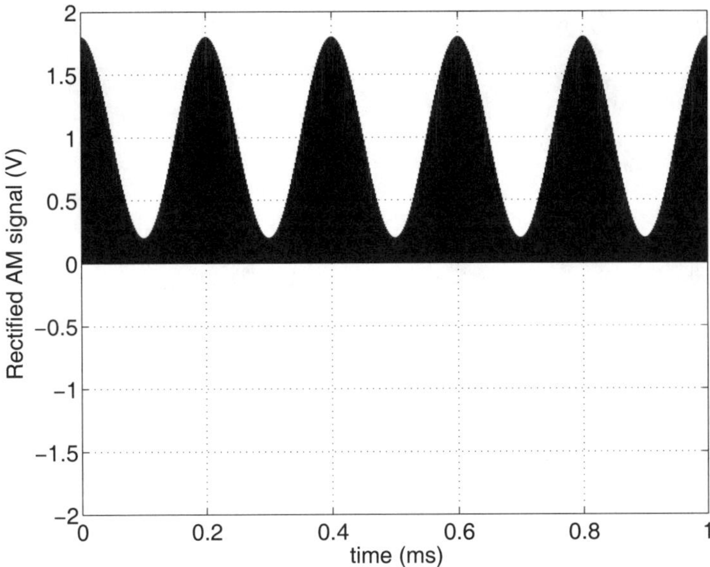

Figure 14.3: A halfwave rectified version of Figure 14.1 in the time domain.

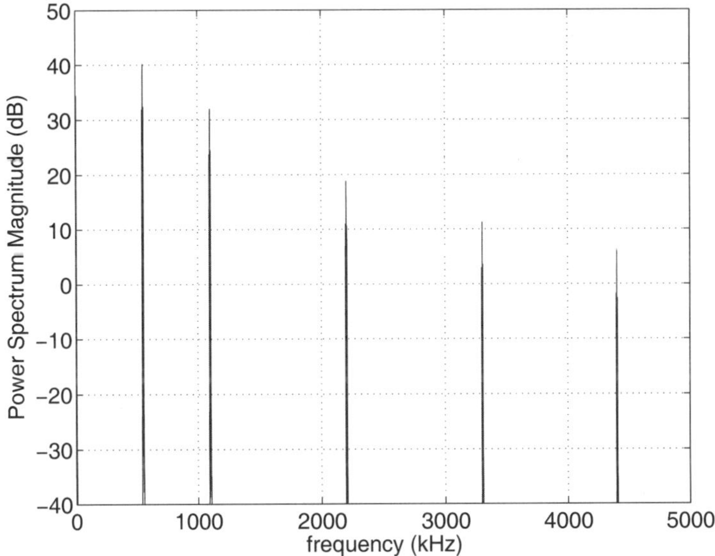

Figure 14.4: A halfwave rectified version of Figure 14.1 in the frequency domain.

AM signal. The nonlinear action of the diode has caused other frequency components to appear in the signal, as is evident in Figure 14.4.

The scale of the plot makes it difficult to distinguish the individual lines of the LSB, the carrier, and the USB; each "spike" on the plot is actually multiple lines. But we can clearly see some components near DC (0 Hz), the fundamental frequency components (centered at $f_c = 550$ kHz), the second harmonic components (centered at $2f_c = 2 \cdot 550 = 1100$ kHz), and the other *even harmonic* components (centered at $nf_c = n \cdot 550$ kHz for $n = 4, 6, 8$). The even harmonics actually continue forever beyond the limits of the plot for $n = 10, 12, 14, \ldots$,

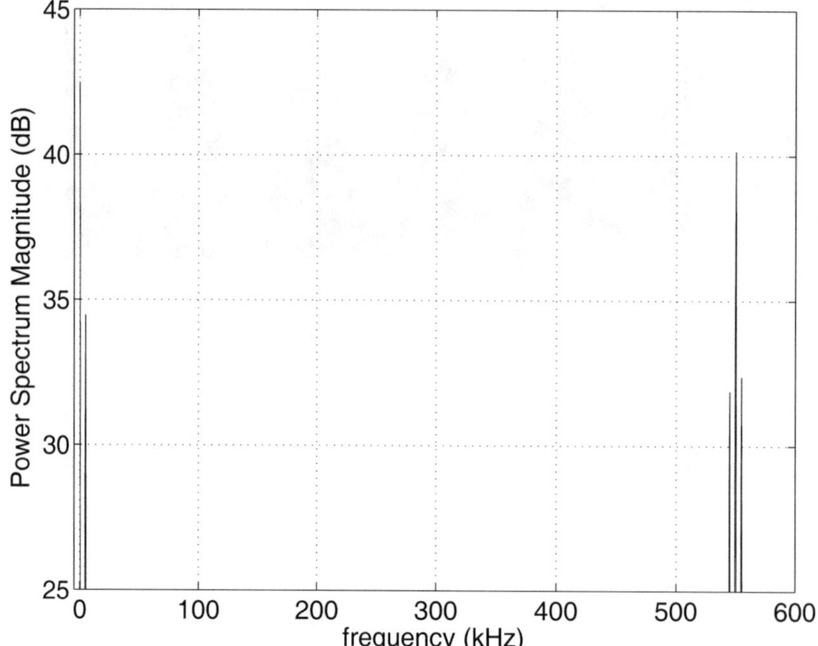

Figure 14.5: An AM signal's spectral content after halfwave rectification.

but the amplitude of the harmonics approaches zero as their frequency increases. Figure 14.5 zooms in on the lower 600 kHz of Figure 14.4; look closely and observe five individual spectral components in Figure 14.5. These individual spectral components occur at the five frequencies that are listed below.

1. D.C., at $f_{DC} = 0$ Hz

2. Message, at $f_{msg} = 5$ kHz

3. Lower sideband (LSB), at $f_c - f_{msg} = 545$ kHz

4. Carrier, at $f_c = 550$ kHz

5. Upper sideband (USB), at $f_c + f_{msg} = 555$ kHz

In the list of spectral components above, the second term is our message and it can be extracted from the spectrum shown in Figure 14.4 by a LP filter. This filter needs to pass the f_{msg} term while providing significant attenuation at the LSB, f_c, and USB frequencies (and all the higher harmonics). These LP filter requirements lead to the design equation

$$BW \ll \frac{1}{\tau} \ll f_c - BW.$$

In this equation, BW is the message signal's bandwidth (in Hz), τ is the LP filter's time constant (in seconds), and f_c is the AM signal's carrier frequency (in Hz). In most radio frequency (RF) systems, $BW \ll f_c$, so the design equation is routinely approximated by

$$BW \ll \frac{1}{\tau} \ll f_c.$$

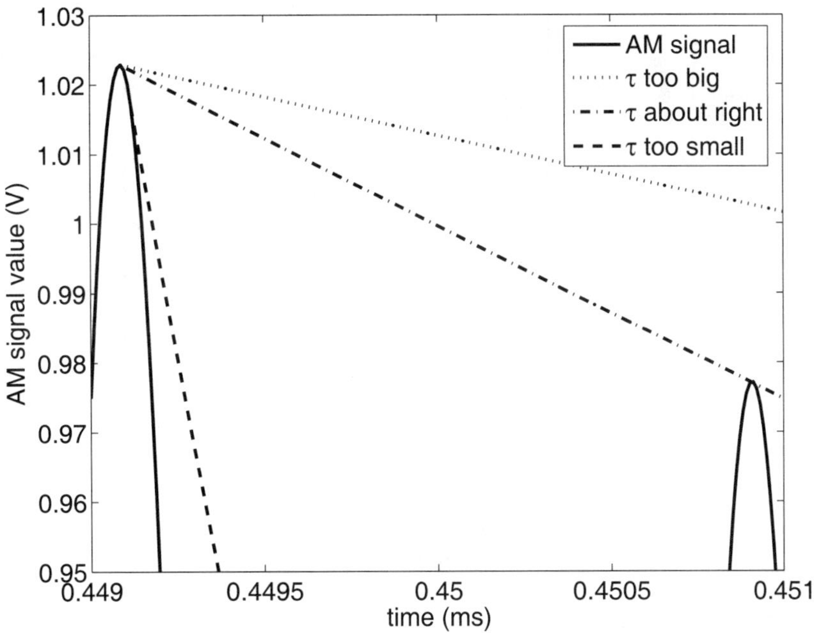

Figure 14.6: The effect of different LP filters on envelope recovery.

Continuing with our example of $f_{msg} = 5$ kHz and $f_c = 550$ kHz requires that

$$5 \text{ kHz} \ll \frac{1}{\tau} \ll 550 \text{ kHz}.$$

Setting $\frac{1}{\tau} \approx 120$ kHz easily satisfies the inequality and allows for direct AM detection/demodulation. Figure 14.6 shows the effect of this LP filter *in the time domain*. This figure shows an extremely magnified portion (showing just the top peaks) of Figure 14.3 with the discharge characteristics of three different LP filters superimposed onto the carrier waveform. The desired effect is for the LP filter to extract the message frequency by connecting the peaks of the halfwave rectified AM signal. Examples of filters that discharge too fast and too slow are also shown in the figure. Setting $\frac{1}{\tau} \approx 120$ kHz is seen to be "about right," as it almost perfectly "connects" the peaks during the filter's discharge process. As we will see below, this may not actually be the optimal choice. In all cases, the discharge rate of an analog filter is controlled by the time constant τ. For a simple first order RC LP filter such as the one shown in Figure 4.1, the time constant is $\tau = RC$.

The previous figures demonstrate the worst-case scenario for using an envelope detector, in which the carrier is at the lowest frequency value allowed (by the FCC) and the message is at the highest frequency allowed. This is the worst case because the relative location of the peaks (in the time domain) drives our LP filter response in one direction, while the separation between the components we want to keep and the components we want to remove (in the frequency domain) drives our LP filter response in the opposite direction. In Figure 14.6, the $\frac{1}{\tau} \approx 120$ kHz filter's discharge characteristic appears to satisfy our needs. However, if we look at the filter response in the frequency domain, this choice does not look nearly as good. This can be seen in Figure 14.7 where the LP filter that seemed to meet our needs in the time domain provides less than 30 dB of attenuation at the carrier frequency of f_c. As previously stated, what is needed is a filter that passes f_{msg} and *significantly* attenuates f_c. From Figure 14.7 it is now clear that a higher performance filter (i.e., higher

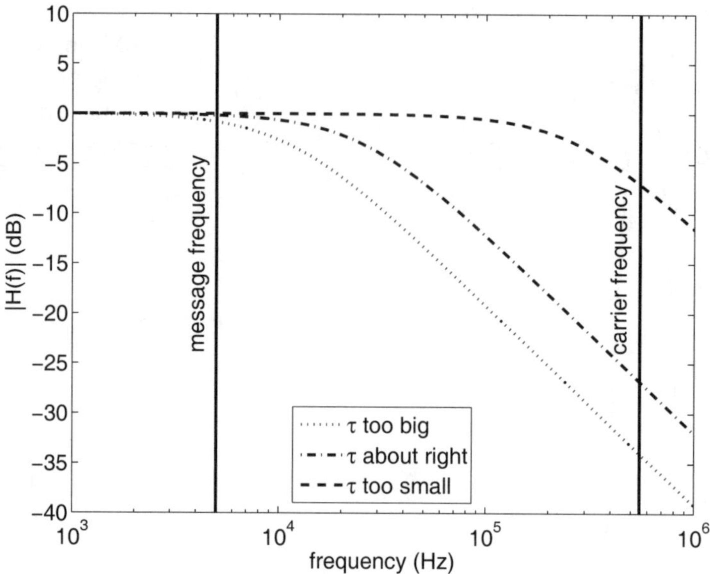

Figure 14.7: The effectiveness of the LP filters used for envelope recovery as viewed in the frequency domain.

order) is required.

Even after an acceptable envelope recovery (LP) filter is designed, there is one last unwanted spectral component at 0 Hz (D.C.) that needs to be removed. In an analog implementation, this can be accomplished with a single D.C. blocking capacitor.

Having made all of these observations, *actual* AM radios typically use a frequency selective, intermediate frequency (IF) based system prior to the envelope detector. The IF-based system provides significant end-to-end gain without amplifier instability/oscillation, as well as providing for much better isolation of the desired frequency channel from any adjacent channels. Better channel isolation is accomplished by the use of a high performance IF filter which then allows the use of lower performance (i.e., less expensive) RF and audio frequency filters.

The previous discussion concerning the importance of the envelope detector's LP filter was necessary to demonstrate that while this technique works very well for commercial AM radio signal recovery, it may not work without severe distortion on our DSK-based AM system, which we limit to audio carrier frequencies to allow the use of the audio codec. This potential filtering problem can be inferred from Figure 14.8, where a 5 kHz message is AM modulated by a 12 kHz carrier. Because the carrier is no longer much higher in frequency than the message frequency (for commercial AM the carrier is more than 100 times higher than the message), it is difficult to see the message envelope in this audio waveform without the "message + D.C." term also being shown in the plot. Despite the appearance of this time domain waveform, the message signal can still be extracted using a high performance LP filter. This can be confirmed by Figure 14.9 where it should be clear that if the LP filter passes the 5 kHz message and if the passband sharply drops off immediately *above* 5 kHz, then the "message + D.C." term can be recovered without distortion from the other frequency components.

We must take great care in choosing frequencies for our audio carrier AM system. We set the carrier frequency to $f_c = 12$ kHz carrier (the center of the codec's alias-free frequency response limit when $F_s = 48$ kHz) to allow "room" for both lower and upper sidebands.

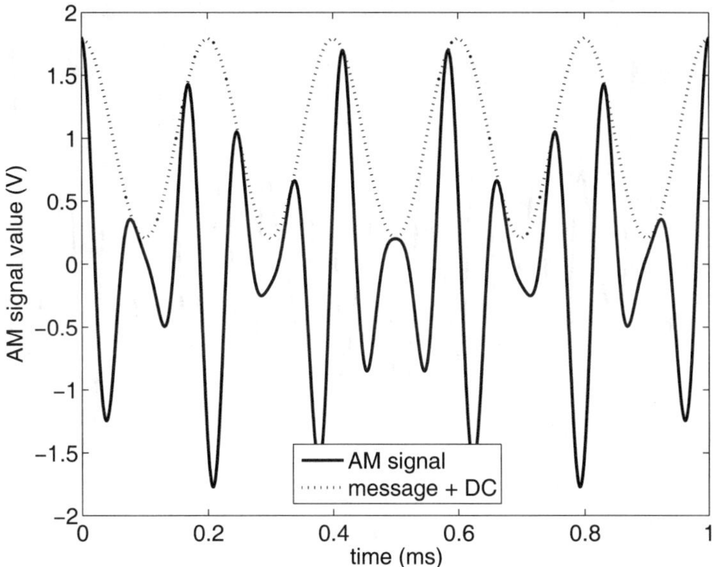

Figure 14.8: AM waveform (f_{msg} = 5 kHz and f_c = 12 kHz).

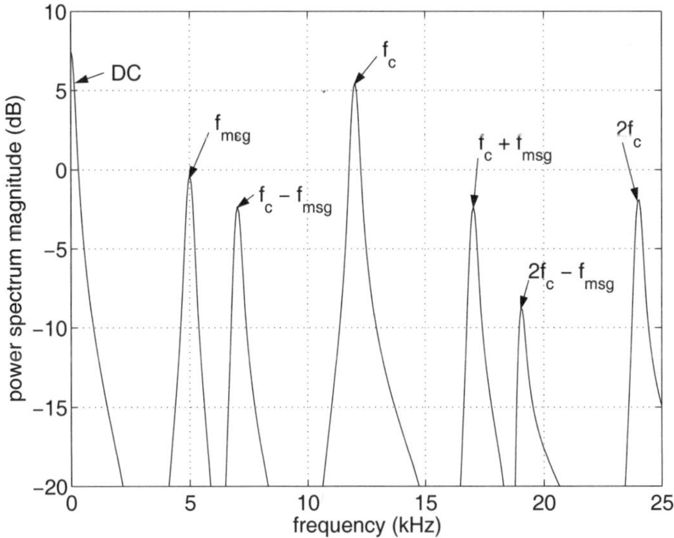

Figure 14.9: AM spectrum (f_{msg} = 5 kHz and f_c = 12 kHz).

But if we *increase* the message frequency too far (i.e., $f_{msg} \geq 6$ kHz), an envelope detector will fail to recover the message properly. At $f_{msg} \geq 6$ kHz, the time domain waveform becomes almost incomprehensible and the spectral components associated with the output of the halfwave rectifier becomes inseparable using the traditional LP filter approach. These concepts can be seen in Figure 14.10 and Figure 14.11. While there are a variety of known solutions to this problem, in the rest of this chapter the focus will be on the Hilbert-based AM receiver.

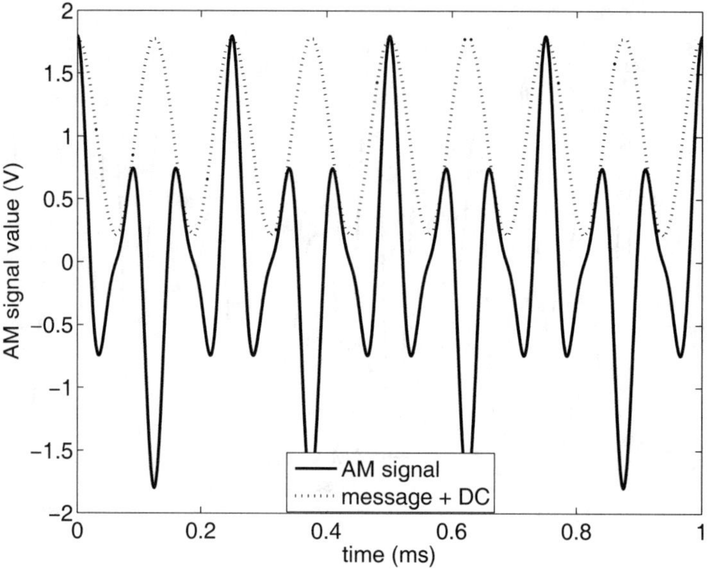

Figure 14.10: AM waveform ($f_{msg} = 8$ kHz and $f_c = 12$ kHz).

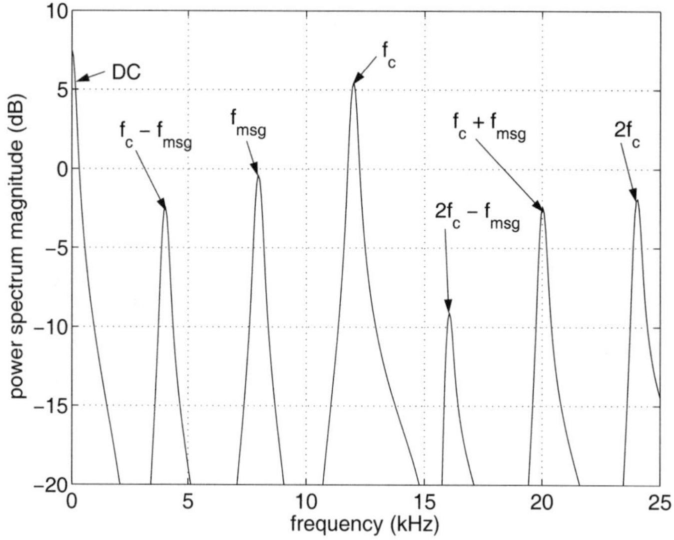

Figure 14.11: AM spectrum ($f_{msg} = 8$ kHz and $f_c = 12$ kHz).

14.1.2   The Hilbert-Based AM Receiver

The Hilbert-based AM receiver extracts the real envelope [63] from the received signal using
the equation

$$r(t) = \sqrt{s^2(t) + \hat{s}^2(t)}.$$

In this equation $r(t)$ is the real envelope of the AM signal. The envelope can be expressed
as $r(t) = m(t) + \text{D.C.}$, where $m(t)$ is the message signal and "D.C." represents the bias that
was added at the AM transmitter to keep $m(t) + \text{D.C.} > 0$. Additionally, $s(t)$ is the received
AM signal and $\hat{s}(t)$ is the Hilbert transform of $s(t)$. Once the real envelope is extracted,

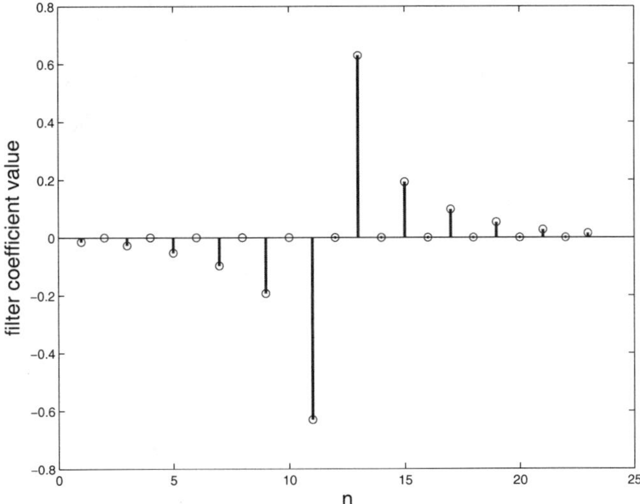

Figure 14.12: Stem plot of the coefficients associated with a 22nd order FIR Hilbert transforming filter.

the D.C. term can be removed using an IIR-based D.C. blocking filter.

To create $\hat{s}(t)$, $s(t)$ must be passed through a system that implements the Hilbert transform. A continuous-time Hilbert transforming filter's impulse response is defined as

$$h(t) = \frac{1}{\pi t}$$

and the frequency response [64, 65] is defined as

$$H(j\omega) = \begin{cases} -j \text{ for } \omega > 0 \\ 0 \text{ for } \omega = 0 \\ +j \text{ for } \omega > 0 \end{cases}$$

The magnitude of the frequency response is one (except it is zero at exactly 0 Hz), and the filter introduces a $-90°$ phase shift for positive frequencies and a $+90°$ phase shift for negative frequencies. This phase shifting filter can either be closely approximated by an FIR digital filter or by using an "FFT/phase shift/IFFT" operation. In this chapter we will use an FIR filter to approximate the Hilbert transform.

To design an FIR-based Hilbert transforming filter in MATLAB we will use the `remez` command.[2] An example of how to use this command is shown below.

Listing 14.1: Using MATLAB to design a Hilbert transforming filter.

```
B = remez(22, [0.1 0.9], [1 1], 'Hilbert');
```

In this listing, variable B will store the 23 filter coefficients, 22 is the filter order, 0.1 and 0.9 are the beginning and ending normalized frequencies over which we want the filter's magnitude response to remain constant, [1 1] represents the magnitudes at those two normalized frequencies, and 'Hilbert' tells MATLAB to design a Hilbert transforming filter. A stem plot of the resulting B coefficients is provided as Figure 14.12. Notice that for Hilbert transforming filters such as this one, specified with a passband that is symmetric

---

[2]At this writing, The MathWorks plans to replace the `remez` function with the `firpm` function.

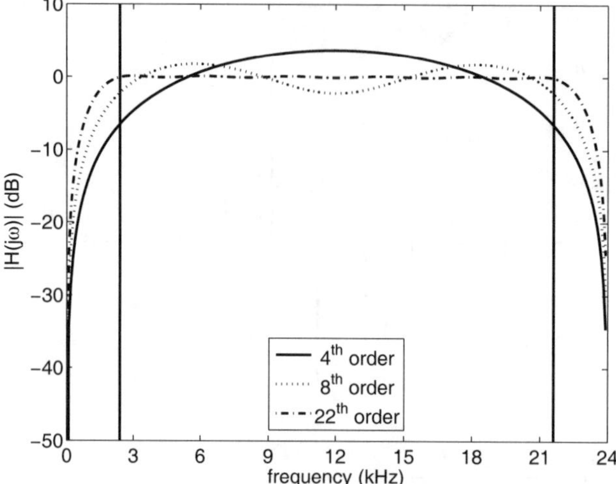

Figure 14.13: Magnitude responses of three Hilbert transforming filters.

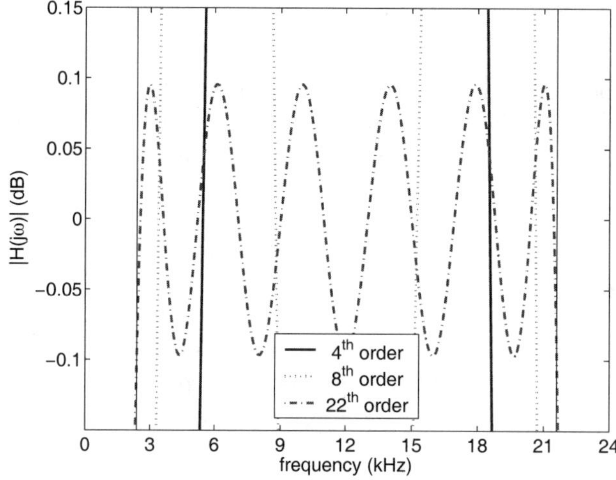

Figure 14.14: Zoomed magnitude response emphasizing the 22nd order Hilbert transforming filter's passband ripple.

on the frequency scale and centered at $F_s/4$, we see that every other filter coefficient has a value of zero. This filter's coefficients are also antisymmetric about the center point (i.e., $h[n] = -h[N-n]$). These traits could be exploited to reduce the computational load of the DSP CPU.[3]

Specifying the Hilbert transforming filter's design parameters depends greatly on the characteristics of the signal that you are trying to transform. In Figure 14.13, three different filters of increasing order are shown. Each filter has the same bandpass (BP) filter specifications. For a given frequency range, the higher the filter order, the flatter the response becomes in the passband. Figure 14.14 shows that as the filter order increases, the filter's passband ripple decreases, but it will never completely disappear.

---

[3]Any linear phase FIR filter will exhibit either symmetrical ($h[n] = h[N-n]$) coefficients or antisymmetrical ($h[n] = -h[N-n]$) coefficients.

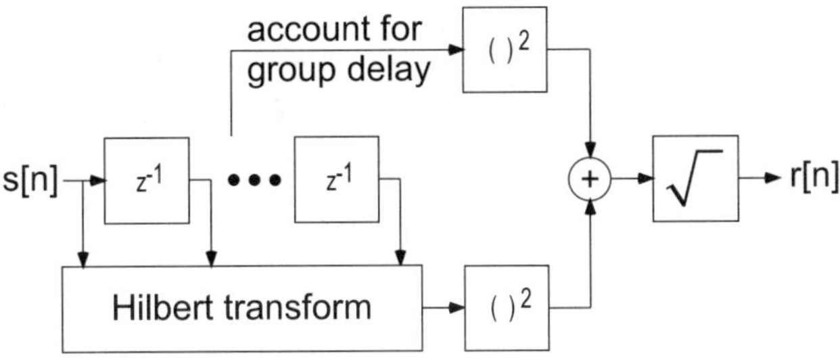

Figure 14.15: Real envelope recovery from and AM signal.

A sampled (i.e., discrete-time) system would replace $s(t)$ by $s[n]$ and $r(t)$ by $r[n]$ in the real envelope extraction equation. The block diagram associated with extracting the real envelope of the signal is shown in Figure 14.15. Note that to "line up" the unfiltered signal $s[n]$ with the filtered signal $\hat{s}[n]$, we need to delay the unfiltered signal by the group delay of the filter. The group delay associated with any linear phase FIR filter, such as the Hilbert transforming filter, is one half of the filter order. For the 22nd order filter in this example, the group delay will be 11 samples. Since the input values to this filter are already being stored for implementation using direct form I (DF-I) techniques, no additional input sample buffering is required for us to delay $s[n]$ as needed.

Finally, $r[n]$ must have its D.C. component removed. This is accomplished with a notch filter where the stopband frequency is set to 0 Hz. This filter places a zero on the unit circle at $z = 1$ to determine the location of the stopband, and a pole on the real axis very near this zero (e.g., $z = 0.95$) to make the slope of the stopband very sharp. The transfer function of this system is,

$$H\left(z\right) = \left(\frac{1+r}{2}\right)\frac{1 - z^{-1}}{1 - rz^{-1}}.$$

In this equation, r represents the location of the pole (we previously used the example where $r = 0.95$) and the $(1 + r)/2$ term normalizes the high frequency (passband) gain to 1.0 (0 dB). The difference equation associated with this system can be derived as follows.

$$H\left(z\right) = \frac{Y\left(z\right)}{X\left(z\right)} = \left(\frac{1+r}{2}\right)\frac{1 - z^{-1}}{1 - rz^{-1}}$$

$$Y\left(z\right)\left(1 - rz^{-1}\right) = X\left(z\right)\left(\frac{1+r}{2}\right)\left(1 - z^{-1}\right)$$

$$y[n] - ry[n-1] = \left(\frac{1+r}{2}\right)\left(x[n] - x[n-1]\right)$$

$$y[n] = ry[n-1] + 0.5\left(1 + r\right)\left(x[n] - x[n-1]\right).$$

We now have enough information to implement the AM demodulator shown in Figure 14.15.

14.2 winDSK6 Demonstration

The winDSK6 program does not provide an equivalent function.

14.3 MATLAB Implementation

The MATLAB listing provided below generates an AM signal, designs a Hilbert transforming filter, uses this filter to recover the real envelope of the signal, filters out the D.C. com-

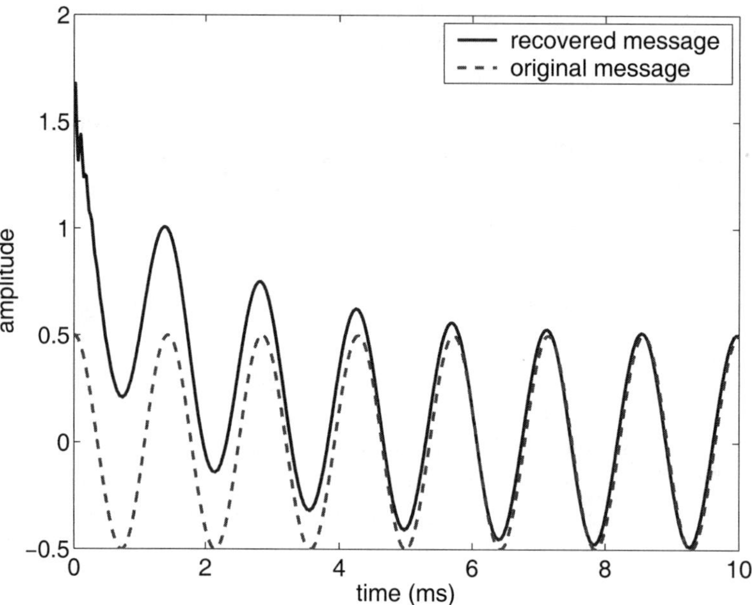

Figure 14.16: MATLAB simulation results of the Hilbert-based AM receiver.

ponent, and displays both the original message and the recovered message. Figure 14.16
shows the output of this simulation. The initial decaying response of the recovered message
is due to the startup transient associated with the D.C. blocking filter. Despite this transient, excellent agreement with the original message signal is achieved in fewer than 10 ms;
thereafter the demodulation would be nearly perfect. Additionally, the very small delay
between the signals is due to the group delay of the D.C. blocking filter.

Failure to properly account for the group delay of the Hilbert transforming filter is a
very common problem that will cause the receiver to fail. Being off by even a single sample
delay while accounting for the group delay of the Hilbert transforming filter will prevent
the proper operation of the system.

Listing 14.2: MATLAB simulation of a Hilbert-based AM receiver.

```
%   Hilbert-based AM receiver simulation
%
% variable declarations
Fs = 48000;            % sample frequency
Ac = 1.0;              % amplitude of the carrier
fc = 12000;            % frequency of the carrier
Amsg = 0.5;            % amplitude of the message
fmsg = 700;            % frequency of the message
HilbertOrder = 62;     % Hilbert transforming filter order
r = 0.990;             % highpass filter pole magnitude
duration = 0.5;        % signal duration in seconds
myFontSize = 16;       % font size for the plot labels

% design the FIR Hilbert transforming filter
B = remez(HilbertOrder, [0.05 0.95], [1 1], 'h');
t = 0:1/Fs:duration;
```

```matlab
17
   % generate the AM/DSB w/carrier signal
19 carrier = Ac*cos(2*pi*fc*t);
   msg = Amsg*cos(2*pi*fmsg*t);
21 AM = (1 + msg).*carrier;

23 % recover the message
   % note the HilbertOrder/2 delay to align the signals
25 % use the MATLAB syntax of "..." to continue a line of code

27 % apply Hilbert transform
   AMhilbert = filter(B, 1, AM);
29
   % get envelope, but account for filter delay
31 OffsetOutput = sqrt((AM(1:2000)).^2 + ...
      (AMhilbert((HilbertOrder/2+1):(2000 + HilbertOrder/2))).^2);
33
   % remove DC component
35 demodOutput = filter([1 -1]*(1 + r)/2, [1 -r], OffsetOutput);

37 % create the desired figure
   plot((0:(length(demodOutput)-1))/Fs*1000, demodOutput)
39 set(gca, 'FontSize', myFontSize)
   hold on
41 plot((0:(length(demodOutput)-1))/Fs*1000, ...
      msg(1:length(demodOutput)), 'r--')
43 ylabel('amplitude')
   xlabel('time (ms)')
45 legend('recovered message', 'original message')
   axis([0 10 -0.5 2.0])
47 hold off
```

14.4 DSK Implementation in C

This version of the Hilbert-based AM receiver is very similar to the MATLAB simulation example. The intention of this first approach is understandability, which may come at the expense of efficiency.

The files necessary to run this application are in the `ccs\Proj_AmRx` directory of Chapter 14. The primary file of interest is the `AMreceiver_ISRs.c`, which contains the interrupt service routines. This file also contains the necessary variable declarations and performs the Hilbert-based AM demodulation.

If you are using a stereo codec (such as the on-board C6713 DSK codec or the PCM 3006-based daughtercard codec for the C6711 DSK), the program could implement independent Left and Right channel demodulators. For clarity, this example program will demodulate only one signal's message, but will output this message to both the Left and Right channels.

In Listing 14.3 shown below, array `x` (Line 1) contains the current and stored (i.e., past) received AM signal values. The Hilbert transforming filter's output value is `y` (Line 2). The present and most recent real envelope values ($r[n]$ and $r[n-1]$) are stored in `envelope` (Line 3) and the output of the D.C. blocking filter is stored in `output` (Line 4). The D.C.

blocking filter's only adjustable parameter is r (Line 5), which controls the location of the pole along the real axis. For proper operation, r should be slightly less than 1.0.

Listing 14.3: Variable declaration associated with a Hilbert-based AM receiver.

```
1  float  x[N];             // received AM signal values
   float  y;                // Hilbert Transforming (HT) filter's output
3  float  envelope[2];      // real envelope
   float  output[2];        // output of the D.C. blocking filter
5  float  r = 0.99;         // pole location for the D.C. blocking filter
   int  i;                  // integer index
```

The code shown below performs the actual AM demodulation operation. The six main steps involved in this operation will be discussed following the code listing.

Listing 14.4: Algorithm associated with Hilbert-based AM demodulation.

```
   /* algorithm begins here */
2  x[0] = CodecDataIn.Channel[LEFT];// current AM signal value
   y = 0;                          // initialize filter's output value

4
   for (i = 0; i < N; i++) {
6      y += x[i]*B[i];              // perform HT (dot-product)
   }

8
   envelope[0] = sqrtf(y*y + x[16]*x[16]); // real envelope

10
   /* implement the D.C. blocking filter */
12 output[0]=r*output[1]+(float)0.5*(r+1)*(envelope[0]-envelope[1]);

14 for (i = N-1; i > 0; i--) {
       x[i] = x[i-1];               // setup for the next input
16 }

18 envelope[1] = envelope[0]; // setup for the next input
   output[1]    = output[0];      // setup for the next input

20
   CodecDataOut.Channel[ LEFT] = output[0]; // setup the LEFT  value
22 CodecDataOut.Channel[RIGHT] = output[0]; // setup the RIGHT value
   /* algorithm ends here */
```

The six real-time steps involved in Hilbert-based AM demodulation

An explanation of Listing 14.4 follows.

1. (Line 2): This code brings the current value of the AM signal into the DSP.

2. (Lines 3–7): These lines of code initializes the filter's output and performs the Hilbert transform of the current and stored input values.

3. (Line 9): This code calculates the real envelope of the AM signal.

4. (Line 12): This code implements the D.C. blocking filter.

5. (Lines 14–19): These lines of code prepare the stored variables to receive the next input value.

6. (Lines 21–22): These lines of code output the message signal to both the Left and Right channels.

Now that you understand the code. . .

Go ahead and copy all of the files into a separate directory. Open the project in CCS and "Rebuild All." Once the build is complete, "Load Program" into the DSK and click on "Run." Your Hilbert-based AM receiver is now running on the DSK. AM modulated signals are available in the `test_signals` directory of the CD-ROM so you can test your AM receiver code. The file names begin with AM.

14.5 Follow-On Challenges

Consider extending what you have learned.

1. Implement the Hilbert transform-based AM receiver using an "FFT/phase shift/IFFT" based Hilbert transform. This should reduce the DSP CPU computational load. See Chapter 8 for details concerning the FFT.

2. Design and implement different bandpass Hilbert transforming filters that are matched in bandwidth to the signal you are trying to recover. Experiment with different filter orders that will still recover the message signal.

3. Implement the Hilbert transform-based AM receiver taking advantage of the fact that every other coefficient for the Hilbert transforming filter is zero. This should reduce the DSP CPU computational load.

4. Implement the Hilbert transform-based AM receiver taking advantage of the fact that Hilbert transforming filters are antisymmetric filters. This should reduce the DSP CPU computational load.

5. Use circular buffering techniques to implement the Hilbert transform-based AM receiver. This should reduce the DSP CPU computational load.

6. Use frame-based techniques to implement the Hilbert-based AM receiver. This should reduce the DSP CPU computational load.

Chapter 15

Project 6: Phase-Locked Loop

15.1 Theory

THE phase locked loop (PLL) is widely used in communications receiver systems. Since even the fundamental theory of phase-locked loops has been the topic of dozens of textbooks, we will only briefly discuss, then implement, a single PLL design: a discretized, second order Costas loop [64, 65]. The simplified block diagram of such a system is shown in Figure 15.1. In this figure, $s[n]$ is the sampled input signal, $m[n]$ is an estimate of the message signal, T is the sample period, and ω_c is the carrier frequency.

The basic operation of the PLL is as follows. The incoming signal $s[n]$ is processed through a filter that implements a Hilbert transforming (HT) operation to create $\hat{s}[n]$. Recall from Chapter 14 that the frequency response magnitude of the HT operation is one (except it is zero at exactly 0 Hz), and it introduces a $-90°$ phase shift for positive frequencies and a $+90°$ phase shift for negative frequencies. A signal commonly called the analytic signal is formed from the sum of the input signal $s[n]$, and the HT filtered version of the input signal $\hat{s}[n]$ that has been multiplied by the imaginary number j (where $j = \sqrt{-1}$). Recall that $jx = |x|\angle 90°$, so multiplication by j is equivalent to a $+90°$ phase shift for all frequencies. Note that you'll need to account for any group delay in the HT filtering operation to make sure the signals "line up" properly, as we did in Chapter 14. In equation

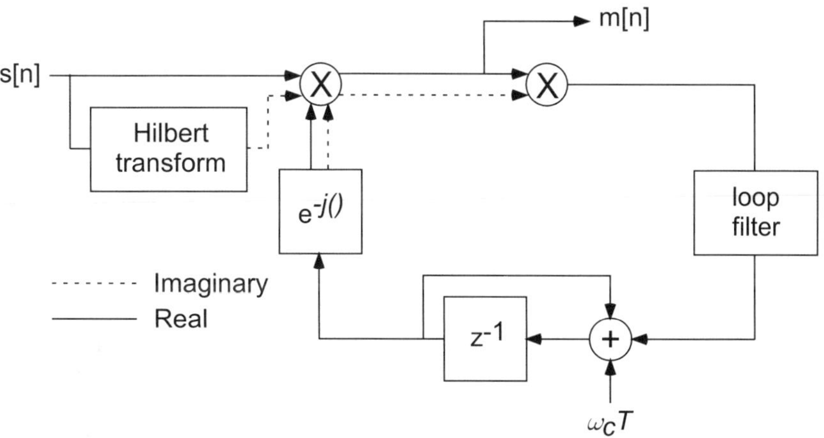

Figure 15.1: The simplified block diagram of a second order Costas loop.

form, the analytic signal is defined as $s[n] + j\hat{s}[n]$. This is why the block diagram uses solid and dashed lines to indicate the real and imaginary signals, respectively. All that remains, at this point, is to strip off the carrier signal.

The first multiplier in Figure 15.1 is actually an approximation of a phase detector. More specifically, the outputs of the complex exponential block $e^{-j(\cdot)}$ are sine and cosine waveforms that, ideally, are at the exact frequency and phase of the incoming signal's carrier. This complex oscillator, in an analog circuit, would be called a local oscillator (LO). The output of the multiplier, which can also be called a mixer, contains both the message signal and an approximation of the LO's phase error.

Various types of coherently detected amplitude modulated (AM) communication signals can be recovered by the real part of this mixer output, $m[n]$. Additional filtering may be required to recover a more accurate estimate of the transmitted message. Binary phase-shift keying (BPSK), the more general M-PSK, and quadrature amplitude modulation (QAM) can all be thought of as special cases of AM. More on this topic can be found in Chapters 16 and 17.

The multiplication of the real and imaginary outputs of the first mixer is the input to the loop filter. This filter is actually a lowpass filter which only allows an estimate of the LO's phase error to be fed back into the complex oscillator. The single input to the complex oscillator (LO) is the output of a 2-input phase accumulator. This accumulator combines the existing phase of the oscillator with the free-running oscillator's next phase increment, $\omega_c T$, and the estimate of the phase error out of the loop filter. The complex oscillator (LO) and the phase accumulator together are routinely referred to as a voltage-controlled oscillator (VCO).

If the LO's rest (or free-running) frequency, is *reasonably* close to the carrier frequency of the input signal, the PLL will almost immediately begin to remove any frequency and phase error between these signals, and "lock" the VCO frequency to the incoming carrier frequency. This is how the PLL got its name. The phase error feedback loop rapidly accomplishes this task without any operator interaction with the system.

15.2   winDSK6 Demonstration

The winDSK6 program does not provide an equivalent function.

15.3   MATLAB Implementation

15.3.1   PLL Simulation

The MATLAB simulation of the PLL is shown in Listing 15.1.

Listing 15.1: Simulation of a PLL.

```
% Simulation of a BPSK modulator and its coherent recovery
%

% input terms
Fmsg = 12000;        % carrier frequency of the BPSK transmitter (Hz)
VCOrestFrequencyError = 200*randn(1);  % VCO's error (Hz)
VCOphaseError = 2*pi*rand(1);          % VCO's error (radians)
Fs = 48000;                            % sample frequency (Hz)
N = 20000;                             % samples in the simulation
samplesPerBit = 20;                    % sample per data bit
```

```matlab
11 dataRate = Fs/samplesPerBit;            % data rate (bits/second)
   beta = 0.002;                           % loop filter parameter "beta"
13 alpha = 0.01;                           % loop filter parameter "alpha"
   noiseVariance = 0.0001;                 % noise variance
15 Nfft = 1024;                            % number of samples in an FFT
   amplitude = 32000;                      % ADC scale factor
17 scaleFactor = 1/32768/32768;            % feedback loop scale factor

19 %  input term initializations
   phaseDetectorOutput = zeros(1, N+1);
21 vcoOutput = [exp(-j*VCOphaseError) zeros(1, N)];
   m = zeros(1, N+1);
23 q = zeros(1, N+1);
   loopFilterOutput = zeros(1, N+1);
25 phi = [VCOphaseError zeros(1, N)];
   Zi = 0;

27

   %  calculated terms
29 Fcarrier = Fmsg + VCOrestFrequencyError; % VCO's rest frequency
   T = 1/Fs;                               % sample period
31 B = [(alpha + beta) -alpha];            % loop filter numerator
   A = [1 -1];                             % loop filter denominator
33 Nbits = N/samplesPerBit;                % number of bits
   noise = sqrt(noiseVariance)*randn(1, N); % AWGN (noise) vector

35

   %  random data generation and expansion (for BPSK modulation)
37 data = 2*(randn(1, Nbits) > 0) - 1;
   expandedData = amplitude*reshape(ones(samplesPerBit,1)*data,1,N);

39

   % BPSK signal and its HT (analytic signal) generation
41 BPSKsignal = cos(2*pi*(0:N-1)*Fmsg/Fs).*expandedData + noise;
   analyticSignal = hilbert(BPSKsignal);

43

   %  processing the data by the PLL
45 for i = 1:N
       phaseDetectorOutput(i+1) = analyticSignal(i)*vcoOutput(i);
47     m(i+1) = real(phaseDetectorOutput(i+1));
       q(i+1) = scaleFactor * real(phaseDetectorOutput(i+1)) ...
49             * imag(phaseDetectorOutput(i+1));
       [loopFilterOutput(i+1), Zf] = filter(B, A, q(i+1), Zi);
51     Zi = Zf;
       phi(i+1) = mod(phi(i) + loopFilterOutput(i+1) ...
53             + 2*pi*Fcarrier*T, 2*pi);
       vcoOutput(i+1) = exp(-j*phi(i+1));
55 end
   %  Plotting commands follow ...
```

An explanation of Listing 15.1 follows.

1. (Line 5): Define the system's carrier frequency as 12 kHz.

2. (Lines 6–7): Define errors in both the carrier's frequency and phase. These errors add realism to the simulation.

3. (Line 8): Defines the system's sample frequency as 48 kHz. This sample frequency matches the rate of the DSK's audio codec.

4. (Line 9): Defines the number of samples in the simulation.

5. (Lines 10–11): Together with the system's sample frequency, these lines of code specify the symbol rate. In this binary phase-shift keying (BPSK) simulation, the symbol rate is equal to the bit rate. It cannot be overemphasized that the requirement specified by Nyquist *must* be met! This requirement can be restated as "you must sample fast enough to prevent aliasing of your input signal." Given the fact that we are implementing the BPSK signal as the multiplication of a 12 kHz carrier with an antipodal pulse train, it should be clear that some aliasing *will* occur since the bandwidth of a perfect antipodal pulse train is infinite.

6. (Lines 12–13): Specify the filter coefficients associated with the loop filter.

7. (Line 14): Like Lines 16–17, adding noise to the system adds realism to the simulation.

8. (Line 16): The full range of a 16-bit ADC and DAC is $+32,767$ to $-32,768$. This amplitude scale factor simulates an input signal near the full range of values allowed into the ADC without exceeding the range of the converter.

9. (Line 17): This term scales the phase error signal *prior* to this signal being fed back into the phase accumulator. As the PLL drives the system's phase error to near zero, we expect the phase error to be a very small number (a small portion of a radian).

10. (Lines 19–26): These lines of code establish the variables for sample-by-sample processing of the PLL simulation. Maintaining a vector instead of only the most recent value will allow for a number of performance plots to be generated.

11. (Lines 31–32): These are the coefficients associated with a single equivalent IIR filter implementation for the loop filter. This is done only for ease of simulation. In [64,65], the loop filter is implemented in parallel form, which can be seen in Figure 15.2.

12. (Lines 37–38): Create the BPSK baseband signal.

13. (Line 41): Mixes the BPSK baseband signal to the carrier frequency and adds noise to the signal.

14. (Line 42): Calculates the analytic signal of the input signal s[n]. Fortunately, MATLAB defines the analytic signal as $s[n] + j\hat{s}[n]$, which is the same definition we introduced earlier. At this point, we must recognize that the variable `analyticSignal` is a complex number.

15. (Line 46): Calculates the output of the first mixer.

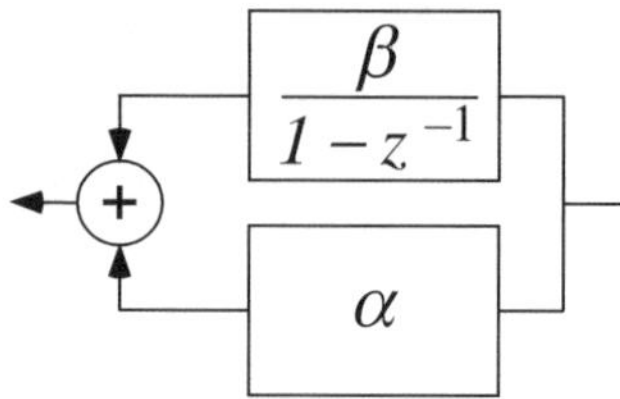

Figure 15.2: The block diagram of the parallel implementation of the loop filter.

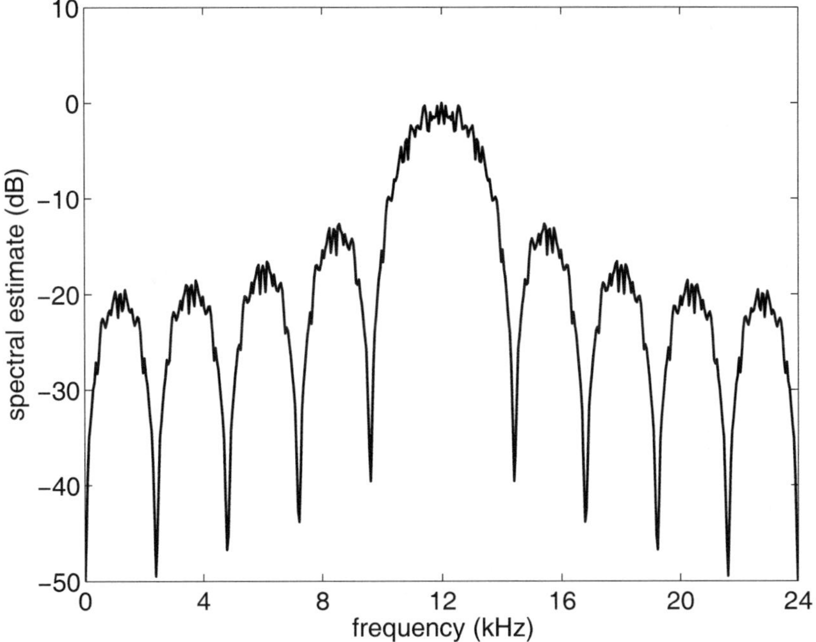

Figure 15.3: The normalized spectral estimate of the BPSK message signal mixed with a carrier frequency of 12 kHz.

16. (Line 47): Calculates the estimate of the message signal.

17. (Lines 48–49): Calculate and scales the output of the second mixer.

18. (Line 50): Performs the loop filtering operation.

19. (Line 51): Saves the filter state by copying the filter's final condition to the filter's initial condition variable.

20. (Lines 52–53): Calculate the next value contained in the phase accumulator. The `mod` command keeps the phase accumulator's output value between 0 and 2π.

21. (Line 54): Calculates the LO's output value. The complex exponential form can also be thought of as sine and cosine terms by using Euler's identity.[1]

Although not shown in the code listing, this simulation results in several plots that allow for a more detailed understanding of the PLL. Given the simulation's ability to randomly initialize the frequency and phase errors associated with the LO and the random nature of the noise that is added to the signal, the behavior of the simulation will be different *every* time the simulation is run. This random behavior can be controlled by either reinitializing the state of the MATLAB random number generator at the beginning of each simulation or by setting the variables in lines 16, 17, and 24 equal to zero. The state of the MATLAB random number generator can be reset using the MATLAB command `randn('state',0)`.

The normalized spectral estimate of the BPSK message signal mixed with a carrier frequency of 12 kHz is shown in Figure 15.3. The pole/zero plot of the loop filter is shown in Figure 15.4. In this figure, a pole is *on* the unit circle and although it may not be

---

[1]Specifically, $e^{-\phi} = \cos(-\phi) + j\sin(-\phi) = \cos(\phi) - j\sin(\phi)$, with the sign simplifications occurring because of the even and odd properties of the cosine and sine functions, respectively.

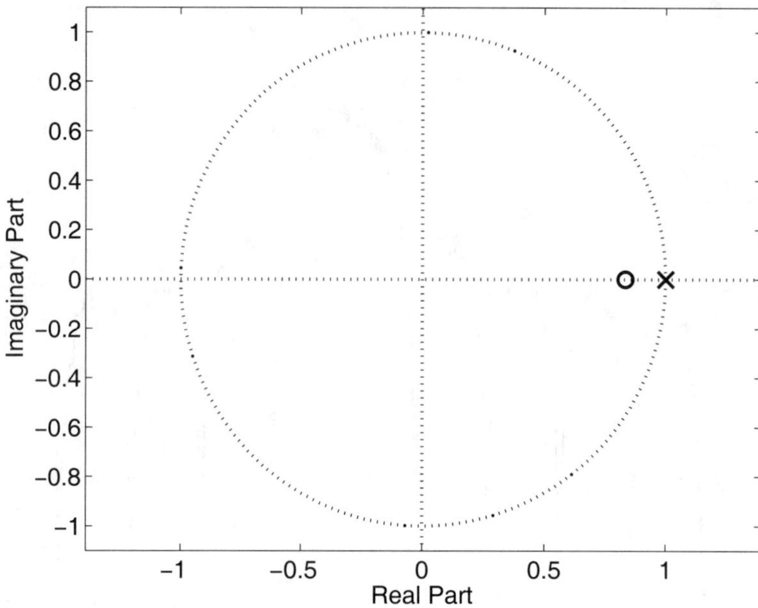

Figure 15.4: The pole/zero plot of the loop filter.

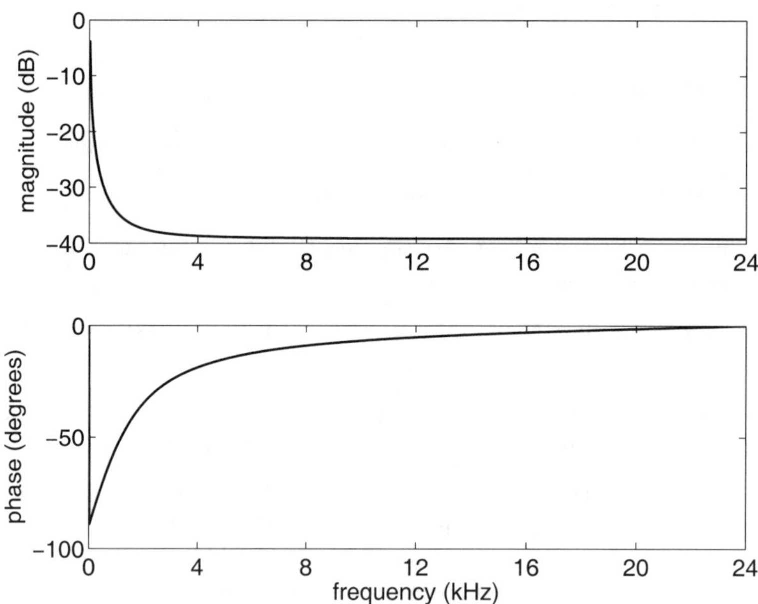

Figure 15.5: The frequency response of the loop filter.

evident from Figure 15.5, this results in infinite gain at 0 Hz (DC). An example of a well behaved startup transient of the PLL is shown in Figure 15.6. While this type of behavior is quite common, the second order Costas loop is blind to 180 degree phase ambiguities and the system's output may be as shown in Figure 15.7. In Figure 15.7, the "recovered message" is clearly 180 degrees out of phase from the "message" signal. This sign error will

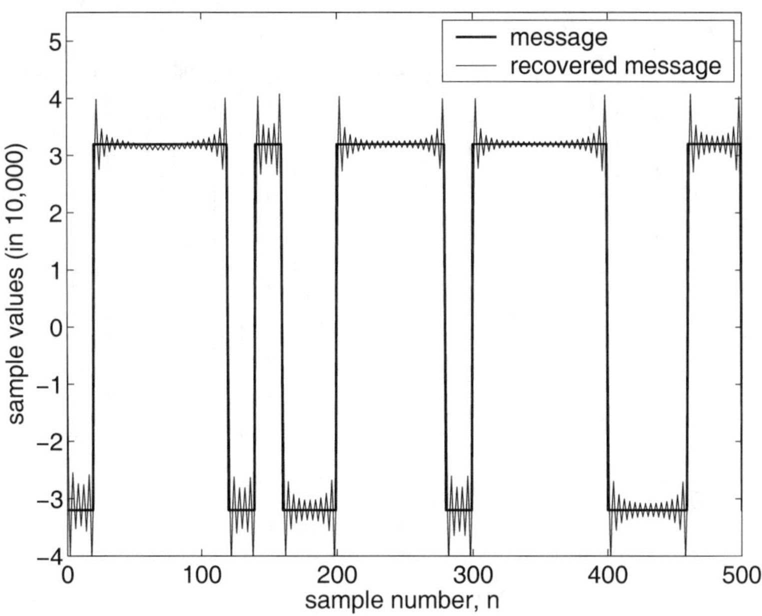

Figure 15.6: Well behaved startup transient of the PLL.

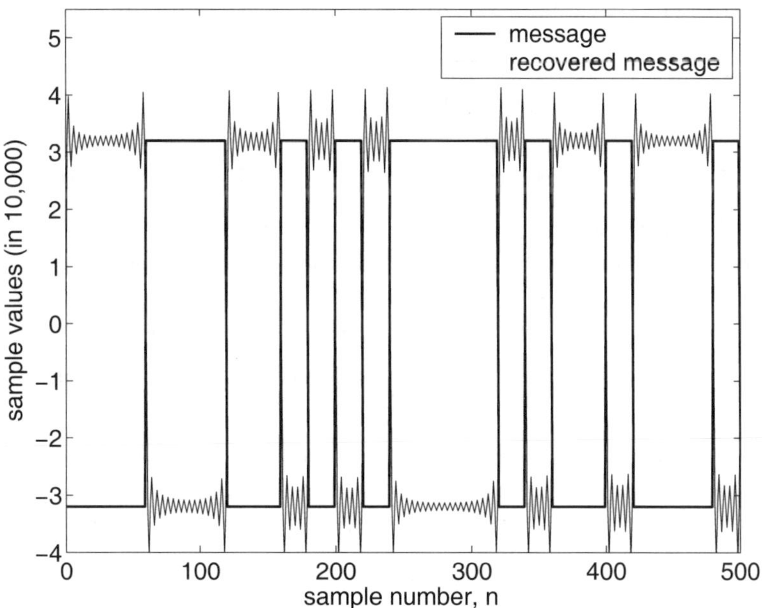

Figure 15.7: Well behaved startup transient of the PLL with a 180 degree phase error.

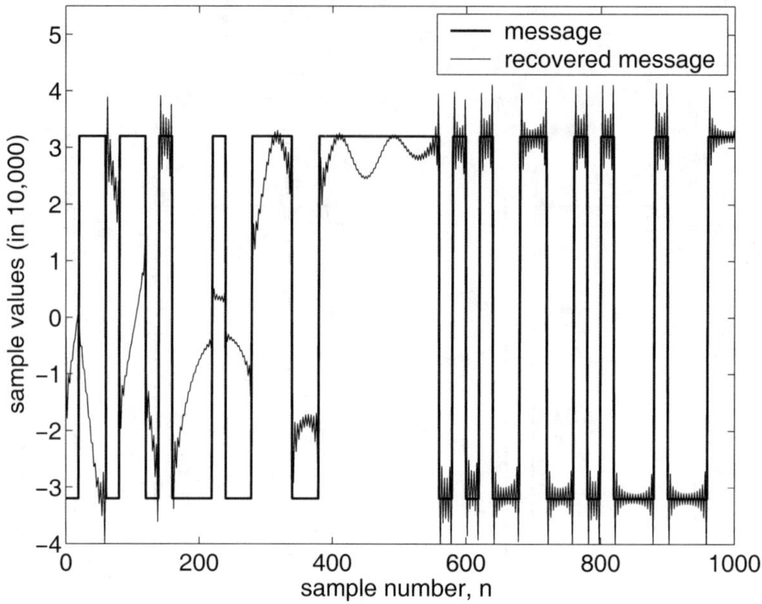

Figure 15.8: A more protracted startup transient of the PLL.

be disastrous (e.g., BPSK communication systems would result in 100% of the bits being received in error).[2] Assuming we are using a two's complement numeric representation, reversing the state of all the bits does *not* just constitute a change in sign! Finally, a very large frequency error was inserted into the LO to shown a more protracted transient during PLL initialization. This is shown in Figure 15.8, where it took somewhere between 300 to 500 samples for the PLL to stabilize and track the incoming signal. In this simulation, there were 20 samples/symbol; this corresponds to 15 to 25 symbols that may not be properly recovered before the PLL stabilizes. Depending on the intended use of the communication system, this transient behavior of the PLL may be disastrous or of little concern.

15.3.2   A Few Updates to the MATLAB Implementation

As mentioned earlier, the simulation we presented was designed to be easily implemented in MATLAB. Since MATLAB is a very high-level language, a few of the specialized commands must be replaced by commands capable of being readily implemented in C/C++. Some issues to consider when transitioning the PLL code from MATLAB to C are given below.

1. The vector-based variables need to be replaced by sample-by-sample variables. This straightforward process will be explained in the next section.

2. There is no longer a need to intentionally offset the LO's frequency or phase and add noise to the incoming signal. While these terms were required to make the simulation more realistic, they are not required for real-time implementation.

3. While MATLAB can easily handle complex numbers, the C language does not have a native complex variable type. A common way to handle complex numbers is to declare

---

[2]If this were an analog PLL performing coherent recovery of an analog voice signal, then this ±180° phase ambiguity would be of little or no concern because the human auditory system will not perceive a difference in the intended message. But for a digital signal we cannot tolerate reversal of all the bits. The use of a known preamble code or using differentially encoded message data can solve this problem. [61, 62, 66].

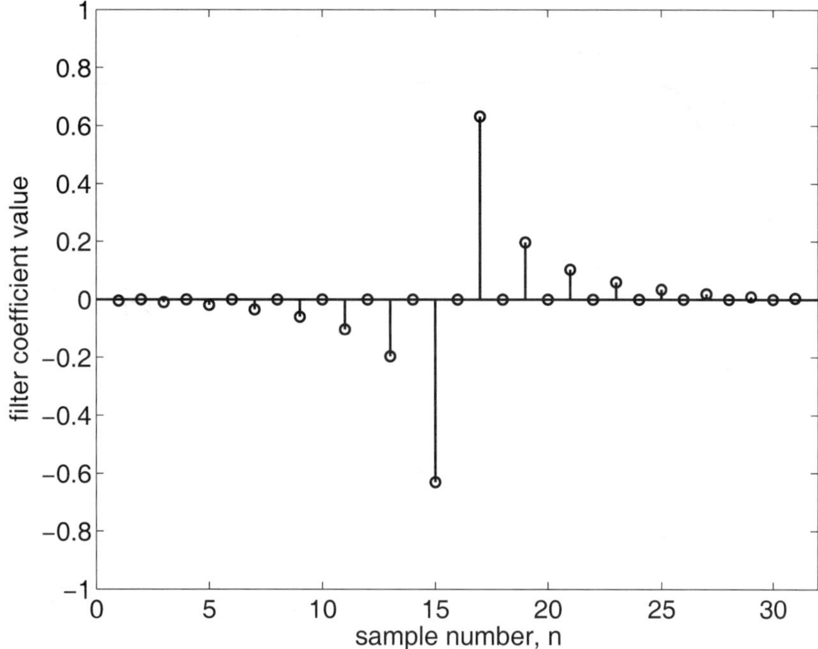

Figure 15.9: Filter coefficients associated with a 30th order FIR Hilbert transforming filter.

real and imaginary variables associated with each of the complex terms. This is just a rectangular implementation of the MATLAB complex terms. Unlike MATLAB, when a rectangular version of the complex terms is used, it is incumbent upon the system programmer to properly code the results associated with all of the complex mathematical operations. For example, if two complex numbers z_1 and z_2 are multiplied together the result would be $z_1 z_2 = (a+jb)(c+jd) = ac+jad+jbc-bd = (ac-bd)+j(ad+bc)$.

4. The Hilbert transform of the incoming signal can be implemented in a number of ways. A sample-by-sample implementation would approximate the transform with a bandpass FIR filter. The FIR filter can be designed in MATLAB, for example, by using the command `B = firpm(30, [0.1 0.9], [1.0 1.0], 'Hilbert')`. In this command, `30` is the order of the filter, `[0.1 0.9]` is the normalized frequency range over which the specified amplitude response `[1.0 1.0]` applies, and `'Hilbert'` tells MATLAB that a Hilbert transforming filter is to be designed. This is the identical syntax used for the **remez** command discussed in Chapter 14. A plot of the resulting filter coefficients is shown in Figure 15.9. Just as we saw with the Hilbert transforming filter we designed in Chapter 14, every other coefficient is zero, and the coefficients exhibit antisymmetry. We can take advantage of the large number of coefficients with a value of zero (and also the antisymmetry) to reduce the computational complexity of the filter. This filter has the frequency response shown in Figure 15.10. While this bandpass implementation of the Hilbert transforming filter may look flat in the passband, a closer inspection of the passband will reveal that some amount of passband ripple is unavoidable. This is clearly shown in Figure 15.11. Increasing the order of the filter or decreasing the normalized frequency range over which the specified amplitude response applies will decrease the passband ripple. The group delay of the Hilbert transforming filter is shown in Figure 15.12. This is an example of a symmetric FIR filter that has a constant group delay (due to its linear phase response). The group

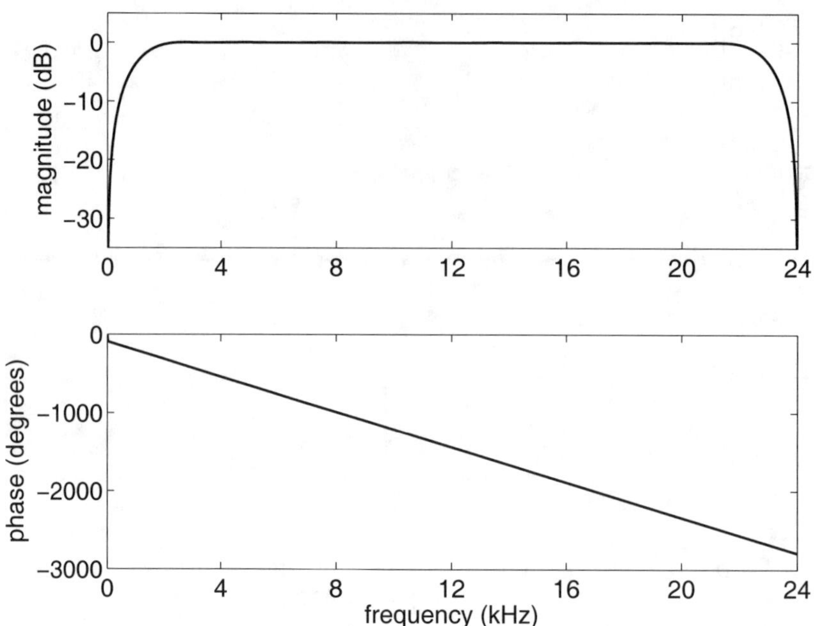

Figure 15.10: Frequency response associated with a 30th order FIR Hilbert transforming filter.

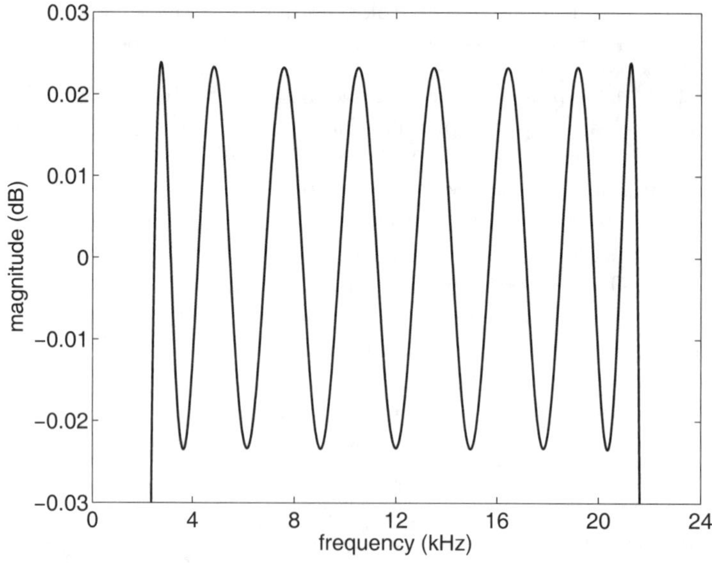

Figure 15.11: Closeup inspection of the passband of the frequency response associated with a 30th order FIR Hilbert transforming filter. The passband ripple is now quite evident.

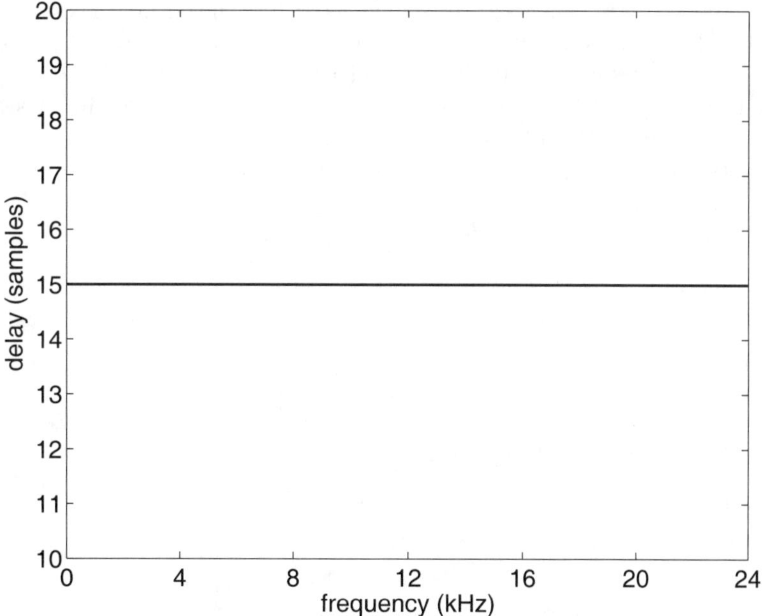

Figure 15.12: The group delay associated with a 30th order FIR Hilbert transforming filter.

delay is equal to half the filter order.

5. Alternatively, an FFT/IFFT can be used to implement the Hilbert transform if a frame-based system is desired. This process can be implemented using the MATLAB code

```
transformedSignal = ifft ( fft ( signal ) .*  ...
    [ -j*ones (1 ,513)  j*ones (1 ,511) ]) ;
```

In this example, there are 1024 samples per frame.

6. The loop filter can alternatively be implemented in its parallel form.

7. The modulus operation should be replaced by more basic C/C++ commands (e.g., an `if` statement followed by the subtraction of 2π whenever the accumulated phase is greater than 2π).

15.4 DSK Implementation in C

15.4.1 Components of the PLL

When you understand the MATLAB code, the translation of the concepts into C is fairly straightforward. The PLL has two major components: a Hilbert transforming FIR filter (any of the techniques discussed in Chapter 3 could be used to implement this filter), and the LO's control loop.

The files necessary to run this application are in the `ccs\PLL` directory of Chapter 15. The primary file of interest is the `PLL_ISRs.c`, which includes the interrupt service routines. This file includes the necessary variable declarations and performs the actual PLL algorithm.

If the DSK codec you're using is a stereo device (such as the on-board codec for the C6713 DSK or the PCM 3006-based daughtercard codec for the C6711 DSK), the program could implement independent Left and Right channel PLLs. For clarity, this example program will implement only one PLL, but will output a delayed version of the input signal and the recovered message signal. If both signals are displayed on a multichannel oscilloscope, the relationship between the modulated signal (input signal) and an estimate of the modulating signal (message) should be clear.

The declaration section of the code is shown in Listing 15.2.

Listing 15.2: Declaration portion of the PLL project code.

```
float  alpha =   0.01;              /* loop filter parameter */
float  beta  = 0.002;              /* loop filter parameter */
float  Fmsg  = 12000;              /* vco rest frequency */
float  Fs = 48000;                 /* sample frequency */
float  x[N+1] = {0,0,0,0,0,0,0,0,0,0,0,0,0,0,0,0,
                0,0,0,0,0,0,0,0,0,0,0,0,0,0,0};  /* input signal
                  [+]*/
float  sReal;          /* real part of the analytic signal */
float  sImag;          /* imag part of the analytic signal */
float  q = 0;          /* input to the loop filter */
float  sigma = 0;      /* part of the loop filter's output */
float  loopFilterOutput = 0;
float  phi = 0;        /* phase accumulator value */
float  pi = 3.14159265358979;
float  phaseDetectorOutputReal;
float  phaseDetectorOutputImag;
float  vcoOutputReal = 1;
float  vcoOutputImag = 0;
float  scaleFactor = 3.0517578125e-5;
```

An explanation of Listing 15.2 follows.

1. (Lines 1–2): Declare and initialize the loop filter coefficients. In this parallel implementation, beta should be much smaller than alpha.

2. (Lines 3–4): Declare the message (or carrier frequency) and the sample frequency, respectively.

3. (Lines 5–6): Declare and initializes the input signal's variable. An FIR filter, even if not initialized, will rapidly *flush* any unintended values from its storage variable. During this transient, it is *possible* that the output of the filter will be QNAN. This "not-a-number" state will be fed into the IIR-based loop filter. The filter cannot recover from this state. For this reason alone, **failure to properly initialize** variables is one of the most common causes of properly written algorithms that will not execute properly.

4. (Lines 7–8): Declare the real and imaginary parts of the analytic signal. These are two of the four inputs to the first mixer.

5. (Lines 9–11): Declare the intermediate variables needed by the loop filter and the loop filter's output.

6. (Line 12): Declares the VCO's current phase, which is the input argument to the LO.

7. (Lines 14–15): Declare the outputs of the first mixer.

8. (Lines 16–17): Declare the final two inputs to the first mixer.

9. (Line 18): Declares a scale factor that brings the product of the second mixer into a reasonable range for an error signal. Remember, the error signal (output of the loop filter) is needed to adjust the phase of the VCO. Assuming that the loop is reasonably close to locking, only small phase adjustments ($<< 1$ radian) should be added to the phase accumulator.

The algorithm section of the code is shown in Listing 15.3.

Listing 15.3: Algorithm portion of the PLL project code.

```
// I added my PLL routine here
x[0] = CodecDataIn.Channel[LEFT]; // current LEFT input value
sImag = 0;                        // initialize the dot-product result

for (i = 0; i <= N; i+=2) { // indexing by 2, B[odd] = 0
    sImag += x[i]*B[i];     // perform the dot-product
}

sReal = x[15]*scaleFactor;  // grpdelay of filter is 15 samples

for (i = N; i > 0; i--) {
    x[i] = x[i-1];          // setup x[] for the next input
}

sImag *= scaleFactor;       // scale prior to loop filter

// execute the D-PLL (the loop)
vcoOutputReal = cosf(phi);
vcoOutputImag = sinf(phi);
phaseDetectorOutputReal=sReal*vcoOutputReal+sImag*vcoOutputImag;
phaseDetectorOutputImag=sImag*vcoOutputReal-sReal*vcoOutputImag;
q = phaseDetectorOutputReal * phaseDetectorOutputImag;
sigma += beta*q;
loopFilterOutput = sigma + alpha*q;
phi += 2*pi*Fmsg/Fs + loopFilterOutput;

while (phi > 2*pi) {
    phi -= 2*pi;    /* modulo 2pi operation */
}

// setup CODEC output values
CodecDataOut.Channel[LEFT]=32768*sReal; // input signal
CodecDataOut.Channel[RIGHT]=32768*phaseDetectorOutputReal; // msg
// end of my PLL routine
```

An explanation of Listing 15.3 follows.

1. (Line 2): Brings the input sample into the ISR.

2. (Lines 3–7 and 11–13): Take the Hilbert transform of the input signal using an FIR filter.

3. (Line 9): Accounts for the group delay of the FIR filter and scales the input signal *prior* to the loop filter.

4. (Line 15): Scales the output of the FIR filter. The outputs of lines 9 and 15 form the analytic signal and represent 2 of the 4 inputs into the first mixer.

5. (Lines 18–19): Calculate the outputs from the complex mixer. These calculations are the last 2 inputs into the first mixer.

6. (Lines 20–21): Calculate the outputs of the first mixer.

7. (Line 22): Calculates the output to the second mixer, which is the input to the loop filter.

8. (Lines 23–24): Calculate the intermediate results and the output of the loop filter.

9. (Line 25): Calculates the value of the phase accumulator, `phi`.

10. (Lines 27–29): Perform a modulus 2π operation on `phi`. The `mod` operator is available in the C implementation of CCS, but it is *only* defined for integer data types. Additionally, this subtraction of 2π technique is much faster than most operators requiring division. The `while` statement can be replaced by an `if` statement since the phase accumulator's value should only increase by ($\pi/2$ plus the loop filter's feedback signal) for each sample. A `while` statement is a more conservative solution to this problem.

15.4.2   System Testing

To test the operation of the PLL, test signals can be created in MATLAB, converted to wave files, and then played back through a computer's sound card. Note that some such signals are included in the `test_signals` directory of the CD-ROM. The filenames of these signals start with AM or with BPSK. MATLAB m-files that create these wave files are contained on the CD-ROM in the `matlab` directory for Chapter 15 and are named `AMsignalGenerator.m` and `BPSKsignalGenerator.m`. With a slight modification to these programs (that is, changing the sample rate to 44,100 Hz), the signals created by these m-files could also be recorded directly on to a CD-R of your choice, and then played back on any a CD player. This effectively turns an inexpensive CD player into an inexpensive communications signal generator.

The response of the system to a 750 Hz message modulated (AM-DSB-SC) with a 12 kHz carrier, as viewed on a multichannel oscilloscope, is shown in Figure 15.13. A typical transient response of the system to a 750 Hz message modulated (AM-DSB-SC) with a 12 kHz carrier, is shown in Figure 15.14. Finally, the response of the system to a 2400 bit per second (bps) BPSK signal with a 12 kHz carrier is shown in Figure 15.15. The BPSK signal was generated by a PC sound card and shows significant distortion due to the bandlimited response of the system.

15.5   Follow-On Challenges

Consider extending what you have learned.

1. Design and implement your own loop filter within the PLL.

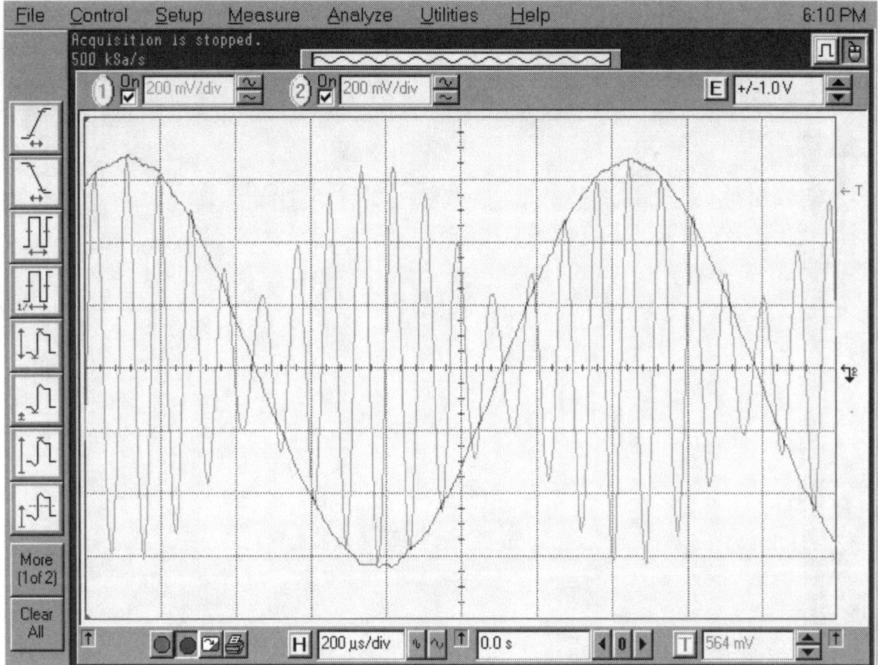

Figure 15.13: The response of the system to a 750 Hz message modulated (AM-DSB-SC) with a 12 kHz carrier, as viewed on a multichannel oscilloscope.

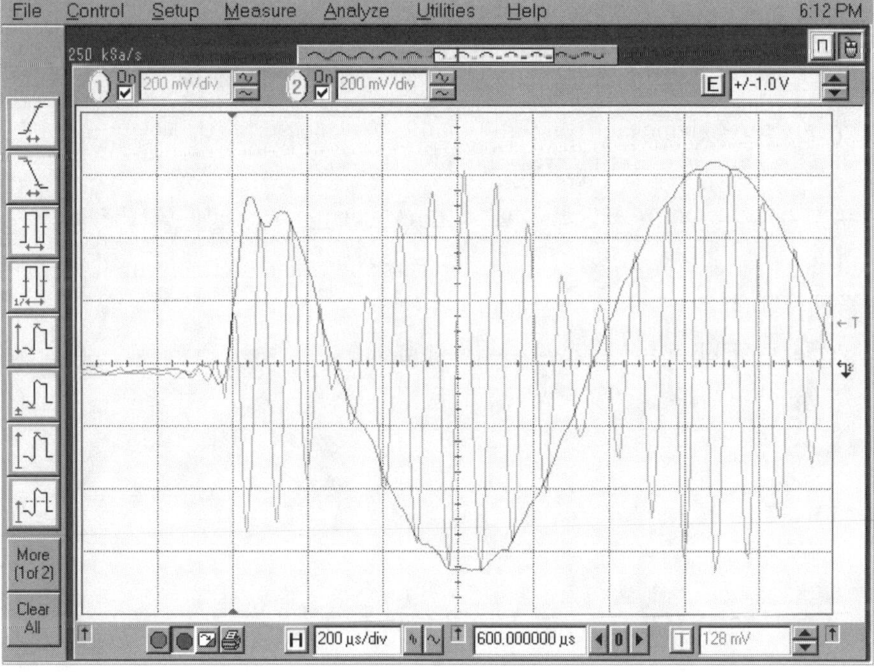

Figure 15.14: A typical transient response of the system to a 750 Hz message modulated (AM-DSB-SC) with a 12 kHz carrier.

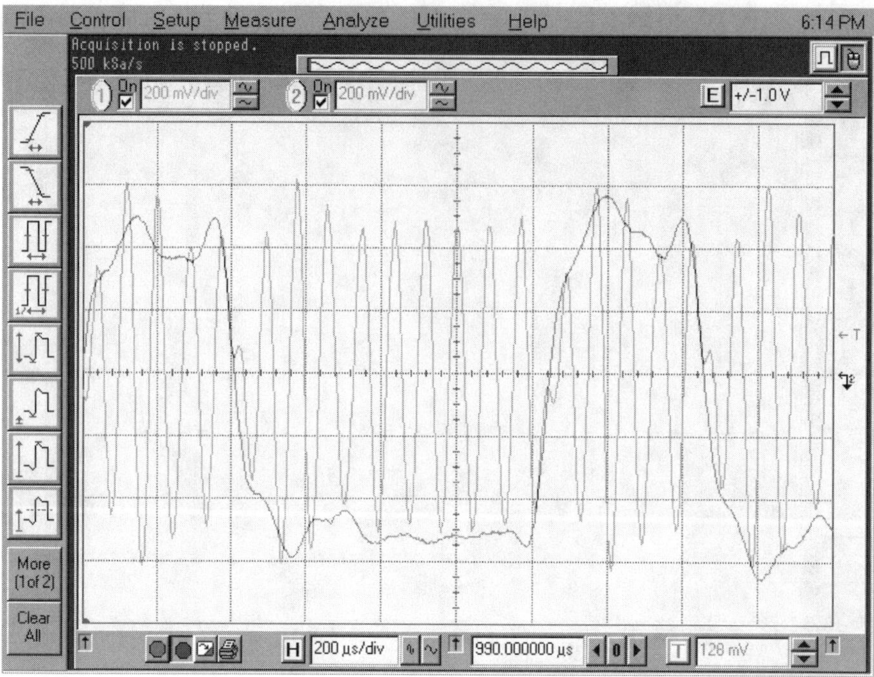

Figure 15.15: A typical transient response of the system to a 750 Hz message modulated (AM-DSB-SC) with a 12 kHz carrier.

2. Design and implement an algorithm that detects, then provides some indication to the user, when the PLL is *locked* and tracking the input signal.

3. There are three significant computational inefficiencies (bottlenecks) in the PLL ISR. Profile the ISR code and identify these bottlenecks.

4. Suggest possible improvements that minimize or remove these bottlenecks.

5. Implement at least one of your improvements and calculate the computational savings of your new code.

6. Implement a PLL using frame-based techniques.

Chapter 16

Project 7: Digital Communications Transmitters

16.1 Theory

IN Chapter 5 we introduced the basic concepts involved in periodic signal generation. Although we mentioned the idea, we intentionally deferred the discussion of aperiodic digital communications signals until now. Since digital communications signals can be generated using an unlimited number of different forms and specifications. We will only introduce one of these forms and a few of the techniques that can be used to produce this type of signal. Specifically, we will discuss

1. Random data and symbol generation.

2. Binary phase shift-keying (BPSK) using antipodal rectangularly shaped bits.

3. BPSK using impulse modulated (IM) raised-cosine shaped bits.

This is a good place to reiterate that this text only gives a brief review of the theory associated with the areas for which we use real-time DSP. For example, in this chapter we can't teach you the in-depth theory of digital communications; you'll need a good communications textbook for that [61, 62, 67].

16.1.1 Random Data and Symbol Generation

In an actual communication system, the data bits that eventually makeup the transmitted symbols would come from an ADC or some other information source. Unfortunately, using real information sources can *greatly* increase the complexity of the system. Additionally, this tends to severely constrain the communication system design. While these constraints are necessary in the design of a real communication system, they tend to overly complicate matters for someone who is just getting started in communication system design and implementation.

To illustrate this point, let's imagine that we want to create a digital communications link capable of sending all of the data generated by the DSK's ADC for a very high fidelity (better than CD-quality) music signal. Assuming a C6713 DSK is being used, this implies stereo (2 channels) of data, with 24 bits per sample, and a nominal sample rate (frequency) of 48,000 samples per seconds. This results in a data rate entering the DSK of $2 \times 24 \times 48,000 = 2,304,000$ bits per second (bps). Remembering that a T-1 data line, which represents the combined data signals from 24 telephone lines, only contains 1,544,000 bps, our desire to transmit such a signal may be a bit aggressive for our first digital communications project!

Instead, we will derive our data bits from a random number generator. A pseudonoise (pn or m-sequence) generator or an array of predeclared data could also be used. Speaking of bits, we need to briefly discuss bits versus symbols in digital communications, since they are often confused. Numeric values are represented in a computer or DSK as *bits*, but what we actually send over a communications link are *symbols*. If our communication system is using a set (called a constellation) of symbols that can take on four different values, then we have two bits per symbol. If our communication system is using a constellation of symbols that can take on sixteen different values, then we have four bits per symbol. Remember that a digital communication system sends and receives symbols, not bits. With that being said, we are discussing BPSK in this chapter, where our symbols can only take on two different values. This means that for BPSK, we have one bit per symbol so the bit rate and the symbol rate are identical.

A relatively straightforward starting point for the design of our digital communication system is to select a data rate such that there are an integer number of samples per symbol. That is, $F_s/R_d = k$, where F_s is the sample frequency, R_d is the data rate, and k is an integer. Using our nominal sample frequency of 48 kHz, and recognizing that we need at least 2 samples per symbol, we can choose from the data rates shown in Table 16.1. For the rest of this chapter, we will use $k = 20$, which implies a *great* number of things about our communication system. This includes, but is not limited to,

1. $F_s/20$ is the symbol rate, which is equal to 2400 symbols per second (sps).

2. The reciprocal of the symbol rate is the symbol period $= 1/2400 = 0.416666$ ms.

3. There are 20 samples associated with each and every symbol period.

While this may all seem very straightforward, a common mistake is to try to change one of these parameters (sample frequency, symbol rate, symbol period, or samples per symbol)

Table 16.1: Data rate (in symbols per second) using an integer number of samples per symbol, where the sample frequency is assumed to be $F_s = 48$ kHz. Note that for BPSK, "symbols per second" is identical to "bits per second."

F_s/k	k	Data Rate (sps)
$F_s/2$	2	24,000
$F_s/3$	3	16,000
$F_s/4$	4	12,000
$F_s/5$	5	9,600
$F_s/6$	6	8,000
$F_s/8$	8	6,000
$F_s/10$	10	4,800
$F_s/12$	12	4,000
$F_s/15$	15	3,200
$F_s/16$	16	3,000
$F_s/20$	20	2,400
$\vdots$	$\vdots$	$\vdots$
F_s/N	N	$48,000/N$

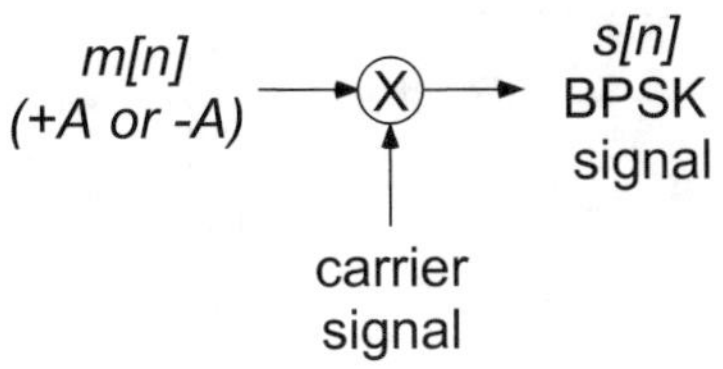

Figure 16.1: The block diagram associated with a rectangularly pulse shaped BPSK transmitter.

without realizing that they are *all* interrelated and therefore cannot be changed independently of one another.

In this project, our modulated message symbols will come out of the DSK's DAC as analog voltage levels. If we were actually going to transmit these symbols, this time varying analog voltage level would then go to other stages such as a power amplifier and an antenna.

16.1.2 BPSK Using Antipodal Rectangularly Shaped Bits

This form of BPSK is infrequently used in actual communication systems, but it is by far the most understandable implementation. Specifically, the BPSK waveform is $s_{BPSK}[nT_s] = m[nT_s]cos[2\pi f_c nT_s]$, where n is a monotonically increasing integer, T_s is the sample period, m is the current bit's value, and f_c is the carrier frequency. Additionally, for antipodal signaling, m will be limited to the values $\pm A$, where A will routinely be an integer that is near the maximum value allowed by the DAC (e.g., $m = 30,000$ for 16-bits per sample). This form of BPSK is a special case of amplitude modulation (AM), in particular the type of AM known as double sideband-suppressed carrier (DSB-SC). A block diagram of this system is shown in Figure 16.1.

16.1.3 BPSK Using Impulse Modulated Raised-Cosine Shaped Bits

Due to its bandlimited nature, the raised-cosine filtered form of BPSK is commonly used in actual communication systems. The signal, however, is more complicated to understand and slightly more difficult to generate. In an impulse modulator, a scaled impulse is used to excite a pulse shaping filter once per symbol period. In our discussions, we will use a raised-cosine FIR filter as the pulse shaping filter. The output of this filter will then be multiplied by the carrier signal. A block diagram of this system is shown in Figure 16.2.

An example of how the impulse modulator functions is shown in Figure 16.3. In this figure, five positive-valued impulses form the input, $x[n]$, of the FIR filter. Each of the five output waveforms is shown individually on the $y[n]$ plot. The *actual* output of the filter would be the sum of the five sinc-shaped output waveforms.

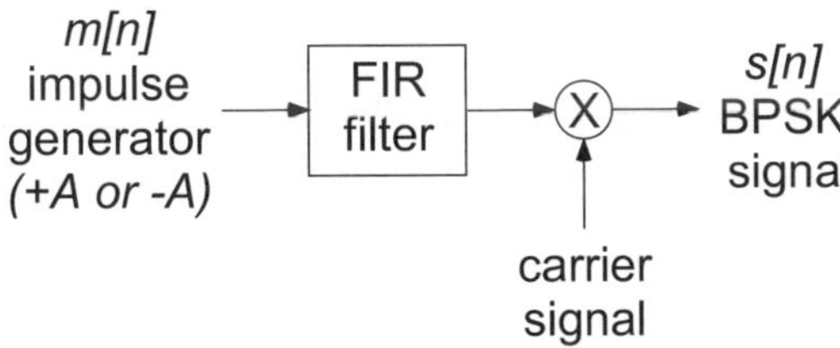

Figure 16.2: The block diagram associated with a impulse modulated BPSK transmitter.

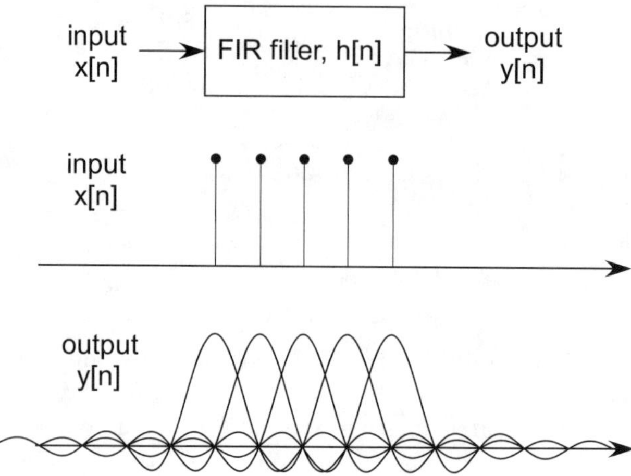

Figure 16.3: An example of how the impulse modulator functions (positive excitation impulses only).

Similarly, if antipodal impulse excitation of the FIR filter is provided, one possible result is shown in Figure 16.4. In both of these figures, notice how *all but one* of the sinc-shaped waveforms have all of their zero crossings in common. This alignment of the zero crossings is by design, and is fundamental to minimizing intersymbol interference (ISI) within the system.

16.2   winDSK6 Demonstration

If you double click on the winDSK6 icon, the winDSK6 application will launch, and a window similar to Figure 16.5 will appear. Before proceeding, be sure the selections in "DSK and Host Configuration" are correct for your setup. Clicking on winDSK6's commDSK button

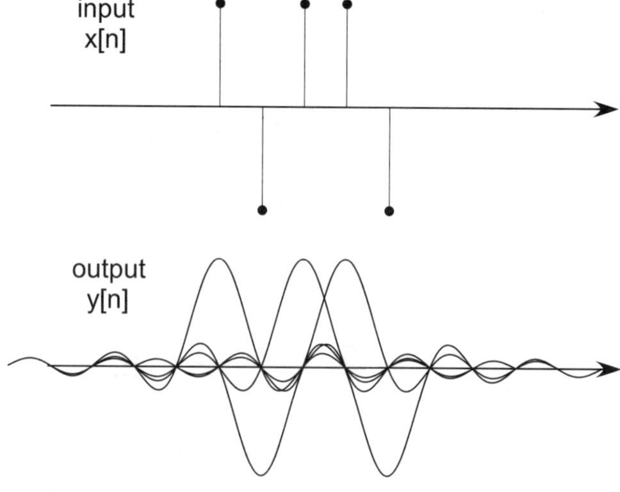

Figure 16.4: Another example of how the impulse modulator functions (antipodal excitation impulses).

will load that program into the attached DSK, and a window similar to Figure 16.6 will appear.

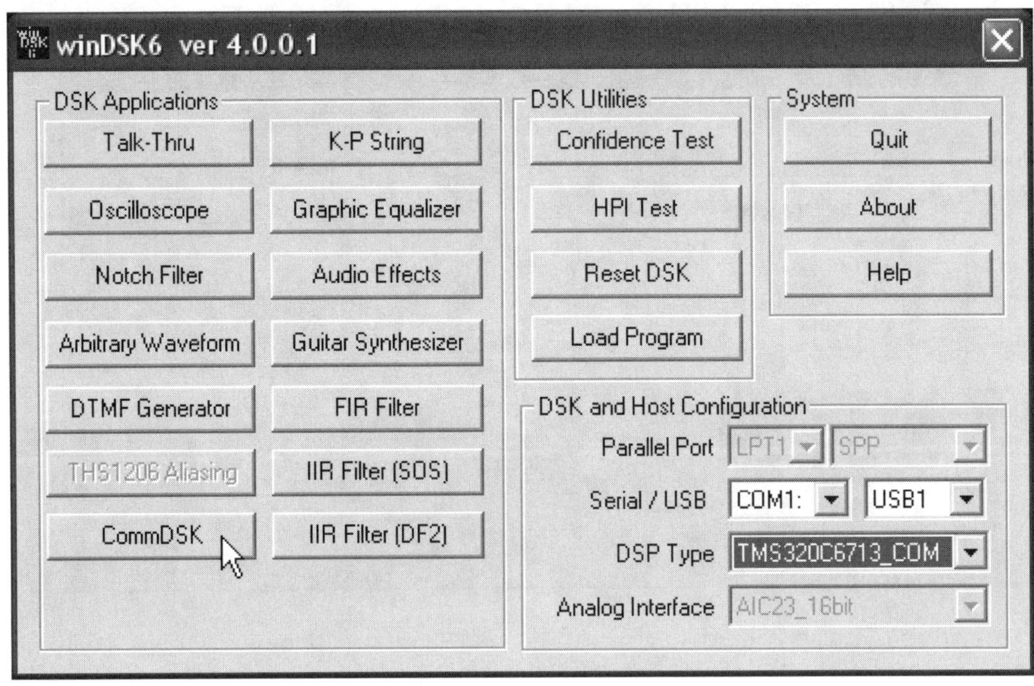

Figure 16.5: winDSK6 ready to load the commDSK application.

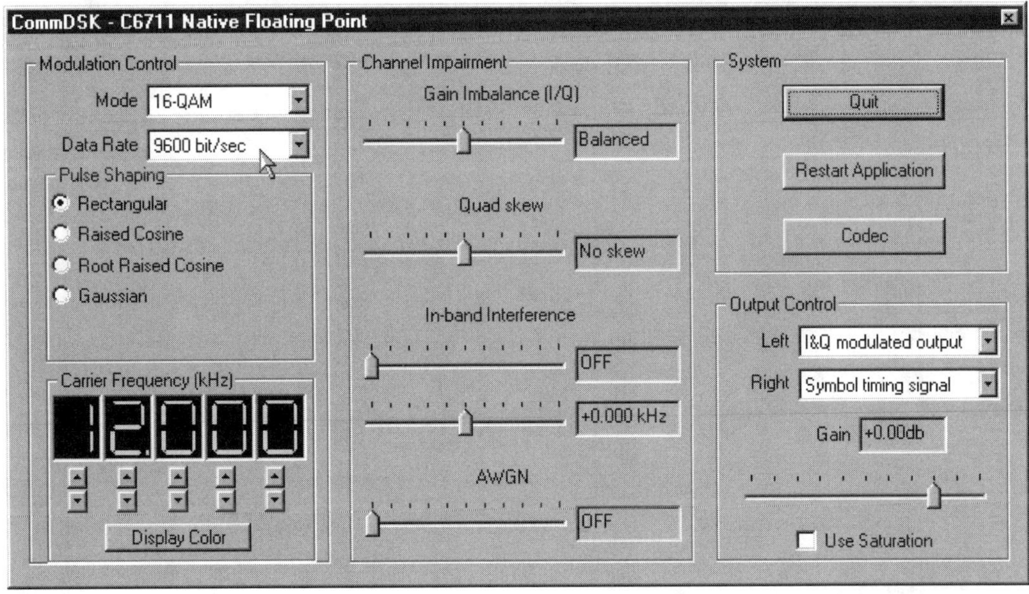

Figure 16.6: winDSK6 running the commDSK application. By default, the modulated signal appears on the Left output channel, and a timing signal appears on the Right output channel. These settings can be changed by the user if needed.

16.2.1   commDSK: Unfiltered BPSK

We need to change a few of the commDSK default settings for now. Change the "Mode" to "BPSK" and the "Data Rate" to "2400 bits/sec" (see Figure 16.7). An example waveform is shown in Figure 16.8.   BPSK has only two possible symbols. In Figure 16.8, you can

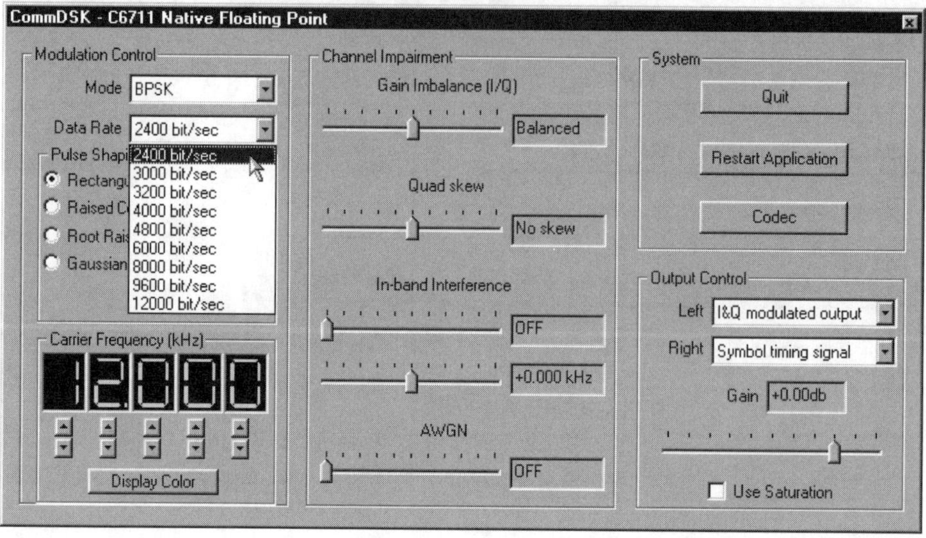

Figure 16.7: commDSK set to generate a rectangular pulse shaped, 2400 bps signal.

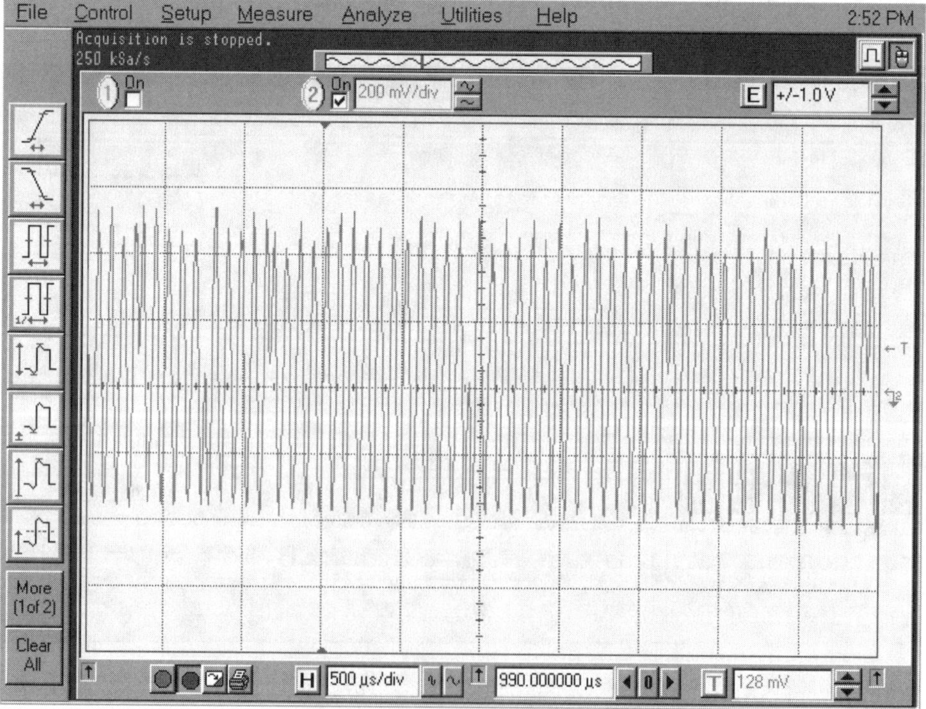

Figure 16.8: commDSK waveform of a rectangular pulse shaped, 2400 bps signal.

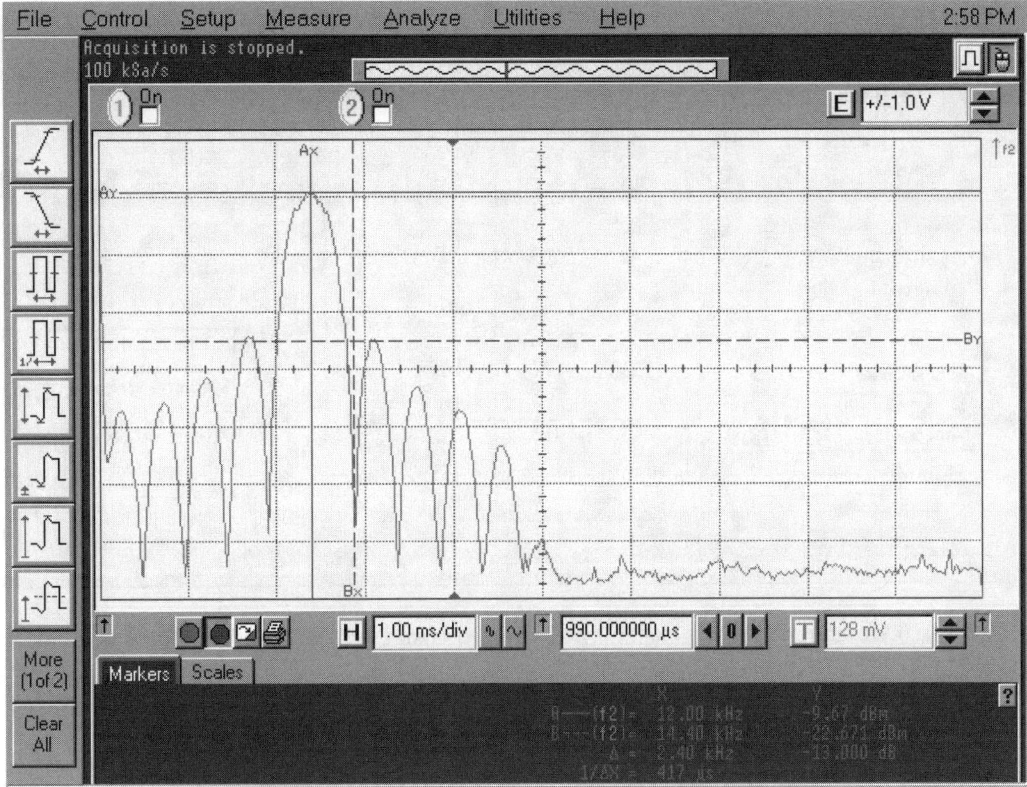

Figure 16.9: An averaged spectrum associated with a rectangular pulse shaped, 2400 bps signal generated by commDSK.

observe the transitions from one type of symbol to the other type of symbol by the sharp phase reversal in the waveform. With the horizontal scale being set to 500 μs per division, the first phase reversal appears to occur at about 350 μs from the left of the display window.

An averaged spectrum associated with this waveform is shown in Figure 16.9. The main lobe is centered at 12 kHz and the first spectral null above 12 kHz occurs at 14.4 kHz. This implies that the main lobe is 4.8 kHz wide (twice the symbol rate), with subsequent nulls occurring at 2.4 kHz spacing. The first side lobe has a relative magnitude that is approximately 13 dB down from the main lobe's peak. These are exactly the expected results associated with band-limited rectangularly-shaped BPSK. Ideally, a perfectly rectangular signal would have a spectrum that would extend to infinity on the frequency axis. But the DSK has a reconstruction lowpass filter built into its codec DAC that removes the majority of the signal energy at frequencies greater than 24 kHz.

16.2.2 commDSK: Raised-Cosine Filtered BPSK

Now we need to adjust commDSK to observe the effect of pulse shaping. We will apply a raised cosine pulse shaping effect, such that we will no longer have rectangular pulses. For the "Pulse Shaping" radio buttons on the commDSK user interface window, select "Raised Cosine." This change is shown in Figure 16.10. An example waveform is shown in Figure 16.11. In this mode, it is much easier to observe the transitions from one type of symbol to the other.

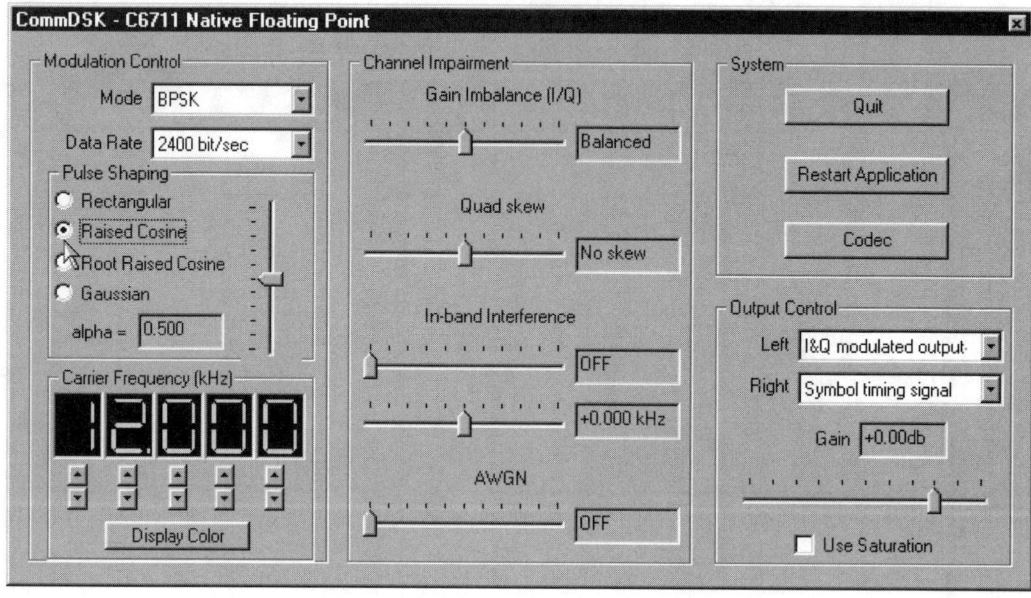

Figure 16.10: commDSK set to generate a raised-cosine pulse shaped, 2400 bps signal.

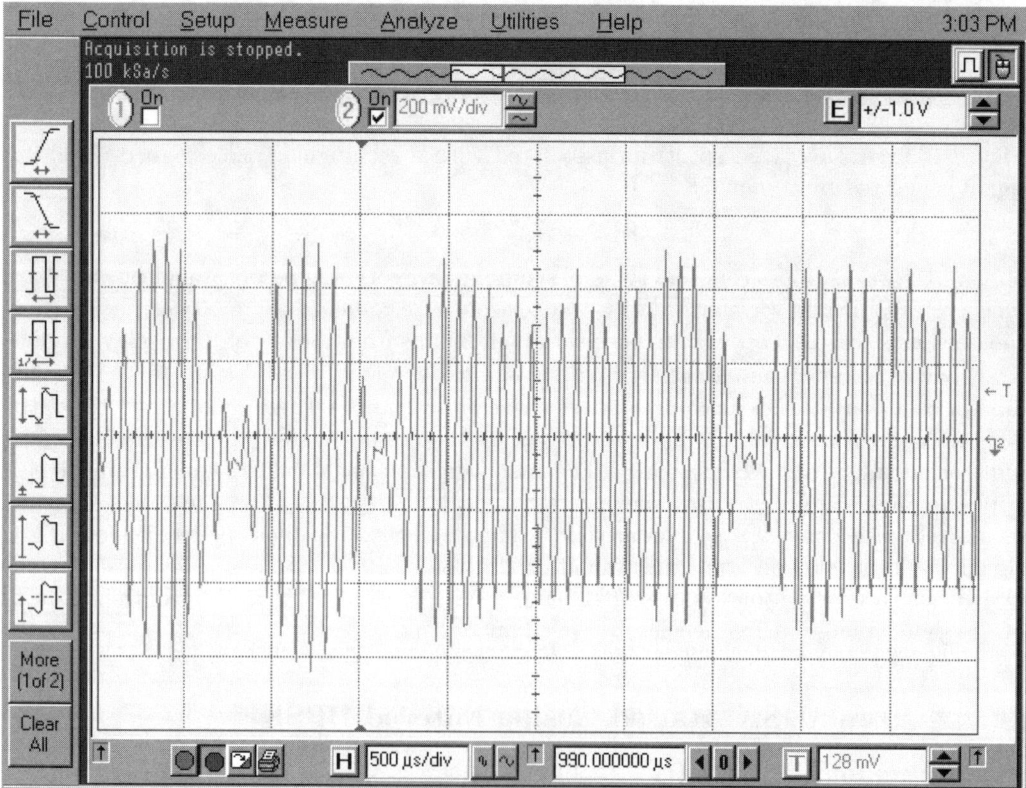

Figure 16.11: commDSK waveform of a raised-cosine pulse shaped, 2400 bps signal.

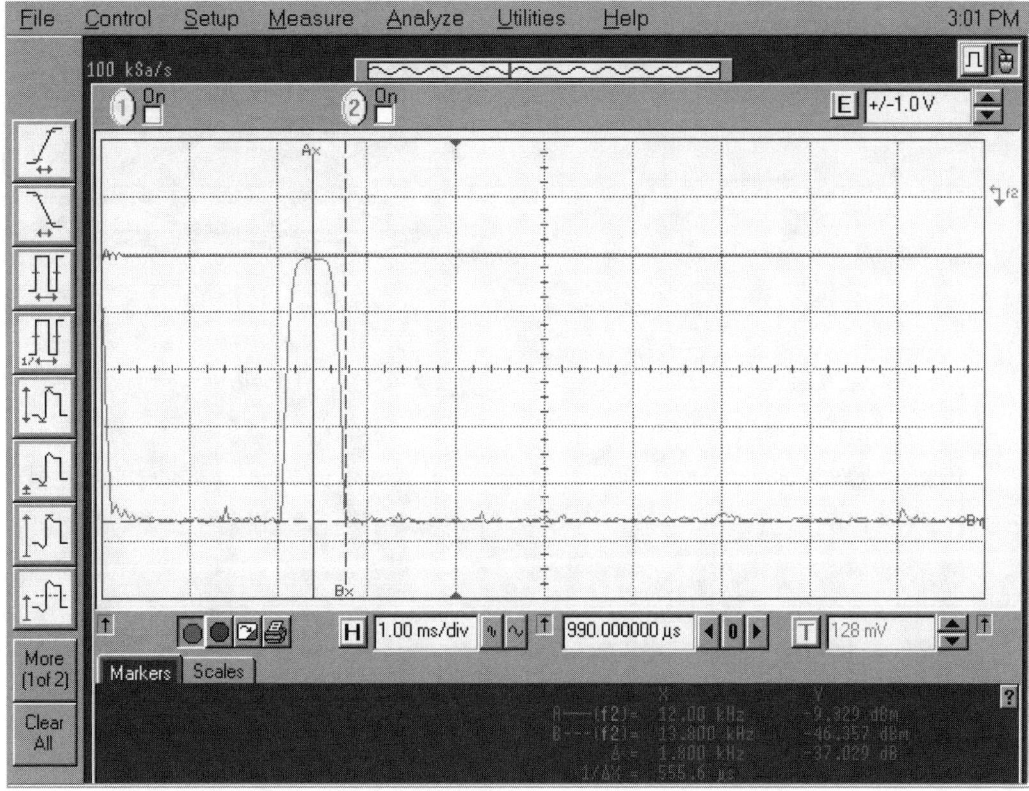

Figure 16.12: An averaged spectrum associated with a raised-cosine pulse shaped, 2400 bps signal generated by commDSK.

An averaged spectrum associated with this waveform is shown in Figure 16.12. Similar to the rectangular shaped signal, the main lobe is centered at 12 kHz but the first spectral null above 12 kHz occurs at 13.8 kHz. This implies that the main lobe is 3.6 kHz wide $(2 \times (13.8 - 12) = 3.6)$. This agrees with the theoretical prediction of $BW = D(1 + \alpha) = 2400 \times (1 + 0.5) = 3600$ Hz, where BW is the signal's bandwidth, D is the symbol rate, and α is the raised-cosine roll-off factor that was used in this case (the roll-off factor must remain between 0 and 1).

In Figure 16.12, there are no sidelobes visible due to the apparent noise floor at approximately -46.357 dBm. With the peak of the main lobe at -9.329 dBm, the expected sidelobe level for raised cosine pulses with a roll-off factor of $\alpha = 0.5$ is lower that the observed noise floor. This noise floor is in fact a limitation of the 8-bit ADC associated with the digitizing oscilloscope that was used to produce this figure.

The commDSK program is capable of generating a number of different digital communications signals at a number of different data rates. These signals may also be distorted using the "Channel Impairment" section of commDSK. These impairments are very helpful if the signal is to be processed by a vector signal analyzer (VSA). An example of a VSA display is shown in Figure 16.13. In this figure, plot A is a trajectory/constellation diagram, plot B is the spectral estimate of the BPSK signal, plot C is the error vector magnitude (EVM), plot D is the eye-pattern, plot E is the EVM's spectral estimate, and plot F reports a number of statistics associated with the performance of the signal being analyzed. These plots can be used to infer a great deal about the communication system performance.

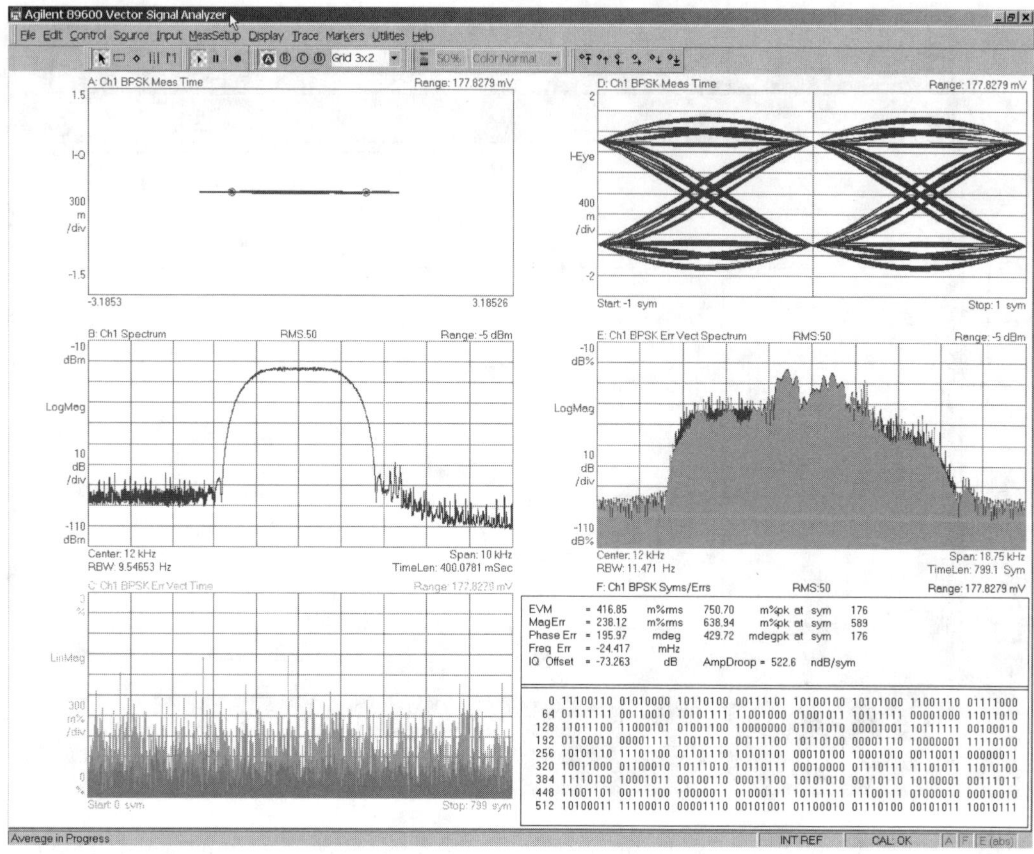

Figure 16.13: An example of a VSA display associated with a raised-cosine pulse shaped, 2400 bps, BPSK signal generated by commDSK.

16.3   MATLAB Implementation

As discussed earlier, we will simulate two types of BPSK signal generation: the rectangular shaped BPSK signal generator and the impulse modulated, raised-cosine BPSK signal generator.

16.3.1   Rectangular Shaped BPSK Signal Generator

The first MATLAB simulation is of a rectangular shaped BPSK signal generator; the code is shown in Listing 16.1.

Listing 16.1: Simulation of a rectangular shaped BPSK signal generator.

```matlab
% input terms
Fs = 48000;            % sample frequency of the simulation (Hz)
dataRate = 2400;       % data rate
time = 0.004;          % length of the signal in seconds
amplitude = 30000;     % scale factor
cosine = [1 0 -1 0];   % cos(n*pi/2) ... Fs/4
counter = 1;           % used to get a new data bit
```

```matlab
   %  calculated terms
10 numberOfSamples = Fs*time;
   samplesPerSymbol = Fs/dataRate;

12
   %  ISR simulation
14 for index = 1:numberOfSamples
       % get a new data bit at the beginning of a symbol period
16     if (counter == 1)
           data = amplitude*(2*(rand > 0.5) - 1);
18     end

20     % create the modulated signal
       output = data*cosine(mod(index,4) + 1)

22
       % reset at the end of a symbol period
24     if (counter == samplesPerSymbol)
           counter = 0;
26     end

28     % increment the counter
       counter = counter + 1;
30 end

32 %  Plotting commands follow ...
```

An explanation of Listing 16.1 follows.

1. (Line 2). Defines the system's sample frequency as 48 kHz. This sample frequency matches the rate of the DSK's audio codec.

2. (Line 3): Defines the data rate as 2400 bits per second (bps).

3. (Line 5): Scales the output signal near the full range of the 16-bit DAC.

4. (Line 6): Defines the output of the local oscillator (LO). This term is defined mathematically as $\cos(n\pi/2)$, for $n = 0, 1, 2$, or 3. This simplifies to $\cos(0\pi/2) = 1$, $\cos(1\pi/2) = 0$, $\cos(2\pi/2) = -1$, or $\cos(3\pi/2) = 0$. Thus the LO will *only* have an output value of $+1, 0, -1$, or 0. For this special case of signal generation, mixing with the LO requires very few computational resources.

5. (Line 7): The `counter` variable is used to determine the current position within a symbol.

6. (Lines 16–18): A new data bit is generated whenever `counter` is equal to 1.

7. (Line 21): Calculates the output value by multiplying (mixing) the data value with the LO's output. The `cosine` variable is accessed by the MATLAB `mod` command which maintains the `index` value between 1 and 4.

8. (Lines 24–26): The `counter` variable is reset at the end of a symbol.

9. (Line 29): The `counter` variable is incremented at the end of the simulated ISR.

An example output plot from this simulation is shown in Figure 16.14.

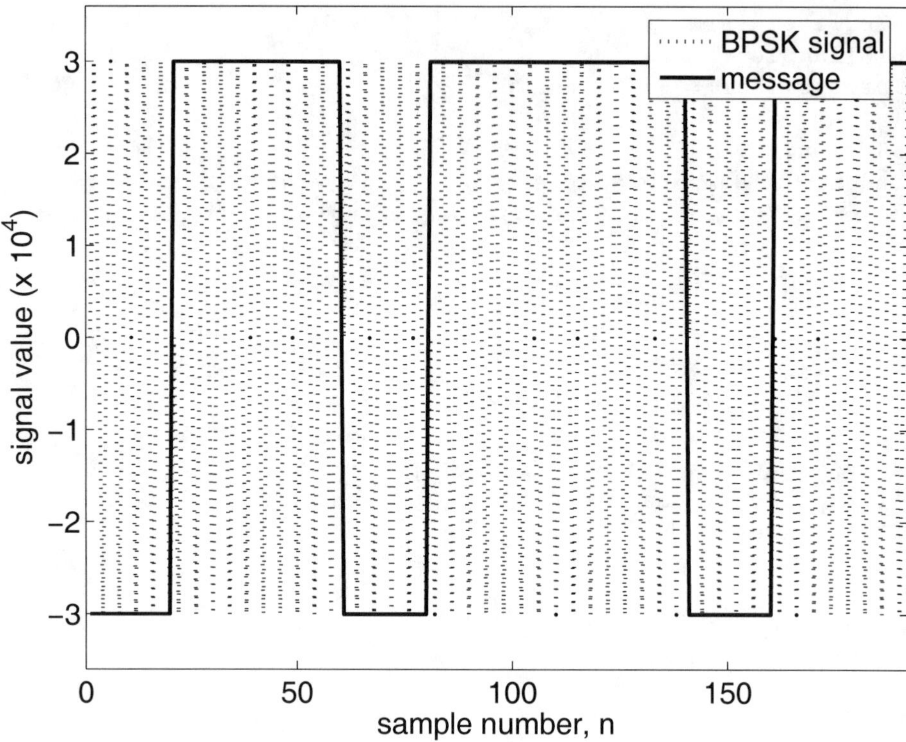

Figure 16.14: An example of the output from the rectangular shaped BPSK simulation. This signal has 2400 bps with a carrier frequency of 12 kHz.

16.3.2 Impulse Modulated Raised-Cosine BPSK Signal Generator

The second MATLAB simulation is of a impulse modulated raised-cosine BPSK signal generator; the code is shown in Listing 16.2. Note the local oscillator (LO) is at $F_s/4$ just like the previous sibilation.

Listing 16.2: Simulation of a impulse modulated raised-cosine BPSK signal generator.

```matlab
% input terms
Fs = 48000;              % sample frequency of the simulation (Hz)
dataRate = 2400;         % data rate
alpha = 0.5;             % raised-cosine rolloff factor
symbols = 3;             % MATLAB "rcosfir" design parameter
time = 0.004;            % length of the signal in seconds
amplitude = 30000;       % scale factor
cosine = [1 0 -1 0];     % cos(n*pi/2) ... Fs/4
counter = 1;             % used to get a new data bit

% calculated terms
numberOfSamples = Fs*time;
samplesPerSymbol = Fs/dataRate;

B = rcosfir(alpha, symbols, samplesPerSymbol, 1/Fs);
Zi = zeros(1, (length(B) - 1));
```

```matlab
18  %  ISR simulation
    for index = 1:numberOfSamples
20      % get a new data bit at the beginning of a symbol period
        if (counter == 1)
22          data = amplitude*(2*(rand > 0.5) - 1);
        else
24          data = 0;
        end
26

        % create the modulated signal
28      [impulseModulatedData, Zf] = filter(B, 1, data, Zi);
        Zi = Zf;
30      output = impulseModulatedData*cosine(mod(index,4) + 1);

32      % reset at the end of a symbol period
        if (counter == samplesPerSymbol)
34          counter = 0;
        end
36

        % increment the counter
38      counter = counter + 1;
    end
40

    %  output terms
42  %  Plotting commands follow ... see the book CD-Rom for the
       [+]details
```

An explanation of Listing 16.2 follows.

1. (Line 2): Defines the system's sample frequency as 48 kHz. This sample frequency matches the rate of the DSK's audio codec.

2. (Line 3): Defines the data rate as 2400 bits per second (bps).

3. (Line 4): Defines the raised-cosine roll-off factor alpha.

4. (Line 5): Defines the length of the raised-cosine FIR filter. The filter length will be equal to 2*`symbols` + 1. This length is in "symbol periods."

5. (Line 7): Scales the output signal near the full range of the 16-bit DAC.

6. (Line 8): Defines the output of the local oscillator (LO). See the LO discussion related to Listing 16.1.

7. (Line 9): The `counter` variable is used to determine the current position within a symbol.

8. (Line 15): Designs the raised cosine filter using MATLAB's `rcosfir` function.

9. (Lines 19–38): These lines of code simulate the real-time ISR.

10. (Lines 19–25): A new data bit is generated whenever `counter` is equal to 1.

11. (Lines 28–29): These lines of code implement the impulse modulator (IM). The IM is just an FIR filter with an input of one or more impulses followed by a large number of zeros. In this simulation, a single impulse with a value of 30,000 is followed by 19 sample values of zero (20 samples per symbol, remember?). The computational savings of this technique will become clear in the next section.

12. (Line 30): Calculates the output value by multiplying (mixing) the data value with the LO's output. The `cosine` variable is accessed by the MATLAB `mod` command which maintains the `index` value between 1 and 4.

13. (Lines 33–35): The `counter` variable is reset at the end of a symbol.

14. (Line 38): The `counter` variable is incremented at the end of the simulated ISR.

An example output plot (with a duration of 4 ms) from this simulation is shown in Figure 16.15. The large order of the FIR filter (120th order) results in a large group delay (of 60 samples). This delay explains why the first "BPSK signal" peaks occur 60 samples after the first "message" impulse.

Even though the amplitude scale factor is set to 30,000, the system's simulation results in output values that approach $\pm 45,000$. These values are much greater than the DAC's maximum output values $(+32,767)$ and $(-32,768)$. This effect re-emphasizes the importance of first using a MATLAB simulation to determine an appropriate scale factor *prior* to implementing the algorithm in real-time hardware. The simulation time was increased to

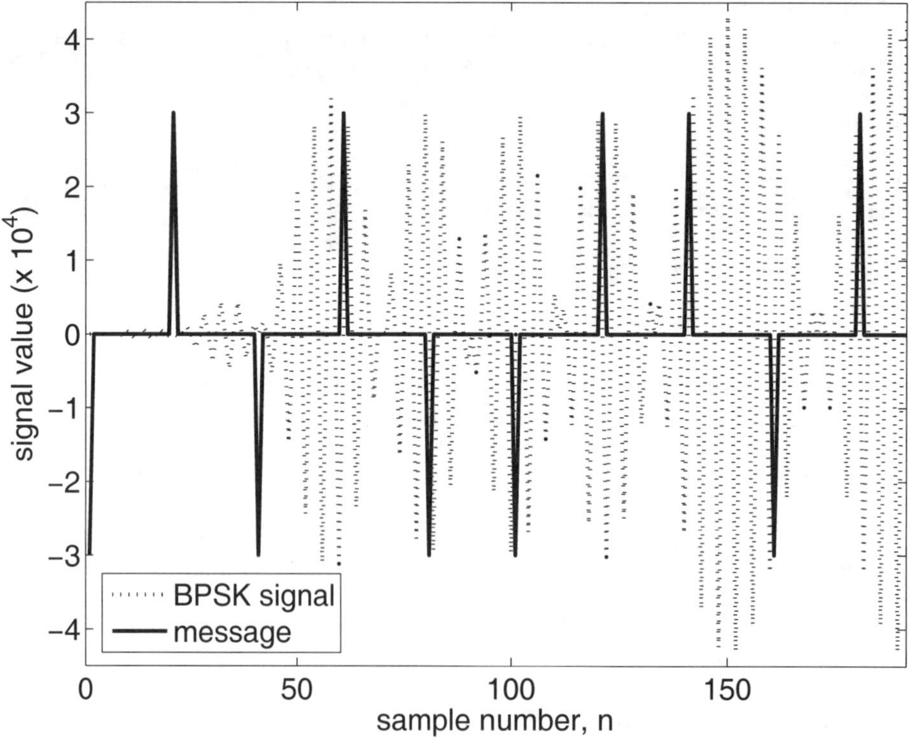

Figure 16.15: An example of the output from the impulse modulated BPSK simulation. This signal has 2400 bps, a roll-off factor of $\alpha = 0.5$, and a carrier frequency of 12 kHz.

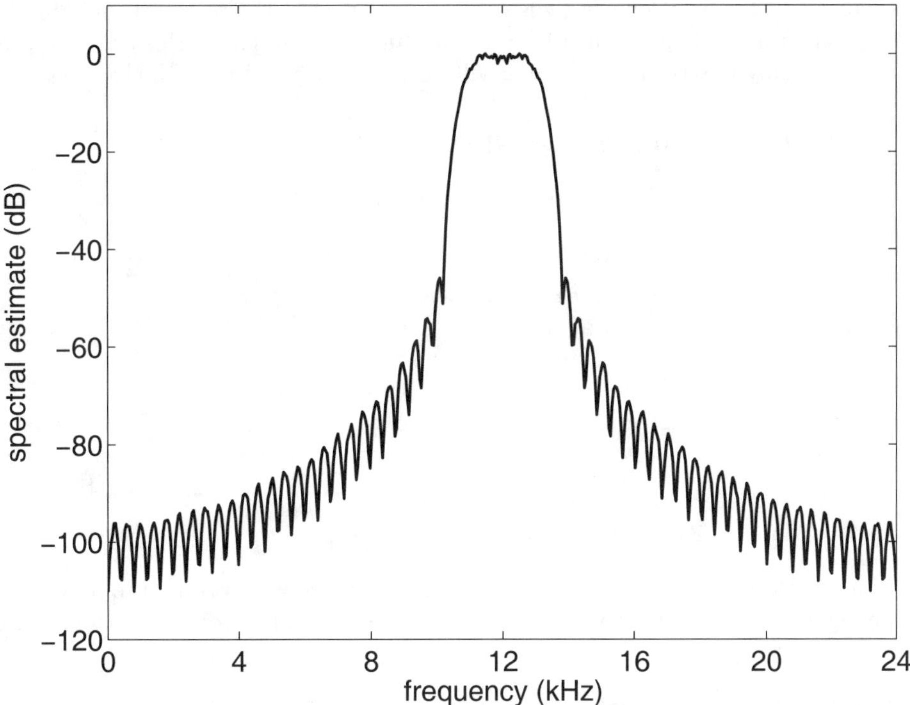

Figure 16.16: An example of the spectral estimate of an impulse modulated BPSK simulation. This signal has 2400 bps, a roll-off factor of $\alpha = 0.5$, and a carrier frequency of 12 kHz.

2 seconds to allow a higher resolution estimate of the resulting normalized power spectral density; this is shown in Figure 16.16.

The benefits of a spectrally compact, raised-cosine pulse shaped signal cannot be overemphasized. As calculated earlier, the null-to-null bandwidth of this raised-cosine signal, as compared to a rectangular signal, is 3600 Hz instead of 4800 Hz. To summarize:

1. The raised-cosine signal is more spectrally compact (smaller effective bandwidth).

2. The raised-cosine signal has a smaller null-to-null bandwidth (more spectrally efficient).

3. The raised-cosine signal is *fairly* easy to generate.

4. The digital communications receiver needed to recover the message from this raised-cosine signal will be more complex than if we used rectangular pulses.

16.4 DSK Implementation in C

When you understand the MATLAB code, the concept translation into C is fairly straightforward. The files necessary to run this application are in the `ccs\DigTx` directory of Chapter 16. The primary file of interest is the `rectangularBPSK_ISRs.c`, which contains the interrupt service routines. This file includes the necessary variable declarations and performs the BPSK generation algorithm.

Assuming the DSK's codec you're using is a stereo device, the program could implement independent Left and Right channel BPSK transmitters. For clarity, this example program will implement only one transmitter, but will output this signal to both channels.

16.4.1   A Rectangular Pulse Shaped BPSK Transmitter

The declaration section of the code is shown in Listing 16.3.

Listing 16.3: Declaration portion of the rectangular BPSK project code.

```
int counter = 0; // counter within a symbol period
int symbol;        // current bit value ... 0 or 1
int data[2] = {-20000, 20000}; // table lookup bit value
int x;                         // bit's scaled value
int samplesPerSymbol = 20;     // number of samples per symbol
int cosine[4] = {1, 0, -1, 0}; // cos functions possible values
int output;                    // BPSK modulator's output
```

An explanation of Listing 16.3 follows.

1. (Line 1): Declares and initializes the `counter` variable which is used to indicate where the algorithm is relative to the beginning of the symbol (0) or the end of the symbol (19).

2. (Line 2): Declares the `symbol` variable, which is the current bit's value (0 or 1).

3. (Line 3): Declares and initializes the antipodal values associated with a 0 or a 1.

4. (Line 4): Declares the `x` variable, which is the current message's value.

5. (Line 5): Declares and initializes the `samplesPerSymbol` variable, which, as its name implies, is the number of samples in a symbol. For BPSK, a symbol and a bit are the same.

6. (Line 6): Declares and initializes the `cosine` variable that contains *all* of the possible LO's values for a 12 kHz cosine carrier.

7. (Line 7) Declares the BPSK modulator's output value.

The algorithm section of the code is shown in Listing 16.4.

Listing 16.4: Algorithm portion of the rectangular BPSK project code.

```
// I added my rectangular BPSK routine here
if (counter == 0) {        // time for a new bit
    symbol = rand() & 1; // equivalent to rand() % 2
    x = data[symbol];      // table lookup of next data value
}

output = x*cosine[counter & 3]; // calculate the output value

if (counter == (samplesPerSymbol - 1)) { // end of the symbol
    counter = -1;
}

counter++;
```

```
15  CodecDataOut.Channel[LEFT]  = output;  // setup the Left  value
    CodecDataOut.Channel[RIGHT] = output;  // setup the Right value
17  // end of my rectangular BPSK routine
```

An explanation of Listing 16.4 follows.

1. (Lines 2–5): Once per symbol, create a random bit, and map that bit to an allowed level. This value remains constant over the next 20 samples.

2. (Line 7): Calculates the product of the current message value with the LO's value. This mixes the BPSK signal to 12 kHz.

3. (Lines 9–11): When `counter` = 19, the algorithm has reached the end of the symbol. At this point `counter` is reset to start the next symbol period.

4. (Line 13): Increments the `counter` variable in preparation for the next ISR call.

5. (Lines 15–16): Output the BPSK transmitter's current value to both the left and right channels.

16.4.2 A Raised-Cosine Pulse Shaped BPSK Transmitter

The declaration section of the code is shown in Listing 16.5. One item that isn't shown here is that the file `coeff.c` is also part of the CCS project. This file contains the coefficients of a 200th order raised cosine FIR filter exported from MATLAB.

Listing 16.5: Declaration portion of the impulse modulation raised-cosine pulse shaped project code.

```
1  int  counter = 0;
   int  samplesPerSymbol = 20;
3  int  symbol;
   int  data[2] = {-15000, 15000};
5  int  cosine[4] = {1, 0, -1, 0};
   int  i;

7
   float  x[10];
9  float  y;
   float  output;
```

An explanation of Listing 16.5 follows.

1. (Line 1): Declares and initializes the `counter` variable, which is used to indicate where the algorithm is relative to the beginning of the symbol (0) or the end of the symbol (19).

2. (Line 2): Declares and initializes the `samplesPerSymbol` variable, which, as its name implies, is the number of sample in a symbol. For BPSK, a symbol and a bit are the same.

3. (Line 3): Declares the `symbol` variable, which is the current bit's value (0 or 1).

4. (Line 4): Declares and initializes the antipodal values associated with the a 0 or a 1.

5. (Line 5): Declares and initializes the `cosine` variable that contains *all* of the possible LO's values for a 12 kHz cosine carrier.

6. (Line 6): Declares the variable `i` that is used as the index in the dot-product.

7. (Line 8): Declares the `x` array that stores the current and past values of the message's value.

8. (Line 9): Declares the variable `y` that is the impulse modulator's current output value.

9. (Line 10) Declares the variable `output` that is the BPSK modulator's output value.

The algorithm section of the code is shown in Listing 16.6.

Listing 16.6: Algorithm portion of the impulse modulation raised-cosine pulse shaped project code.

```
// I added IM BPSK routine here
if (counter == 0) {
    symbol = rand() & 1; // faster version of rand() % 2
    x[0] = data[symbol]; // read the table
}


// perform impulse modulation based on the FIR filter, B[N]
y = 0;


for (i = 0; i < 10; i++) {
    y += x[i]*B[counter + 20*i];    // perform the dot-product
}


if (counter == (samplesPerSymbol - 1)) {
    counter = -1;


    /* shift x[] in preparation for the next symbol */
    for (i = 9; i > 0; i--) {
        x[i] = x[i-1];              // setup x[] for the next input
    }
}


counter++;


output = y*cosine[counter & 3];


CodecDataOut.Channel[LEFT]  = output; // setup the LEFT  value
CodecDataOut.Channel[RIGHT] = output; // setup the RIGHT value
// end of IM BPSK routine
```

An explanation of Listing 16.6 follows.

1. (Lines 2–5): Once per symbol, create a random bit, and map that bit to an allowed level. This value remains constant over the next 20 samples.

2. (Lines 8–12 and 18–20): Perform the FIR filtering associated with the impulse modulator. The vector `B` contains the coefficients of the raised cosine filter designed with and exported from MATLAB, using the `rcosfir` function. Even though a 200th order filter is being implemented, only 10 multiplies are required. This is because *all* of the other multiplies have a zero in the operation. These other multiplies are therefore *not* required since the outcome is already known! This is one of the most important advantages of an impulse modulator.

3. (Lines 14–15): When `counter` is equal to 19, the algorithm has reached the end of the symbol. At this point `counter` is reset to start the next symbol period.

4. (Line 23): Increments the variable `counter` in preparation for the next ISR.

5. (Line 25): Calculate the product of the current message value with the LO's value. This mixes the BPSK signal up to 12 kHz.

6. (Lines 27–28): Output the BPSK transmitter's current value to both the left and right channels.

16.4.3 Summary of Real-Time Code

Thus we have created two real-time implementations of a BPSK transmitter. The second version, while more complicated, is closer to what is used in actual communication systems due to its superior spectral characteristics. The analog voltage output from the DSK's DAC represents message symbols at a data rate of 2,400 bps (remember that for BPSK, bits and symbols are equivalent). This data stream could be used with a second DSK that is configured as a digital receiver, a concept covered in Chapter 17.

16.5 Follow-On Challenges

Consider extending what you have learned.

1. In the MATLAB impulse modulator simulations we used a 120th order FIR filter. In the real-time implementation we used a 200th order FIR filter. What is the effect of using a smaller order filter on the system's performance?

2. Design a bandpass raised-cosine pulse shaping FIR filter/system that when excited by a impulse, creates the modulated waveform directly (no separate mixer required).

3. Given your design in Challenge 2 above, what are the advantages and disadvantages of this approach?

4. How do the approaches discussed in this chapter compare to the analog filter approach? That is, where you would generate a rectangular shaped BPSK signal and then filter the signal to the desired bandwidth using traditional analog filters.

Chapter 17

Project 8: Digital Communications Receivers

17.1 Theory

IN Chapter 16 we introduced a few of the basic techniques that can be used to generate a binary phase shift-keying (BPSK) signal. Since there are an unlimited number of different forms and specifications associated with the generation of a BPSK signal, it should not be surprising that there are just as many variations of the receiver. In this chapter, we will only introduce one of these forms and a few of the techniques that can be used to recover the message contained within the BPSK signal.

Specifically, we will discuss a simplified BPSK receiver that must

1. Recover the carrier and remove its effects from the incoming signal. This will be accomplished using a phase-locked loop (PLL) (see Chapter 15). This process will recover a baseband version of the BPSK signal.

2. Process the recovered baseband signal through an FIR-based matched filter (MF). Assume our BPSK signal was generated at the transmitter using impulse modulated (IM) "root raised-cosine" shaped pulses at 2,400 symbols per second. This is similar to, but an interesting and common variation of, the normal "raised cosine" shaped pulses we explored in Chapter 16. As is required for MF operation, the receiver will use an identical root raised-cosine filter to what was used in the transmitter. We selected a MF-based receiver because it results in the optimal signal to noise ratio (SNR) of the decision statistic in the presence of additive white Gaussian noise (AWGN) [67].

3. Finally, symbol timing must be recovered from the signal that comes from the receiver's MF. We will use a maximum likelihood (ML) based timing recovery loop to determine when to sample the matched filter's output. This sampling/decision process is equivalent to determining where, on average, the eye-pattern is the most "open." We will discuss eye-patterns in more detail later in the chapter. This sampling/decision process also converts a series of filtered sampled signal values back into a message bit (0 or a 1). Recall that for BPSK, bits and symbols are equivalent.

At this point in our discussion of a BPSK receiver, it cannot be emphasized enough that carrier recovery, matched filtering, and symbol timing recovery must *all* occur properly within the receiver in order for the individual message bits to be recovered correctly. The combination of these three required processes makes this the most challenging project that we will attempt in this text. It may reduce your anxiety a bit when we tell you that this project actually consists of stringing together two concepts you've already seen and developing only one new concept. Specifically, the phase locked loop (PLL) was developed in Chapter 15 and the matched filter is just an FIR filter, which was developed in Chapter 3. The new portion of this project is the ML-based timing recovery system, which is where we will direct the majority of our discussion. Note that we will use the acronym NCO for "numerically controlled oscillator" to be consistent with the literature, but this is really just another name for a direct digital synthesizer (DDS) such as we discussed in Chapter 5.

The simplified block diagram of the end-to-end BPSK transmitter-channel-receiver is shown in Figure 17.1. In this figure, the receiver block diagram is contained within the dashed box. The timing recovery portion of the receiver is a simplified version derived from Figure 17.2. See [66] for more details.

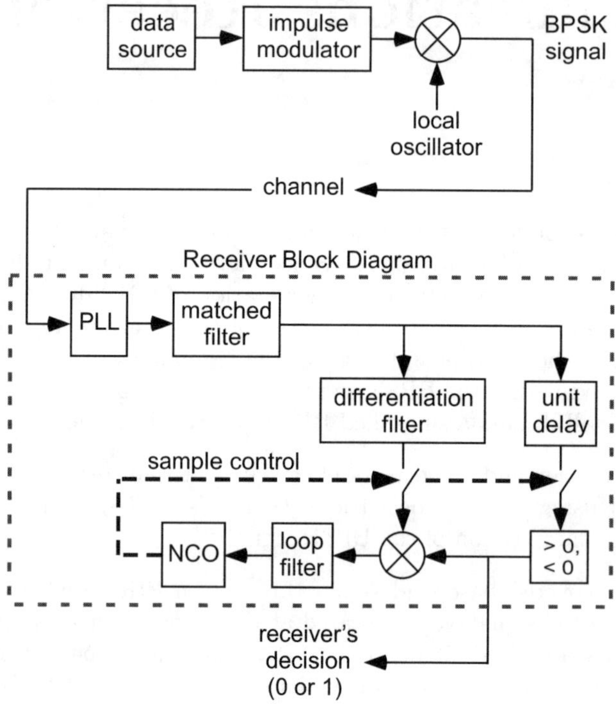

Figure 17.1: A simplified block diagram of a BPSK communications system (transmitter/channel/receiver). PLL: phase locked loop. NCO: numerically controlled oscillator.

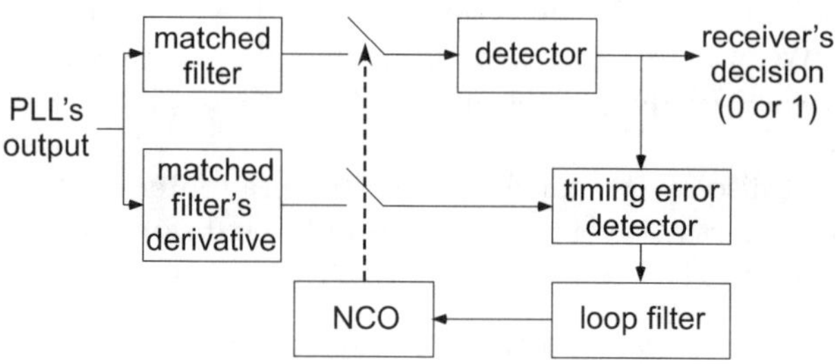

Figure 17.2: A maximum likelihood based timing recovery scheme. NCO: numerically controlled oscillator.

A few key simplifications of Figure 17.2 were made to arrive at Figure 17.1. Specifically,

1. Rather than implement two high-order FIR filters (both a matched filter and a "matched filter derivative" filter) as shown in Figure 17.2, we will implement only the matched filter, then *approximate* the derivative of the matched filter output.

2. The derivative operation will be implemented with a very simple second order FIR filter that has a group delay of one sample. This unit delay must be accounted for to properly align the two signals.

3. The detector shown in Figure 17.2 will be implemented with a slicer. That is, positive signals will be mapped to +1 (message bit value of 1) and negative signals will be mapped to −1 (message bit value of 0). The detector block can be thought of as being the point where the receiver *decides* the value of the received message bit.

4. The timing error detector (TED) as shown in Figure 17.2 will be approximated by the multiplication operation.

17.1.1 The Output of the Matched Filter

After the PLL removes the effects of the transmitter's carrier, the signal is passed through a matched filter. As previously mentioned, this filter optimizes the signal-to-noise ratio of the decision statistic in the presence of AWGN. An example of the output of the matched filter is shown in Figure 17.3. In this figure the time scale is 5.00 ms/div, and there are 10

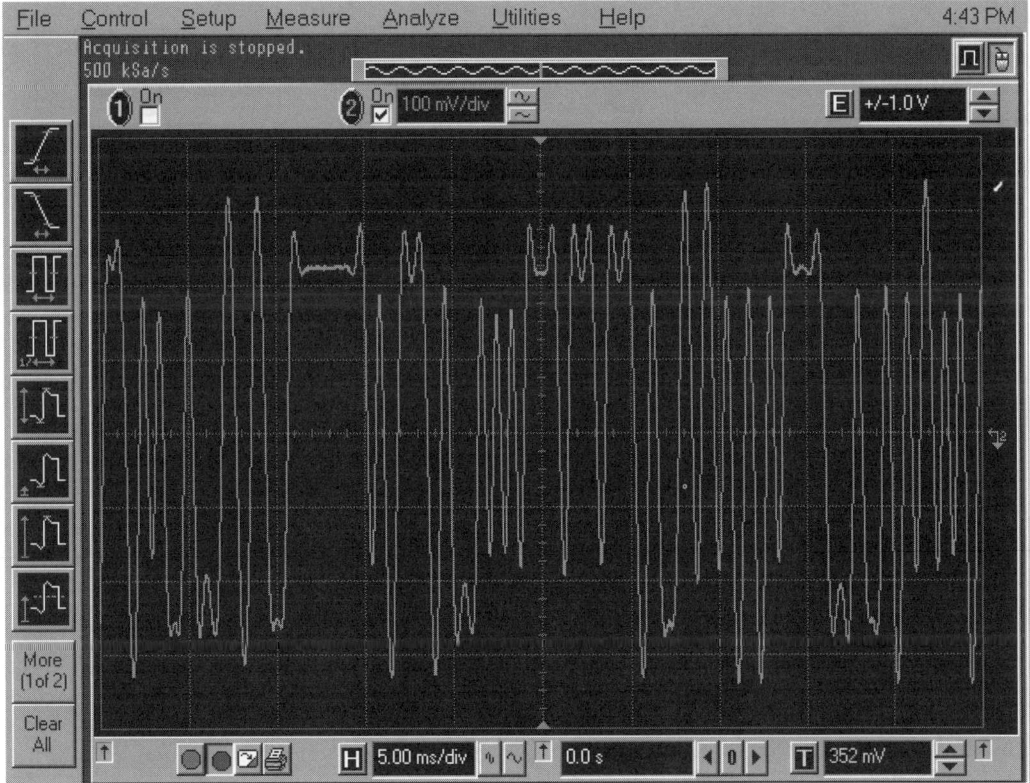

Figure 17.3: Output of the receiver's matched filter (120 bits).

horizontal divisions shown, so we see a 50 ms duration of data in the figure. At the given data rate of 2400 symbols/second (sps), this means the 50 ms "snapshot" is showing us 120 symbols of data. Since this is BPSK, 120 symbols is equivalent to 120 bits.

There was no noise intentionally added to the signal that resulted in Figure 17.3. How difficult is it for you to visually determine the values of all 120 symbols? Can you imagine trying to determine the message symbol values if a significant amount of noise had been added to the signal? Any real-world signal will unavoidably have noise added to it. Perhaps you can begin to appreciate how difficult a task the receiver must accomplish. The entire message symbol detection process is made possible by knowing when is the "best" time to sample the matched filter's output.

17.1.2   The Eye-Pattern

Before we commence our discussion of the timing recovery loop, we need to briefly review the concept of the eye-pattern. As previously mentioned, the BPSK receiver must remove the effects of the carrier (also called "down conversion"), filter the down-converted signal, and then sample the resulting signal at just the right time (timing recovery process) to convert (in this case) 20 samples (the number of samples in a transmitted symbol) back into a message symbol ($+1$ or -1). This process is equivalent to creating what is commonly called an eye-pattern and sampling this pattern at the point of maximum average eye opening. An example of an eye-pattern created using an oscilloscope and a recovered symbol timing signal to trigger the display is shown in Figure 17.4. In this figure, 100 ms of the

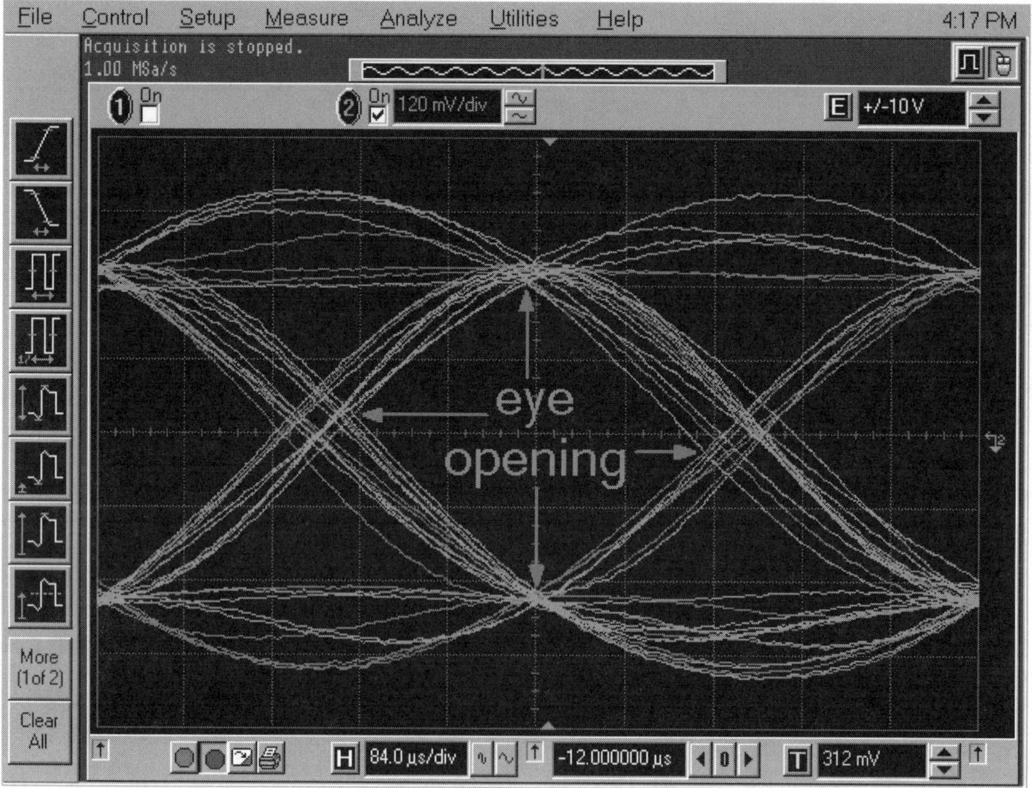

Figure 17.4: BPSK eye-pattern (100 ms of data from MF output).

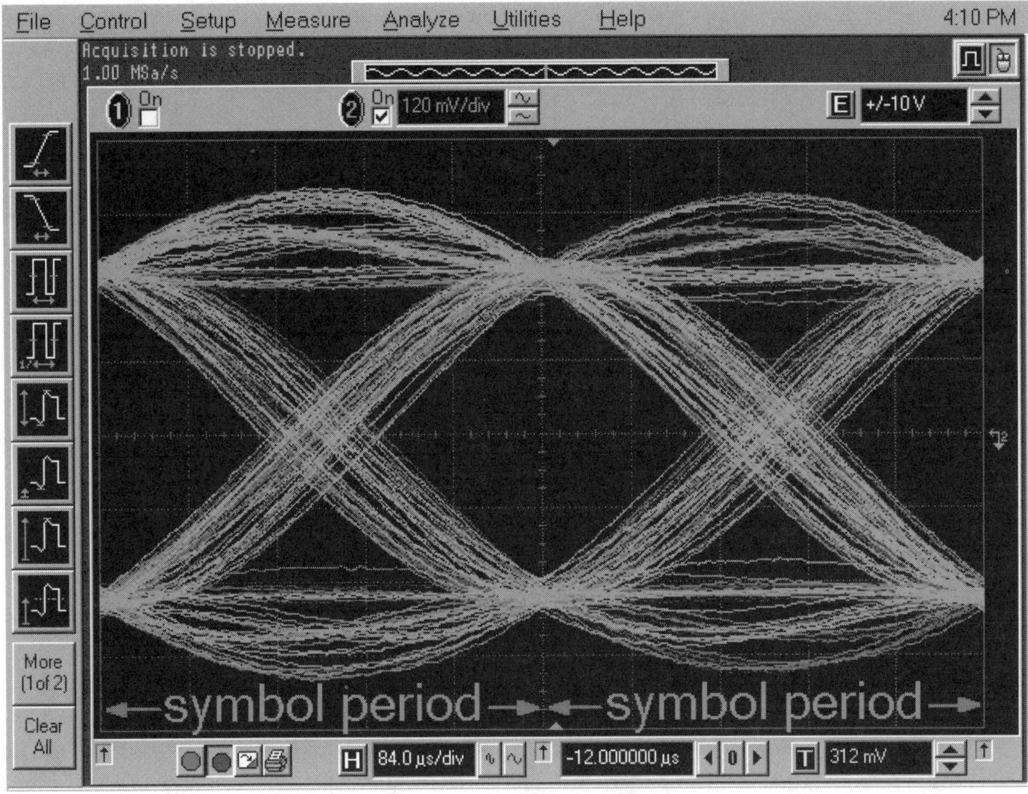

Figure 17.5: BPSK eye-pattern (1000 ms of data from MF output).

matched filter's output is displayed. The eye opening is labeled, but this opening is not symmetric. To achieve a symmetric eye opening, considerably more data (gathered over time) is required to be displayed. In Figure 17.5 a full second of matched filter output is displayed. The eye is now symmetric and the symbol period is labeled on the horizontal axis. The symbol rate is 2400 symbols per second, which results in a symbol period of $1/2400 = 416.67$ μs. Considering that we wish to display three eye openings (i.e., two symbol periods), the time scale (the horizontal axis of the oscilloscope) needs to be set to $(1/2400) * 2/10 = 83.33$ μs/div, so the oscilloscope time-base was set to the nearest value of 84 μs/div.

At this point it should be clear that eye-pattern displays of short periods of time allow for individual traces to be seen but that the characteristics of the eye opening can be misleading due to insufficient data. On the other hand, if a much longer time period is displayed, the individual traces on the eye-pattern will be obscured, but the average characteristics of the eye opening will be more clearly visible.

17.1.3 Maximum Likelihood Timing Recovery

As stated earlier, the approximation of the optimum ML timing recovery loop that we will implement is shown in the receiver block diagram portion of Figure 17.1. This diagram is a modified version of the optimal ML timing recovery loop discussed in Chapter 7 of the text by Mengali and D'Andrea [66]. The first step in this process is to multiply properly sampled outputs from the matched filter's decision ($+1$ or -1) by the derivative of the

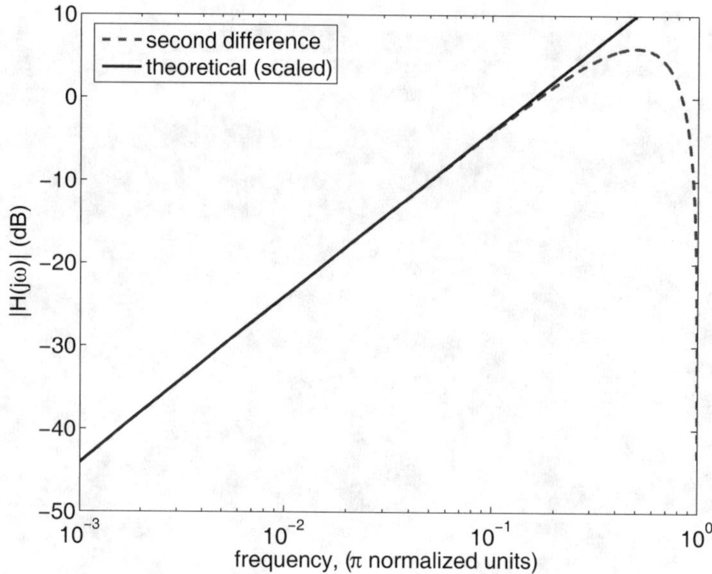

Figure 17.6: The frequency response of a second order FIR difference filter compared with a scaled version of the theoretical response.

output of the matched filter (this multiplication takes the place of the timing error detector in Figure 17.2). Therefore we need to somehow implement the derivative operation. The derivative can be obtained using a number of different techniques, but we will approximate it with a simple difference operation using a second order FIR filter, $y[n] = x[n] - x[n - 2]$, which has a group delay of one sample. As shown in Figure 17.6, this implementation is more than adequate since the difference operation very closely approximates the derivative operation for highly over-sampled signals. As is almost always the case, the group delay associated with this filter must be accounted for prior to continuing with the algorithm.

An intuitive understanding of why this multiplication results in a signal that is proportional to the timing error can be seen in Figure 17.7. Recall that for this BPSK signal, a symbol value of -1 is equivalent to a bit value of 0, and a symbol value of $+1$ is equivalent to a bit value of 1. The eye-pattern traces labeled as "A_{01}" are associated with message bit transitions from 0 to 1 and the traces labeled as "A_{10}" are associated with message bit transitions from 1 to 0. For a random message of equiprobable bit values, the traces labeled "A_{01}" and "A_{10}", when taken together, should occur with a probability of 0.5.

The eye-pattern traces labeled as "B" represent extended strings of either ones or zeros (i.e., the bit value doesn't change). The eye-pattern traces labeled as "C" represent an extend string of either ones or zeros followed by a message bit change, or a message bit change followed by an extended string of either ones or zeros. For a random message of equiprobable bit values, the traces labeled "B" and "C", when taken together, should also occur with a probability of 0.5.

Only the "A" region traces result in a proper error signal being generated by the timing error detector (TED). In the "B" region, the slope (i.e., the derivative) of the traces is very close to zero, which results in almost no error signal being generated. In the "C" region, the traces result in an error signal of the "wrong" polarity. This wrong polarity signal can be viewed as either providing positive feedback or as a reinforcement of the system's error, which is not what we want. Fortunately, the effects of the "A" region are stronger than that of the combined "B" and "C" regions. Therefore, with the NCO set to free-run at the

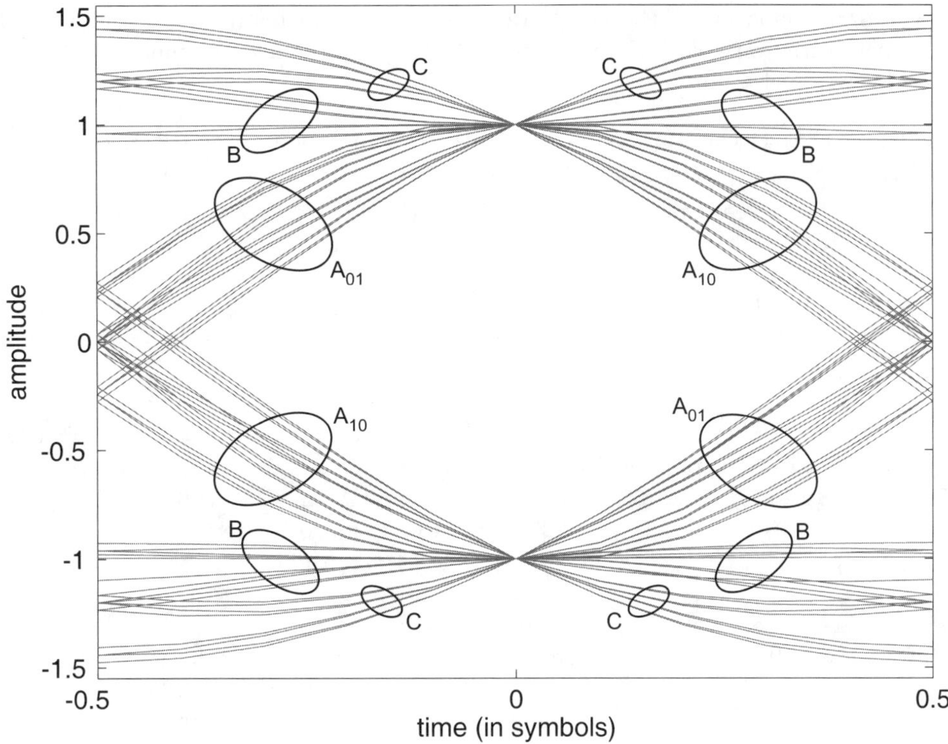

Figure 17.7: The timing recovery scheme for an ideal eye-pattern.

2400 symbols per second rate, the system's error signal should only need to account for very small differences between the transmitter's and the receiver's symbol clocks. This allows for proper operation of the timing recovery loop because, on average, the system tracks the maximum eye opening.

Finally, the sampling and decision at the output of the matched filter MF extracts the transmitted data symbols ($+1$ or -1). The slicer, which implements an even simpler decision, actually decides between $x > 0$ or $x < 0$. Based upon the symbol recovered, the data bits (1 or 0) are obtained from the received BPSK signal.

Important: At this point in the book, we hope you've caught on to the need to go beyond the text and to read the *full* program listings associated with a given chapter on the accompanying CD-ROM. While many partial code listings and explanations have been given in previous chapters, the following discussion is intentionally not as detailed. This an attempt to help you further develop the ability to *teach yourself* how read and understand real-time DSP code; this is a necessary step toward creating your own original code. Try to go beyond what we've written.

17.2 winDSK6 Demonstration

The winDSK6 program does not provide an equivalent receiver function. The commDSK application of winDSK6 has transmitter functions only.

17.3 MATLAB Implementation

The MATLAB simulation of the BPSK receiver is shown in Listings 17.1 and 17.2. Listing 17.1 details the variable declarations while Listing 17.2 details the simulated ISR portion of the MATLAB script file.

Listing 17.1: Declarations associated with the simulation of the BPSK receiver.

```
1  % generate the BPSK transmitter's signal
   [BPSKsignal, dataArray] = impModBPSK(0.1);
3
   % simulation inputs ... PLL
5  alphaPLL = 0.010;              % PLL's loop filter parameter "alpha"
   betaPLL  = 0.002;              % PLL's loop filter parameter "beta"
7  N  = 20;                       % samples per symbol
   Fs = 48000;                    % simulation sample frequency
9  phaseAccumPLL = randn(1);      % current phase accumulator's value
   VCOphaseError = 2*pi*rand(1);  % selecting a random phase error
11 VCOrestFrequencyError = randn(1);  % error in the VCO's rest freq
   Fcarrier = 12000;              % carrier freq of the transmitter
13 phi = VCOphaseError;           % initializing the VCO's phase

15 % simulation inputs ... ML timing recovery
   alphaML = 0.0040;   % ML loop filter parameter "alpha"
17 betaML  = 0.0002;   % ML loop filter parameter "beta"
   alpha   = 0.5;      % root raised-cosine rolloff factor
19 symbols = 3;        % MATLAB "rcosfir" design parameter

21 phaseAccumML = 2*pi*rand(1);% initializing the ML NCO
   symbolsPerSecond = 2401;     % symbol rate w/ an offset from 2400
```

An explanation of Listing 17.1 follows.

1. (Line 2): Generation of the transmitter's signal.

2. (Lines 5–6): Loop filter parameters associated with the PLL.

3. (Lines 7–8): The sample frequency (samples per second) divided by the number of samples per symbol determines the number of symbols per second (symbol rate or baud).

4. (Lines 9–11, 13, 21, and 22): Add realism to the simulation by ensuring that errors exist in the PLL's initial frequency and phase and also in the ML timing recovery loop's initial phase and symbol rate.

5. (Line 12): Carrier rest frequency is set to match the transmitter's carrier frequency (12 kHz).

6. (Lines 16–17): Loop filter parameters associated with the ML timing recovery loop.

7. (Line 18): Root raised-cosine roll-off factor. This should match the roll-off factor of the transmitter.

8. (line 19): A MATLAB filter design parameter associated with the length (order) of the FIR filter that was designed using the `rcosfir` function.

Listing 17.2: ISR simulation of the BPSK receiver.

```matlab
% commencing ISR simulation
for i = 1:length(BPSKsignal)
    % processing the data by the PLL
    phaseDetectorOutput = analyticSignal(i)*vcoOutput;
    m = 6*real(phaseDetectorOutput);    % scale for a max value
    q = real(phaseDetectorOutput)*imag(phaseDetectorOutput);
    [loopFilterOutputPLL,Zi_pll]=filter(B_PLL, A_PLL, q, Zi_pll);
    loopFilterOutputPLLSummary = ...
        [loopFilterOutputPLLSummary loopFilterOutputPLL]; % plot
    phi = mod(phi + loopFilterOutputPLL + 2*pi*Fcarrier*T, 2*pi);
    vcoOutput = exp(-j*phi);


    % processing the data by the ML-based receiver
    [MFoutput, Zi_MF] = filter(B_MF, 1, m, Zi_MF);
    [diffMFoutput,Zi_diff]=filter([1 0 -1], 1, MFoutput,Zi_diff);


    phaseAccumML = phaseAccumML + phaseIncML;
    if phaseAccumML >= 2*pi
        phaseAccumML = phaseAccumML - 2*pi;
        decision = sign(delayedMFoutput);
        [error, Zi_ML_loop] = filter(B_ML, A_ML, ...
            decision*diffMFoutput, Zi_ML_loop);
        phaseAccumML = phaseAccumML - error;
        errorSummary = [errorSummary error];          % plot
        decisionSummary = [decisionSummary decision]; % plot
    else
        errorSummary = [errorSummary 0];              % plot
    end


    delayedMFoutput = MFoutput; % accounts for group delay

    % state storage for plotting ... not part of the ISR
    delayedMFoutputSummary = ...
        [delayedMFoutputSummary delayedMFoutput];
    decisionSummaryHoldOn = [decisionSummaryHoldOn decision];
end

% output terms
% Plot commands follow ...
```

An explanation of Listing 17.2 follows.

1. (Lines 2–36): This "for" loop simulates the sample-by-sample processing of the received data.

2. (Lines 4–11): This block of code implements the PLL discussed in Chapter 15.

3. (Line 14): Performs the matched filtering of the data out of the PLL.

4. (Line 15): Performs the second difference of the output of the matched filter. This is a very good approximation of differentiation for this over-sampled signal.

5. (Line 17): Updates the value of the phase accumulator associated with the ML timing recovery loop. For a free-running frequency of 2400 symbols per second, every time the ISR is called the accumulator's value must increment by $2\pi/20 = \pi/10$ radians.

6. (Lines 18–28): Ensure that whenever the phase accumulator reaches 2π, the accumulator's value is reduced by 2π. For reasonable values of the ML timing loop error this is equivalent to a modulo 2π operation. On average, this block of code should only run once every 20 ISR calls. This effectively is a decimation by 20 operation which results in only these samples being passed to the ML timing recovery loop filter.

7. (Line 30): Accounts for the group delay of the second difference FIR filter.

Remember that the simulation randomly initializes the frequency and phase errors associated with the PLL and ML timing recovery loops; this (plus the random nature of any noise that may be added to the input signal) means that the behavior of the simulation will be different *every time* the program is run. This random behavior can be controlled by either reinitializing the state of the MATLAB random number generator at the beginning of each simulation or by setting the variables in lines 9, 10, 11, and 21 equal to zero. The state of the MATLAB random number generator can be reset using the MATLAB command `randn('state',0)`.

A simulation output example exhibiting excellent behavior of the PLL and ML timing recovery loops is shown in Figure 17.8. In this figure the first (top) subplot shows how the PLL's error damps toward zero error. The second subplot shows the ML timing error also settling towards an average value of zero. Remember that the phase angle between adjacent samples as measured by the ML timing recovery NCO is $\pi/10 \approx 0.314$ radians. In

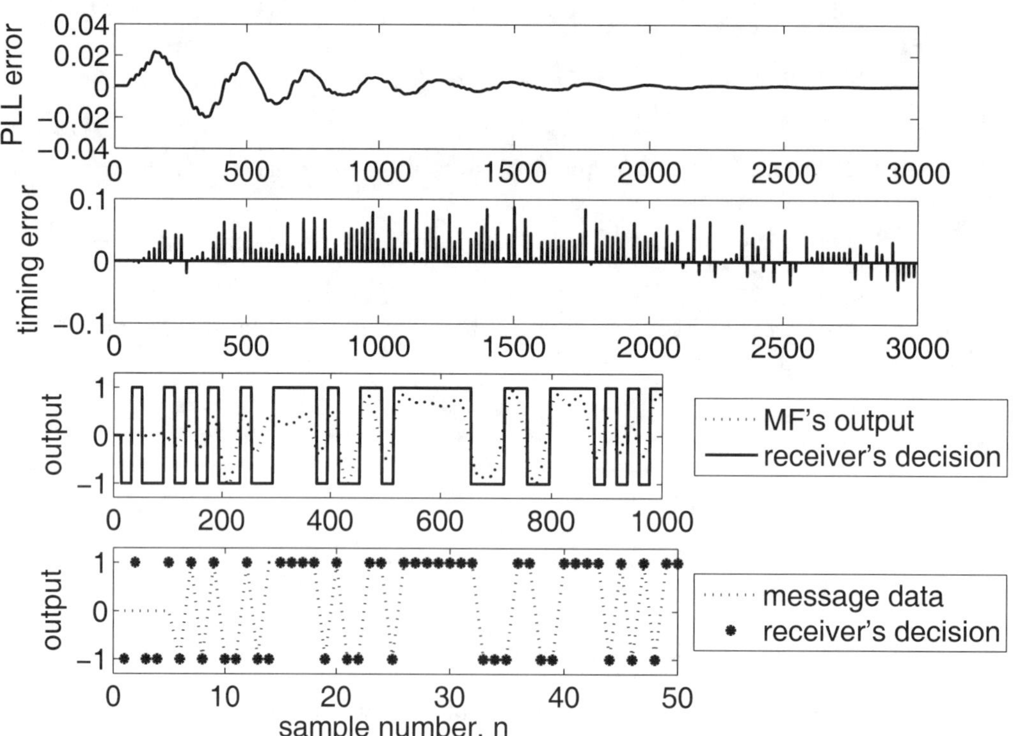

Figure 17.8: BPSK receiver exhibiting excellent PLL and ML timing recovery performance.

this example the timing error never exceeds 0.1 radians. Keep in mind that the timing error is *only* defined when a sample is taken. So the subplot is showing an average of 19 zero error values for every non-zero value. The third subplot shows the output of the matched filter and the receiver's decision. This effectively turns the analog output of the MF back into a digital signal (data symbols). The fourth (bottom) subplot compares the transmitted and received symbols. In this simulation, the receiver processed about 15 symbols before the errors stopped occurring. Remember the symbol to bit mapping for this BPSK example is defined as $(+1 \rightarrow 1)$, $(-1 \rightarrow 0)$.

In Figure 17.9 everything in the first, second, and third subplots indicates that the algorithm is working quite well. The fourth subplot, however, clearly shows that the received symbols have been flipped (a transmitted -1 was interpreted as a $+1$, and a transmitted $+1$ was interpreted as a -1)! When we map the received symbols to bits, this will result in message bit flipping, and our bit error rate will be nearly 100%. This is unacceptable; a reasonable error rate is more on the order of one error for every million bits received! How did this catastrophe occur?

As we discussed in the PLL chapter, the Costas loop PLL is *blind* to 180 degree phase ambiguities. But there are ways to get around this problem. Sending a known preamble signal at the beginning of a transmission would allow you to identify the phase inversion so you could remove it. Another popular technique is to use differential encoding of the transmitted data (with the accompanying differential decoding after message recovery); this makes your data impervious to phase inversion at the cost of more complexity in the transmitter and receiver.

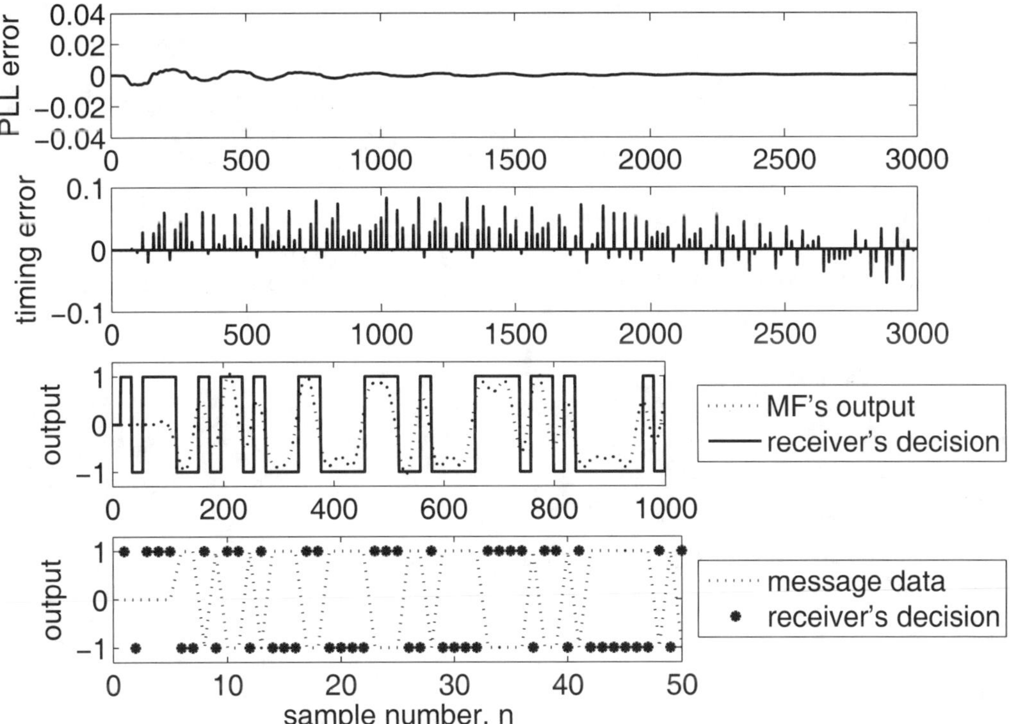

Figure 17.9: BPSK receiver exhibiting excellent PLL and ML timing recovery performance. However, the Costas loop locked up 180 degrees out of phase. This will result in message bit flipping, and a bit error rate of nearly 100%.

17.4   DSK Implementation in C

17.4.1   Components of the Digital Receiver

The real-time implementation of the BPSK receiver is shown in Listings 17.3 and 17.4. Listing 17.3 provides the declaration of most of the variables used, and Listing 17.4 implements the algorithm. The files necessary to run this application are in the ccs\DigRx directory of Chapter 17. The primary file of interest is the BPSK_rcvr_ISRs.c, which contains the interrupt service routines. This file includes most of the necessary variable declarations and performs the BPSK receiver operation.

Listing 17.3: Declaration portion of the BPSK receiver project code.

```
1  float  alpha_PLL = 0.01;  // loop filter parameter
   float  beta_PLL = 0.002;  // loop filter parameter
3
   float  alpha_ML = 0.1;    // loop filter parameter
5  float  beta_ML = 0.02;    // loop filter parameter

7  float  twoPi = 6.2831853072;      // 2pi
   float  piBy2 = 1.57079632679;     // pi/2
9  float  piBy10 = 0.314159265359;  // pi/10
   float  piBy100 = 0.0314159265359;  // pi/100
11 float  scaleFactor = 3.0517578125e-5;
   float  gain = 3276.8;
13
   float  x[N+1] = {0,0,0,0,0,0,0}; // input signal
15 float  sReal = 0;  // real part of the analytical signal
   float  sImag = 0;  // imag part of the analytical signal
17
   float  phaseDetectorOutputReal[M+1] = {0,0,0,0,0,0,0,0,
19     0,0,0,0,0,0,0,0,0,0,0,0,0,0,0,0,0,0,0,0,0,0,
       0,0,0,0,0,0,0,0,0,0};  // real part of the analytical signal
21 float  phaseDetectorOutputImag = 0;
   float  vcoOutputReal = 1;
23 float  vcoOutputImag = 0;
   float  q_loop = 0;        // input to the loop filter
25 float  sigma_loop = 0; // part of the PLL loop filter
   float  PLL_loopFilterOutput = 0;
27 float  phi = 0; // phase accumulator value

29 float  *pLeft = phaseDetectorOutputReal;
   float  *p = 0;
31 float  matchedfilterout[3] = {0,0,0};

33 float  diffoutput = 0;
   float  sigma_ML = 0;  // part of the ML loop filter
35 float  ML_loopFilterOutput = 0;
   float  accumulator = 0;
37 float  adjustment = 0;
   float  data = 0;
39 volatile float ML_on_off = 1;
```

```
41  int i = 0;
    int sync = 0;
```

An explanation of Listing 17.3 follows.

1. (Lines 1–2): Declare the PLL's loop filter parameters.

2. (Lines 4–5): Declare the ML timing recovery loop's filter parameters.

3. (Lines 7–12): Declare various constants used in the algorithm.

4. (Lines 14–27): Declare the variables necessary to implement the PLL.

5. (Lines 29–31): Declare the variables necessary to implement the MF.

6. (Lines 33–38): Declare the variables necessary to implement the ML timing recovery loop.

7. (Line 39): Declares a flag to turn the ML timing recovery loop "ON" or "OFF". This can be used to demonstrate the effect of not having an error control process in the timing recovery loop.

8. (Line 41): Declares an integer index used in a loop.

9. (Line 42): Declares an integer synchronization signal that is used to send a timing pulse through the second codec channel. This timing pulse can be used to trigger an oscilloscope to create an eye-pattern.

Listing 17.4: Algorithm portion of the BPSK receiver project code.

```
    // my algorithm starts here ...
2   // bring in input value
    x[0] = CodecDataIn.Channel[LEFT];   // current LEFT input value

4
    // execute Hilbert transform and group delay compensation
6   sImag = (x[0] - x[6])*B_hilbert[0] + (x[2] - x[4])*B_hilbert[2];
    sReal = x[3]*scaleFactor; // scale and account the group delay

8
    // setup x for the next input
10  for(i = N; i >0; i--) {
        x[i] = x[i-1];
12  }

14  sImag *= scaleFactor;

16  // execute the PLL
    vcoOutputReal = cosf(phi);
18  vcoOutputImag = sinf(phi);
    phaseDetectorOutputReal[0] = sReal*vcoOutputReal ...
20      + sImag*vcoOutputImag;
    phaseDetectorOutputImag = sImag*vcoOutputReal ...
22      - sReal*vcoOutputImag;
    q_loop = phaseDetectorOutputReal[0] * phaseDetectorOutputImag;
24  sigma_loop += beta_PLL*q_loop;
```

```c
   PLL_loopFilterOutput = sigma_loop + alpha_PLL*q_loop;
26 phi += piBy2 + PLL_loopFilterOutput;

28 while(phi > twoPi) {
       phi -= twoPi;  // modulo 2pi operation for accumulator
30 }

32 // execute the matched filter (MF)
   *pLeft  = phaseDetectorOutputReal[0];
34
   matchedfilterout[0] = 0;
36 p = pLeft;
   if(++pLeft > &phaseDetectorOutputReal[M])
38     pLeft = phaseDetectorOutputReal;
   for (i = 0; i <= M; i++) {  // do LEFT channel FIR
40         matchedfilterout[0] += *p— * B_MF[i];
           if(p < phaseDetectorOutputReal)
42               p = &phaseDetectorOutputReal[M];
   }
44
   // execute the differentiation filter
46 diffoutput = matchedfilterout[0] - matchedfilterout[2];

48 // execute the ML timing recovery loop
   sync = 0;
50 if(accumulator >= twoPi) {
       sync = 20000;
52     data = -1;
       if(matchedfilterout[0] >= 0) { // recover data
54         data = 1;
       }
56     adjustment = data*diffoutput;
       sigma_ML += beta_ML*adjustment;
58     // prevents timing adjustments of more than +/- 1 sample
       if (sigma_ML > piBy10) {
60         sigma_ML = piBy10;
       }
62     else if (sigma_ML < -piBy10) {
           sigma_ML = -piBy10;
64     }
       ML_loopFilterOutput = sigma_ML + alpha_ML*adjustment;
66
       // prevents timing adjustments of more than +/- 0.1 sample
68     if (ML_loopFilterOutput > piBy100) {
           ML_loopFilterOutput = piBy100;
70     }
       else if (ML_loopFilterOutput < -piBy100) {
72         ML_loopFilterOutput = -piBy100;
       }
74     if (ML_on_off == 1) {
           accumulator -= (twoPi + ML_loopFilterOutput);
```

```
76        }
       else {
78            accumulator -= twoPi;
          }
80 }

82 // increment the accumulator
   accumulator += piBy10;

84
   // setup matchedfilterout for the next input
86 matchedfilterout[2] = matchedfilterout[1];
   matchedfilterout[1] = matchedfilterout[0];

88
   CodecDataOut.Channel[LEFT] = sync; // O-scope trigger pulse
90 CodecDataOut.Channel[RIGHT] = gain*matchedfilterout[1];
   // ... stop adding code here
```

1. (Line 3): Brings the input value into the ISR.

2. (Line 6): Performs the Hilbert transform on the incoming signal. An inspection of the B_hilbert coefficients in the `hilbert.c` file included in this CCS project reveals an odd symmetry with the three odd numbered terms all equal to zero. This greatly simplifies the dot-product required to implement this FIR filter. A zoomed-in comparison of the passband frequency response of the filter as designed in MATLAB and this filter with every other coefficient set equal to zero is shown in Figure 17.10. Notice that this filter is *very* close to "flat" in the frequency band of interest (12 kHz ± 2 kHz. This is not immediately obvious until you notice that the vertical axis units are milli-dB (mdB). The figure shows the very small but unavoidable passband ripple in both filters, but they are *very* close to being the same.

3. (Line 7): Applies a scale factor to the signal and accounts for the group delay of the Hilbert transforming FIR filter prior to the PLL.

4. (Lines 10–12): Update the buffered values of x for the next ISR call.

5. (Line 14): Applies a scale factor to the signal prior to the PLL.

6. (Lines 17–30): Implement the PLL.

7. (Lines 33–43): Implement the FIR-based matched filter.

8. (Line 46): Implements the FIR-based second difference filter which is a very close approximation to differentiation.

9. (Line 49): Turns the codec's channel synchronization pulse off.

10. (Lines 50–80): Implement the ML timing recovery loop (the details follow).

11. (Line 50): When the total phase of the ML timing loop's NCO is $\geq 2\pi$, the sampling operation that supplies the next 2 inputs to the ML timing loop is activated.

12. (Line 51): Drives the codec's output synchronization pulse high.

13. (Lines 52–55): Implement the detector (slicer) to decide which symbol was received.

14. (Line 56): Performs the timing error detector operation (multiplication) of the properly aligned matched filter's output and its derivative.

15. (Lines 57–73): Implement the ML timing recovery loop's loop filter (the details follow).

16. (Lines 57–64): Implements the IIR portion of the filter. Note that a non-linear element (a limiter) prevents this filter's output from exceeding $\pm\pi/10$ radians (± 1 sample).

17. (Lines 65–73): Finishes the implementation of the loop filter and adds another limiter to prevent the total error from exceeding $\pm\pi/100$ (± 0.1 samples).

18. (Lines 74–79): Apply the ML timing recovery loop's error to the NCO if the ML loop is turned "ON," otherwise the loop free-runs at the nominal symbol rate (2400 sps).

19. (Line 83): Increments the ML timing loop NCO's phase by $\pi/10$ each time the ISR is called. This sets the NCO's free-running speed to the the symbol rate (2400 sps).

20. (Lines 86–87): Update the buffered values of the matched filter's output in preparation for the next second difference operation.

21. (Lines 89–90): Output the synchronization signal to the codec's Left channel and the group-delayed matched filter output to the codec's Right channel.

17.4.2   System Testing

Creating an eye-pattern is an excellent way to determine how well your receiver is working. The DSK system should be set up as follows.

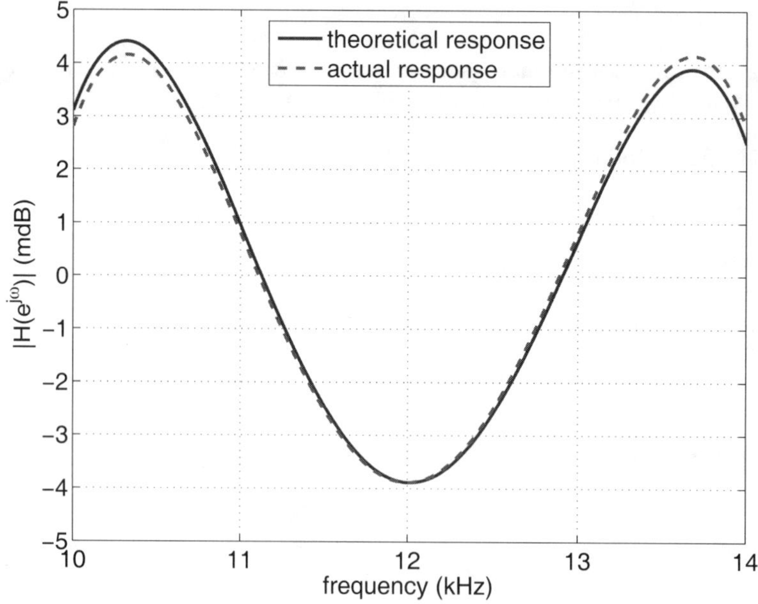

Figure 17.10: Comparison of the MATLAB designed Hilbert transform filter's passband frequency response and the same filter with every other coefficient set equal to zero. Note that the vertical axis scale is in milli-dB.

1. The CD-ROM for the book contains wav-files (in the `test_signals` directory, where the file name starts with "BPSK RRC" for BPSK root raised cosine) that can be played through any wav-file playback device (CD player, PC sound card, etc....) to "stand in" for the BPSK transmitter. This signal should be connected as the input to the DSK's ADC. Alternatively, a second DSK could serve as the transmitter, running code similar to the C code for impulse modulated BPSK transmitter in Chapter 16 (but using a *root* raised cosine filter, not a raised cosine filter). Or a second DSK could serve as the transmitter, running the `commDSK` application of `winDSK6`.

2. Use winDSK6 or an oscilloscope to verify that the input to the DSK is near, but does not reach, the maximum range of the DSK's codec (typically $-1 < x < +1$ volt).

3. Connect the signal source to the DSK's input.

4. Attach the two output channels (Left and Right) of the DSK's codec to two input channels of an oscilloscope in the following manner. Set the oscilloscope display to trigger off of the oscilloscope channel to which the Left output of the DSK (the sync pulse) is connected. Display both input channels of the oscilloscope on the screen.

5. Ensure that the BPSK input signal is playing, load the BPSK receiver's `.out` file into the DSK, and start the DSK (run).

With the oscilloscope's display persistence turned up, an image similar to Figure 17.11 should be visible.

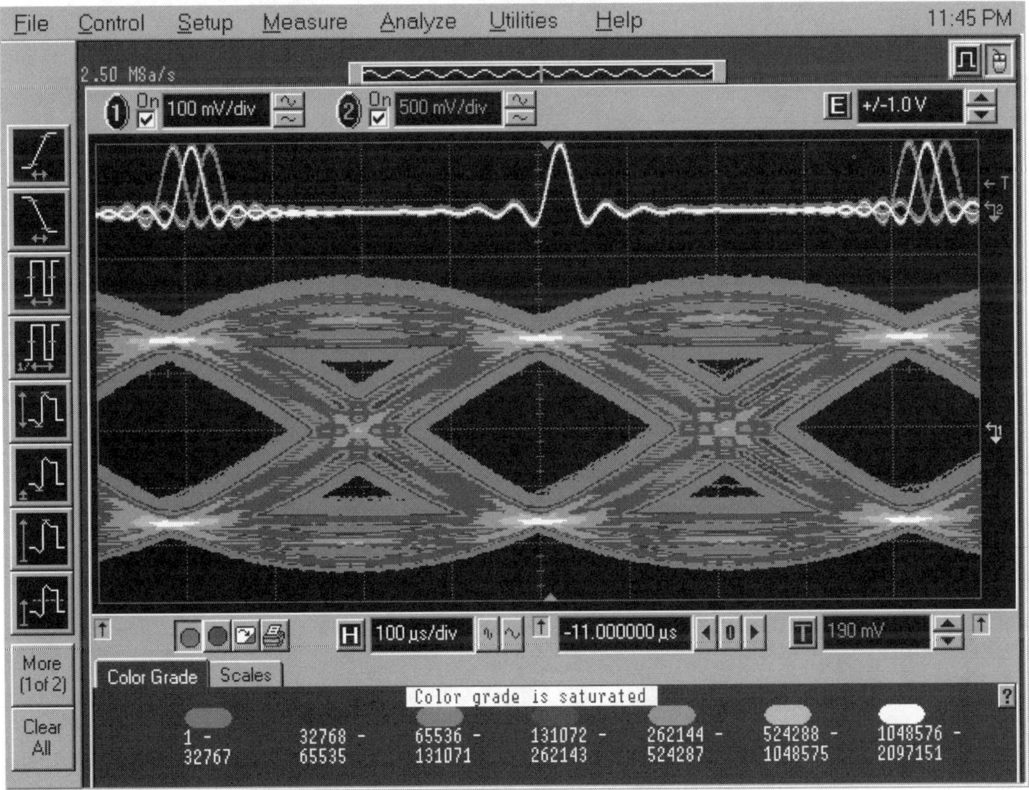

Figure 17.11: BPSK eye-pattern histogram with recovered timing pulses.

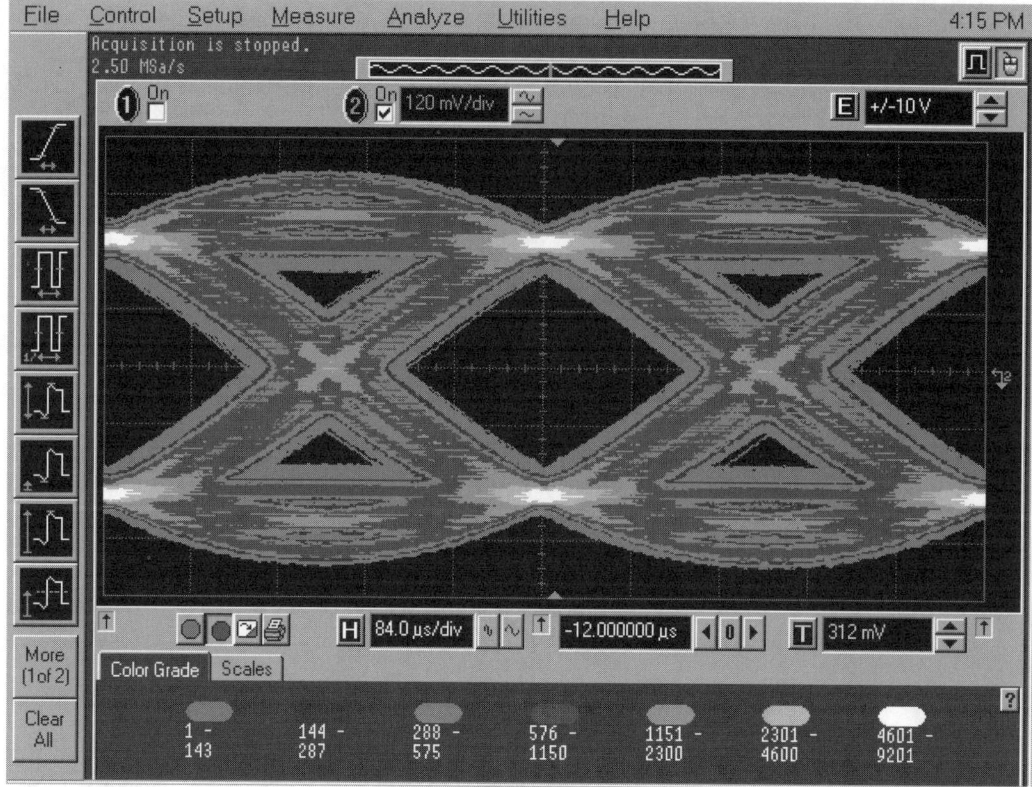

Figure 17.12: BPSK eye-pattern histogram.

In this figure a color-coded histogram (shown here in the text as shades of gray) has been produced that indicates where on the screen the MF output trace is most often located. Synchronization pulses are shown at the top of the display screen. The center synchronization pulse appears to be very stable since it is being used to trigger the oscilloscope. The synchronization pulse immediate before and after this center pulse show clear indications that the ML timing recovery loop occasionally tracked one sample to either side of the location of maximum eye opening. This very slight timing error results in a timing jitter that causes the eye opening to close slightly in both the vertical and horizontal directions. As the eye opening gets larger, the receiver is performing better.

Without the synchronization pulses displayed, a traditional eye-pattern should be displayed such as the one shown in Figure 17.12.

17.5   Follow-On Challenges

Consider extending what you have learned.

1. Design and implement your own loop filter within the ML timing recovery loop.

2. Design and implement an algorithm that detects, then provides some indication to the user, when the ML timing recovery loop is *locked* and tracking the symbol rate.

3. Profile the ISR code and identify any bottlenecks.

4. Suggest possible improvements that minimize or remove these bottlenecks.

5. Implement at least one of your improvements and calculate the computational savings of your new code.

6. Implement the BPSK receiver using frame-based techniques.

Appendices

Appendix A

Code Composer Studio: A Brief Tutorial

A.1   Introduction

CODE Composer Studio™ (CCS) is Texas Instruments' integrated development environment (IDE) for developing routines on a wide variety of their DSPs. In CCS, the editing, code generation, and debugging tools are all integrated into one unified environment. You can select the target DSP, adjust the optimization parameters, and set the user preferences as you desire.

An application is developed based on the concept of a project, where the information in the project file (*.pjt) determines what source code is used and how it will be processed. Learning to use Code Composer Studio is a necessary step in bridging the gap between DSP theory and real-time DSP. We recommend that you devote some time to getting to know CCS. You can do that on your own, or you can use this appendix to get you started.

At the end of this appendix is a description of a typical set of files used on a typical project for this text.

A.2   Starting Code Composer Studio

This tutorial assumes that CCS is properly installed on a Windows 98SE, 2000, ME, or XP-based computer. As of this printing, the latest version of CCS (version 3.1) only supports Windows 2000 and Windows XP. As long as you are running CCS version 2.0 or higher, this tutorial will sequence you through the basic steps involved in creating, compiling, loading, and running (executing) a project. If CCS is not installed, please install it now. The following discussion assumes you are using a C6713 DSK. There will be only minor differences if you are using a C6711 DSK. When you are ready...

1. Power up the C6713 DSK by plugging in its power supply and connecting the power supply cable to the DSK. Verify that the DSK "boots up" properly by observing the power on self test (POST) which takes about 15 seconds. After the POST is complete verify that the +5V LED is "ON," all four user LEDs (D7, D8, D9, and D10) are "ON," and the "USB IN USE" LED is "ON." All of these LEDs are green.

2. Plug the USB cable into the DSK and then plug the other end into either the host PC or a USB hub that is already connected to the host PC.

3. Launch CCS by double clicking on its icon, which after a standard CCS install will be located on the computer "desktop." Icons from two different CCS versions are shown below; your installation should be similar to one of these.

As CCS starts up, a splash screen similar to one of the two screens shown below will appear.

Shortly after the splash screen appears, a "Waiting for USB Enumeration" dialog box, as shown below, will appear.

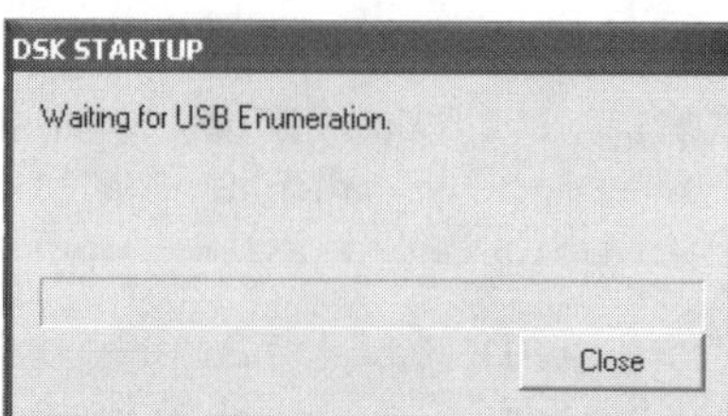

Finally, the CCS integrated development environment (IDE) will finish loading and an interface similar one of the two screens shown below will appear.

If your version of CCS displays the "No target connected" icon, as shown next, in the lower left-hand corner of the CCS IDE screen,

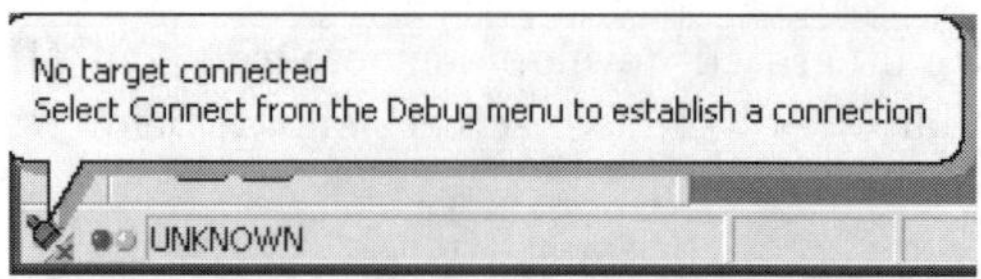

you will need to connect your DSP to CCS by selecting the "Debug" pull down menu and then select "Connect" as shown below.

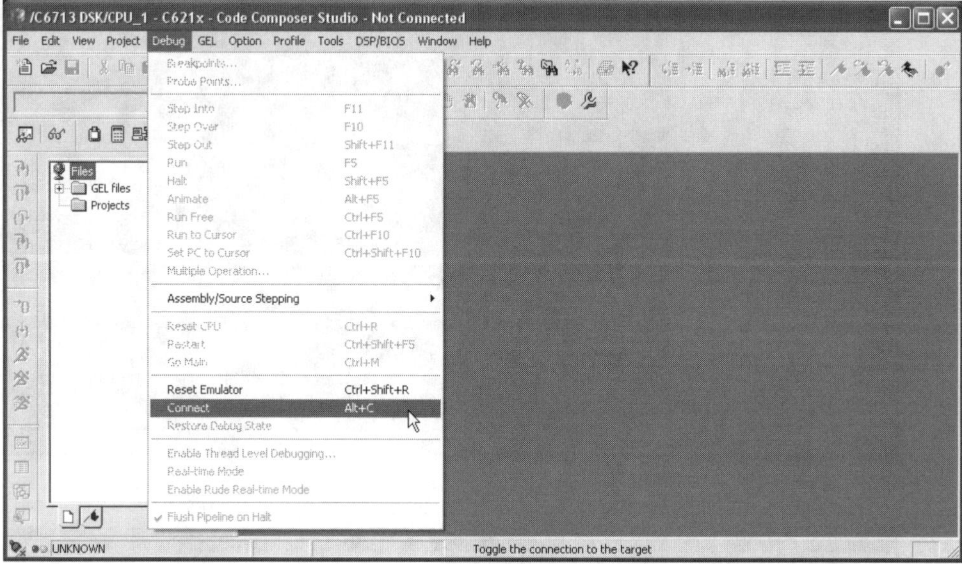

If your DSP is not connected to CCS, this fact will also be indicated in the title bar of the CCS window.

Once your DSP is connected to CCS, the lower left-hand corner of the CCS IDE screen will update as shown below. Notice that the "The target is now connected" bubble will only be present for a few seconds. Additionally, the title bar no longer indicates "Not Connected."

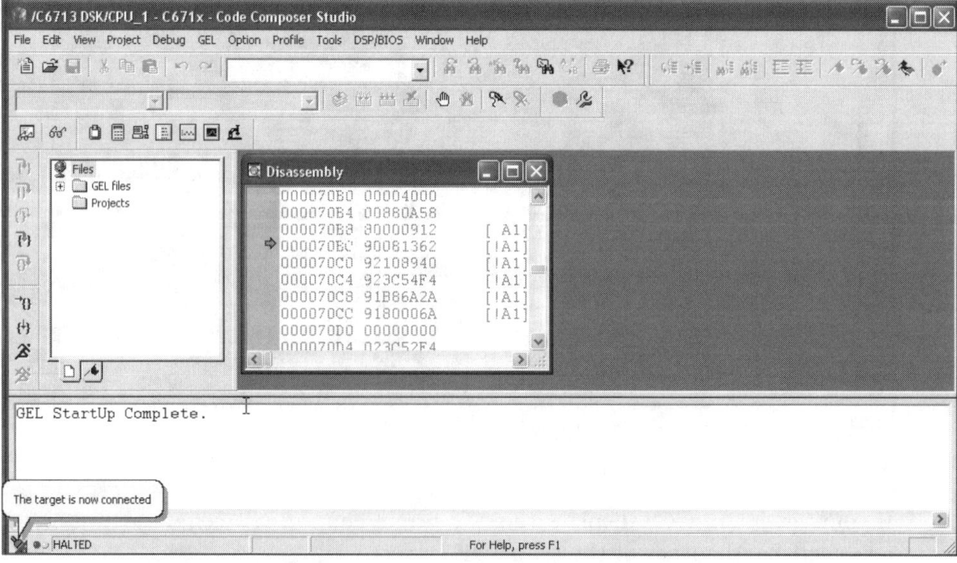

A.3 Needed Files

4. **The following sections of the tutorial assumes that you have installed the text's software that is included on the enclosed CD-ROM.**

5. Use a file management program such as Windows Explorer to verify that the directory `common_code` exists at `C:\CD\common_code`, where `C:\CD` may be some other name or disk location that you specified when you installed the text's software. For simplicity, we just assume `C:\CD` in our discussion here. If this directory structure does not exist, you either need to create it or install the text's software. There should be several files in the `common_code` directory. If these files are not already in the directory, they can be copied from the `common_code` directory of the text CD-ROM.

6. Some form of organization is required for your projects and the dozens of files you will eventually create. The directory structure that is created when you install the text's software is shown below.

A.4   Creating a New Project

7. During the CCS installation procedure, the `C:\CCStudio_3.1\MyProjects` directory was created. This appendix, however, assumes that you intend to use the text's software and its associated directory structure. In other words, we will not be using the `C:\CCStudio_3.1\MyProjects` directory.

8. We will now create a project and its associated directory by adding a new project. To create a new CCS project, click on the "Project" pull down menu as shown below.

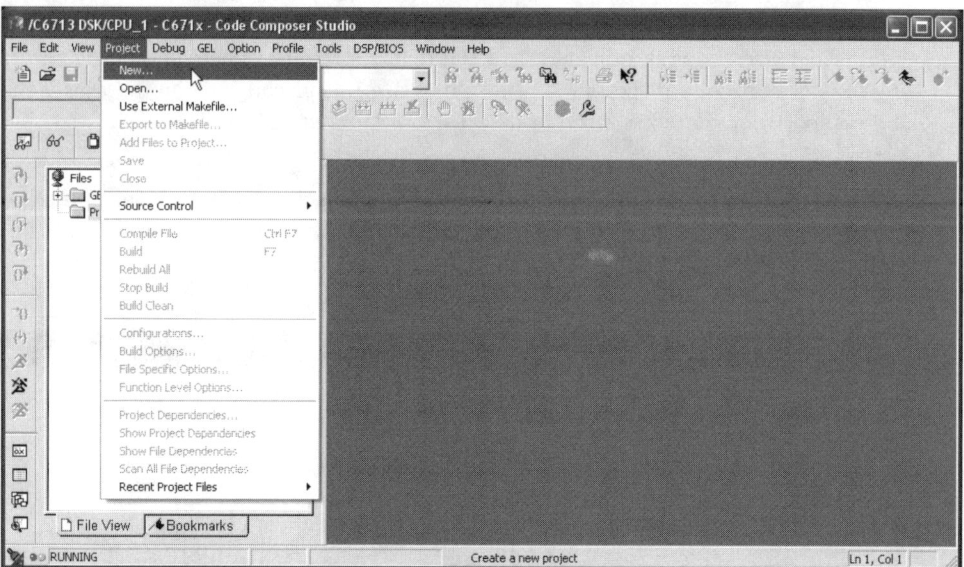

9. Clicking on "New..." brings up a dialog box similar to the one shown below.

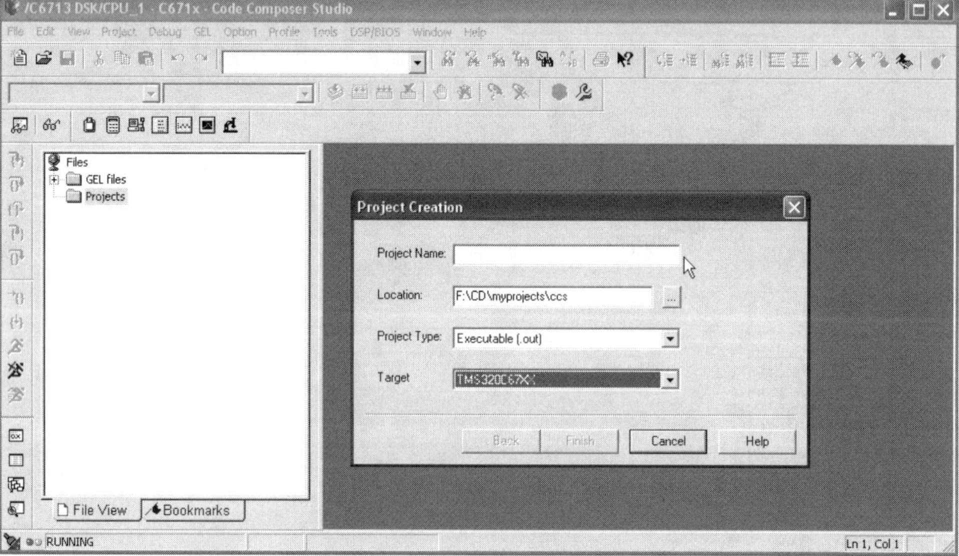

If the "Location:" box does not point to `C:\CD\myprojects\ccs`, use the ⬚ button at the right of the "Location" box and navigate to the correct directory. Also ensure that the TMS320C67xx target is selected.

10. Notice that CCS "looks" to the directory that it was last using. We almost always will create a new directory for each new project. In the "Project Name:" box, type in `myFirstProject`. Did you notice that as you typed in the project name, the "Location:" box was automatically updated with the new project's name? When you click on "Finish," this process will create a new directory called `myFirstProject`.

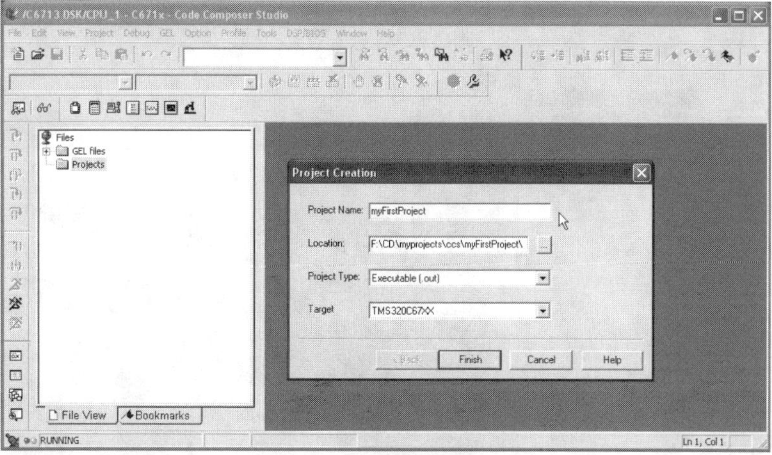

11. Once you "Refresh" the directory structure, you will see something similar to the figure shown below. The `myFirstProject` sub-directory has now been added to the `myprojects` directory and within this new directory, the file `myFirstProject.pjt` now exists. CCS version 2 (and later) use a `.pjt` extension for project files.

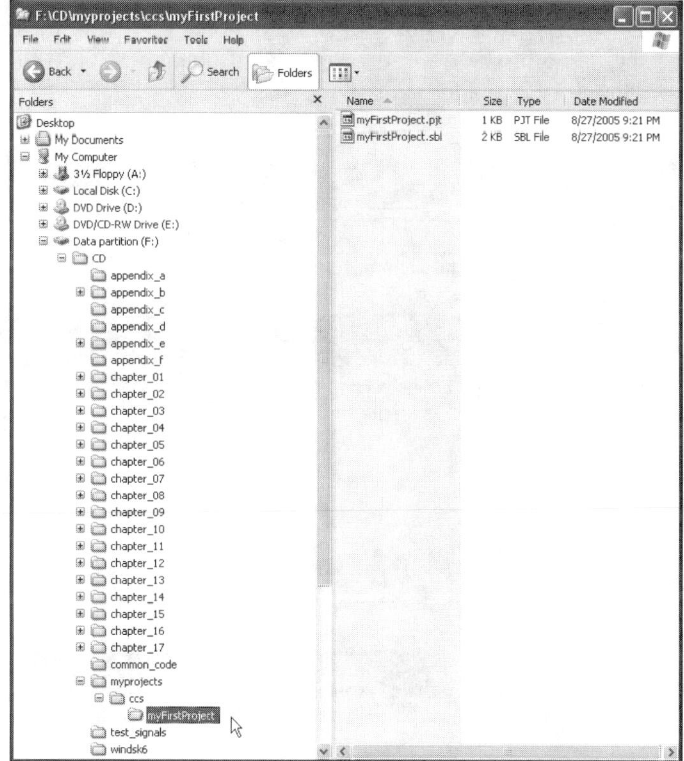

Your new project is now open in CCS.

A.5   Opening Existing Projects

12. In the future, when you need to open an existing project, click on "Project," then "Open...." The display should look similar to the one shown below.

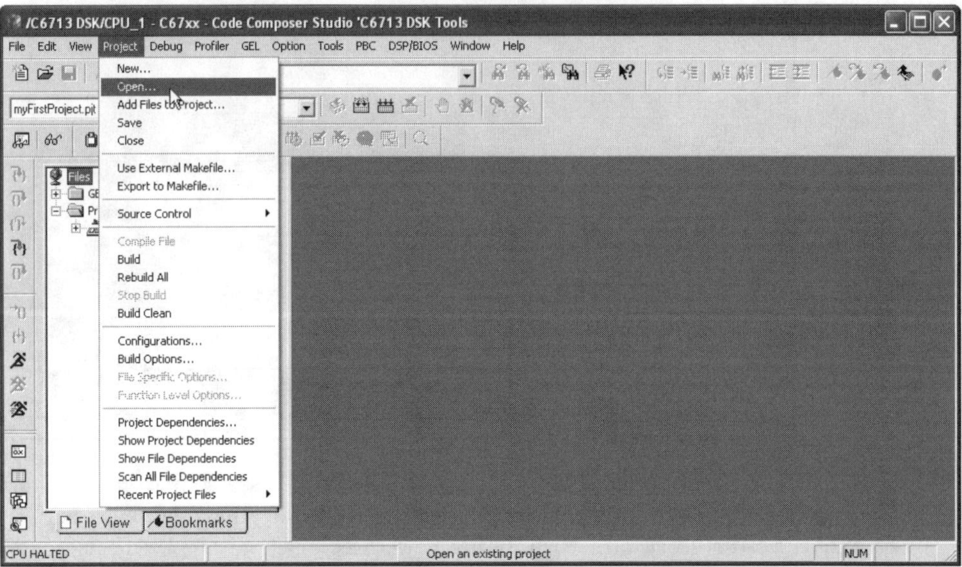

13. Find your project's directory or folder and open the "myFirstProject.pjt" file, or whatever the name of the desired project is. Remember, CCS always looks initially in the last directory or folder that it was using.

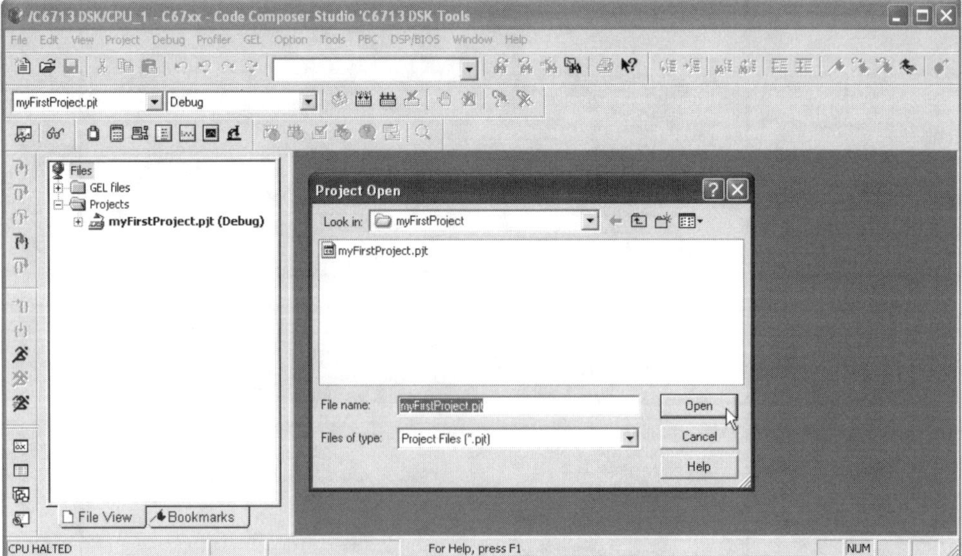

14. Your Code Composer window should now look similar to the figure shown next.

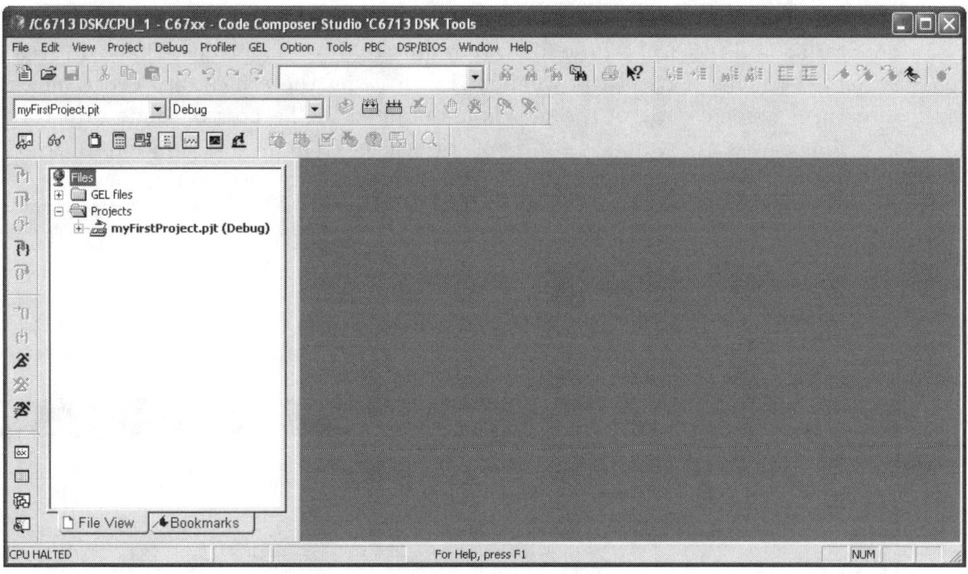

A.6   Adding Files to a Project

15. The directory structure on the left side of the CCS window now has a "+" next to the "myFirstProject.pjt" heading. Clicking on the "+" next to "myFirstProject.pjt" results in an expanded view of your project files. There are currently no files associated with this project.

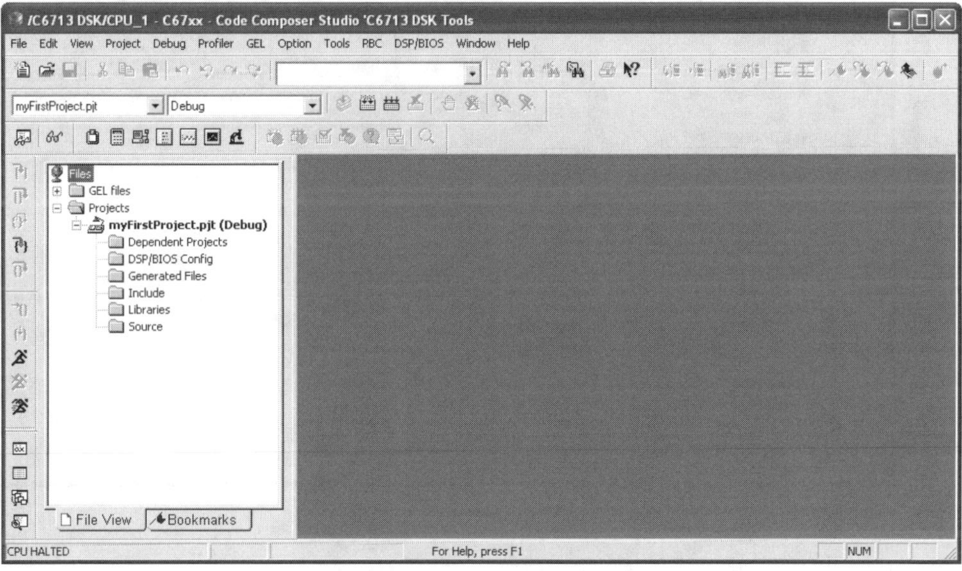

16. To add files to a project, the files must already exist somewhere on your computer. Adding a project doesn't move the location of a file; it just links the project to the file wherever it happens to be located. Most of your files that are specific to a particular project related to this text should be kept in the project directory under `C:\CD\myprojects\ccs`. Adding all of the required files to the project is a multi-step

process. There are four files, as shown in the figure below, that are contained in the **common_code** directory. These files are required for every project that you create that is related to this text. To prevent having to copy these files into every new project directory, a single copy of the files is maintained in the **common_code** directory. This technique saves work, makes your software easier to maintain, and minimizes the files in your own project's directory. It's also good software engineering practice.

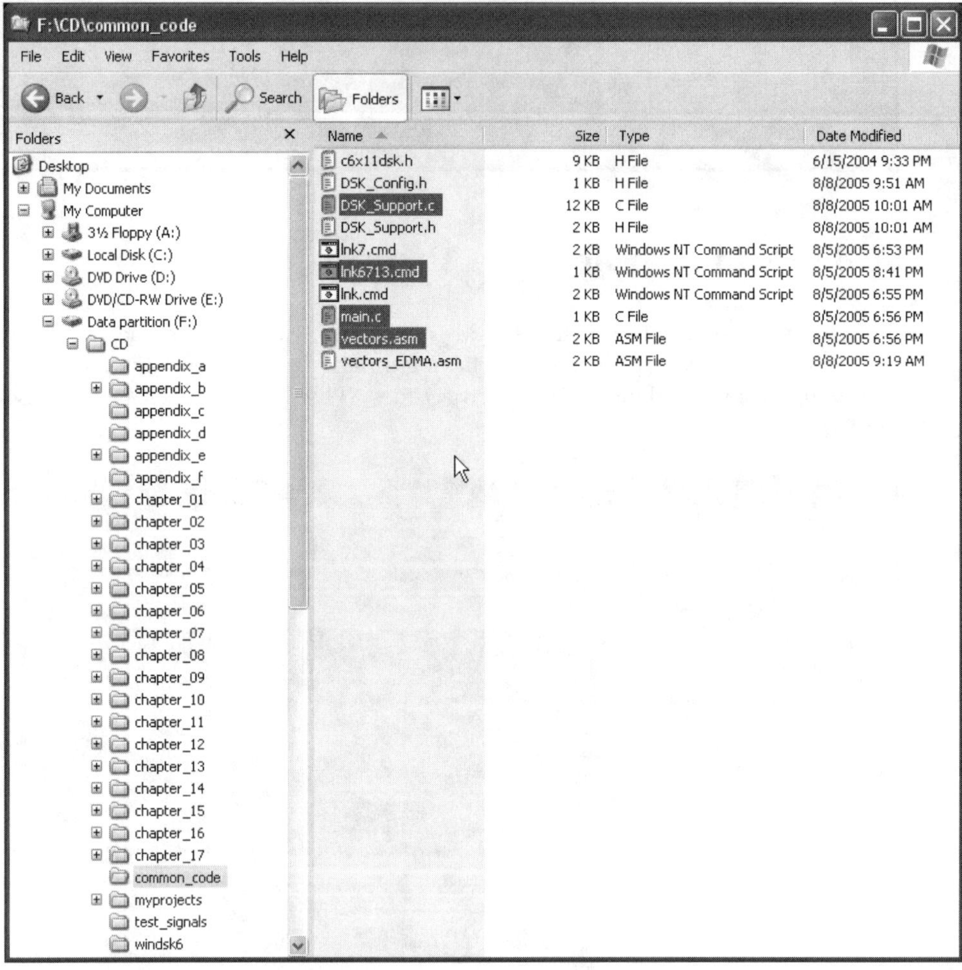

17. These files can now be added to the project by clicking on "Add Files to Project..." as shown in the next figure.

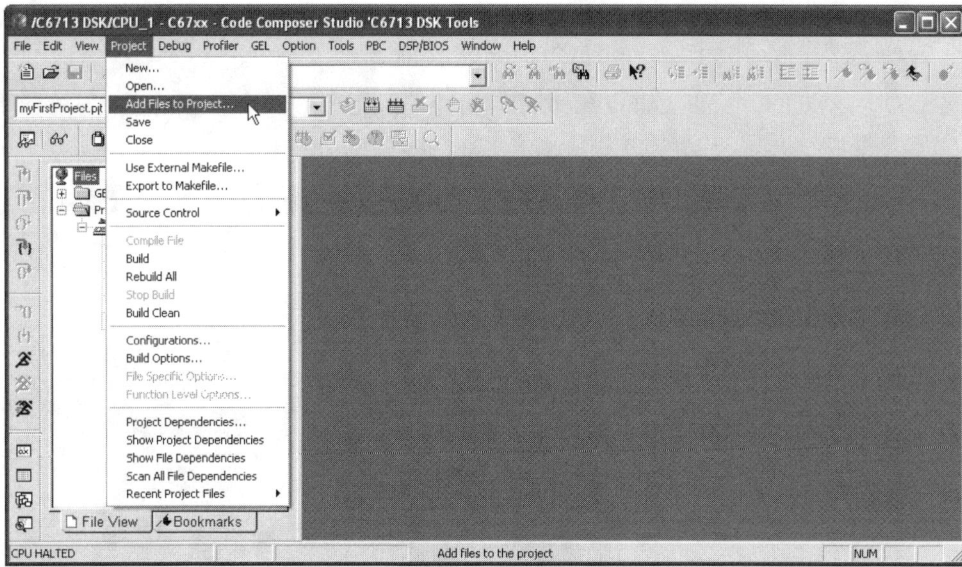

18. In a given directory, all of the files having the selected file extension for a given project are displayed in the dialog box. Alternatively, you can select "All Files (*.*)" in the "Files of type" box. Notice that the "All Files (*.*)" option was selected in the figure shown below.

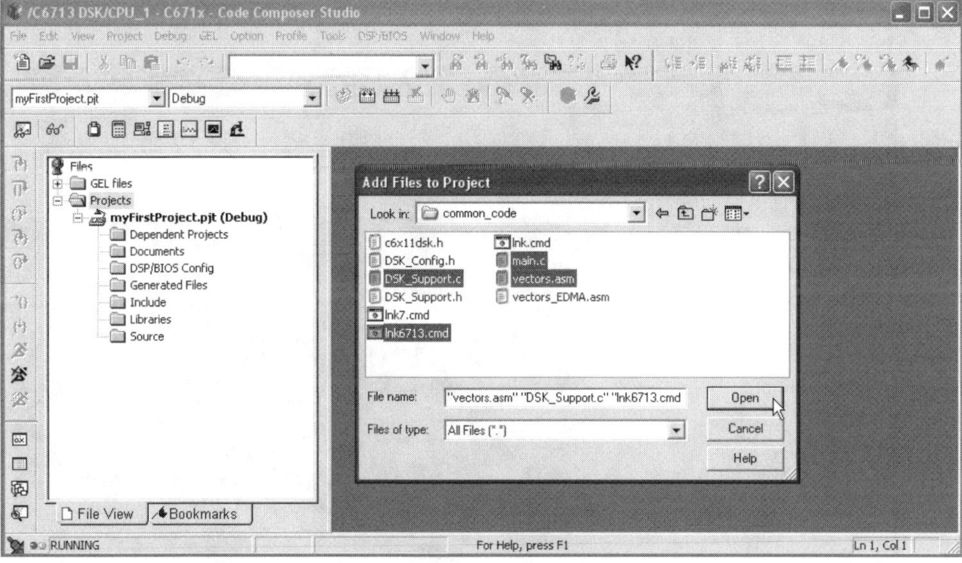

19. You can add the displayed files to the project by selecting each of the files and clicking on the "Open" button. The standard Windows shortcut keys ("Ctrl" to add multiple files or "Shift" to add all of the files between any two points) are also available.

20. Use a file management program such as Windows Explorer to copy the files `ISRs.c` and `Startup.c` to your newly created `myFirstProject` directory. The files can be copied from the text CD-ROM (look for `chapter_02\ccs\MyTalkThrough`) or from the equivalent directory is installed on your hard drive.

21. Use the techniques that you used to add the `common_code` files to the project to add these two files to the `myFirstProject` project.

22. Once you have added all of these files to the project you may "Scan All Dependencies" as shown in the figure below.

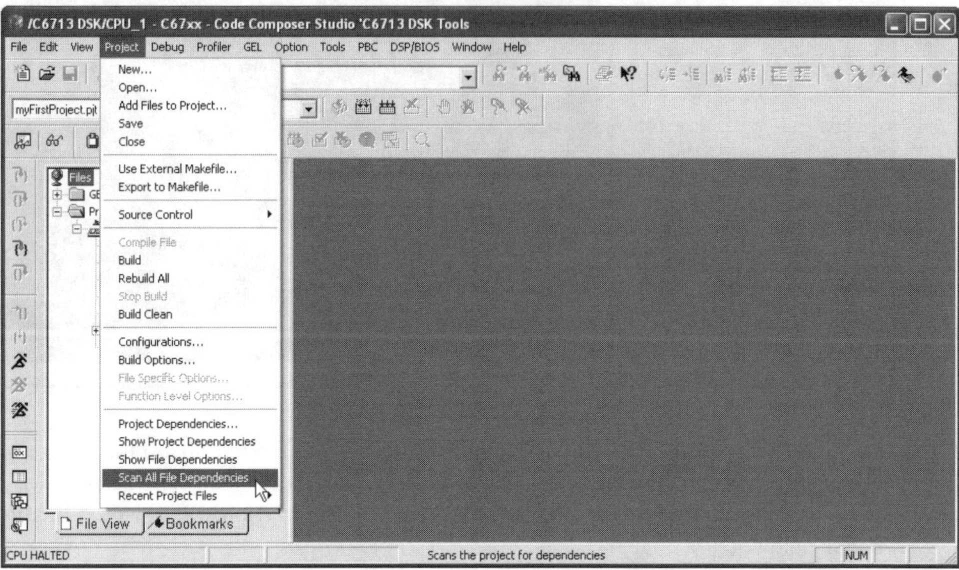

23. This scan will result in a display similar to the figure shown below.

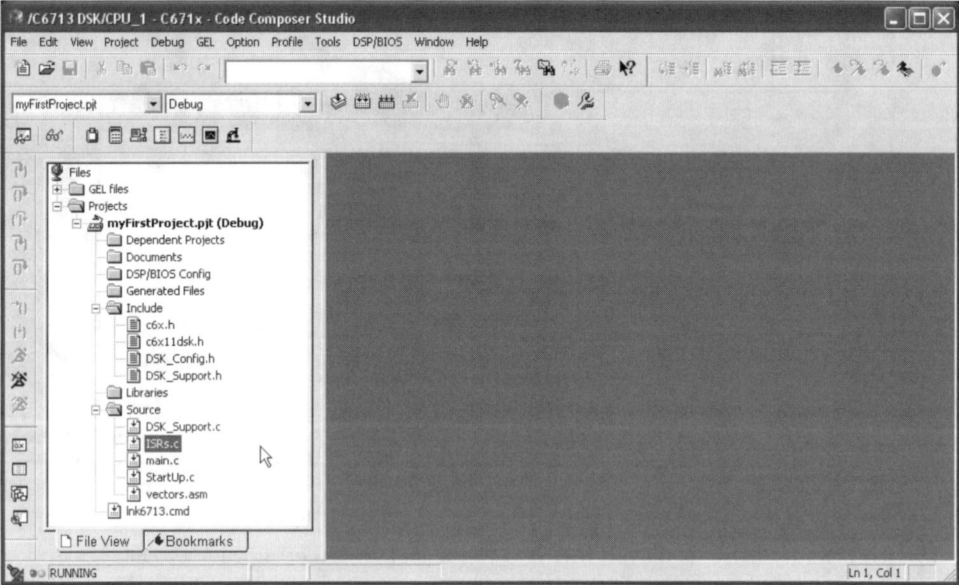

A.7   Project Options

24. A number of default options were automatically assigned to your new project. These options can be viewed by clicking on "Project," "Options...," as shown in the figure

below. For our first project, the default settings work well except for one setting. The default "Target Version" in the Basic Compiler settings assumes a 670x processor is being used. You should select the "C671x" target as shown in the figure below.

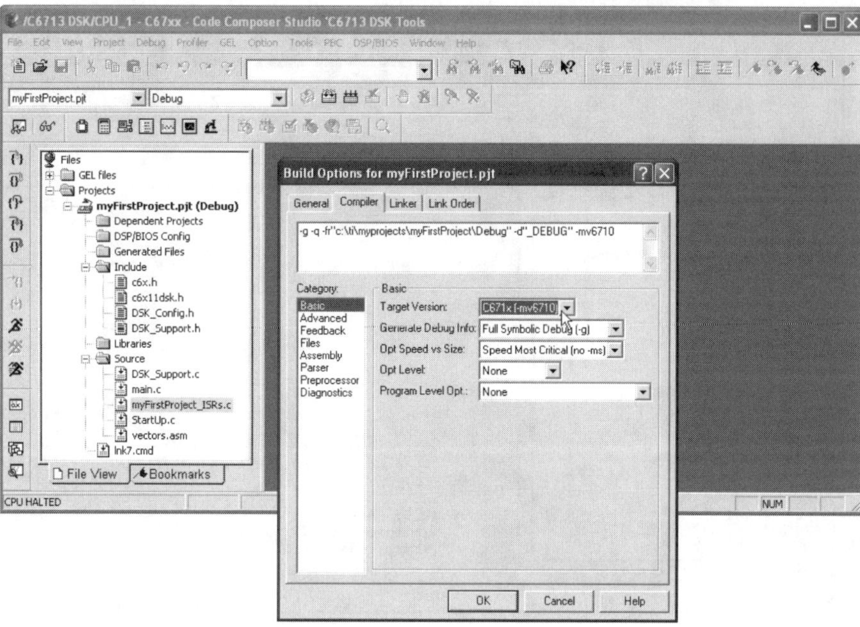

25. Verify that you have selected the correct `CodecType` in the `DSK_Config.h` file, which is located in the `common_code` directory. The available options are shown below.

```
// uncomment the line for the codec in use
// #define CodecType        TLC320AD535
// #define CodecType        eDSP_PCM3006
// #define CodecType        TI_PCM3003_16bit
   #define CodecType        DSK6713_16bit
```

The `DSK6713_16bit` (otherwise known as the TLV320AIC23) is the native codec built onto the C6713 DSK's printed circuit board. The `eDSP_PCM3006` is on the stereo-capable daughter card that attaches to the DSK at J3. The `TI_PCM3003_16bit` is TI's version of the `eDSP_PCM3006` board, and the `TLC320AD535` is the native monaural codec built onto the C6711 DSK's printed circuit board.

Comment out all the `CodecType` lines that do not apply to your current project; only one `CodecType` line should selected. While you can have multiple codecs attached to the DSK, the software that you are using will only allow you to access one codec at a time. This will limit your projects to either one or two channels (i.e., monaural or stereo).

A.8 Building the Project

26. The buttons near the top and along the left of the CCS window are shortcuts for a number of the more commonly used features of CCS. The "Rebuild All" process accomplishes compiling and linking and results in an output file that you can load

into the DSP target (DSK). Resting the pointer on a button will cause the program to display the button's intended purpose or function. Also notice that in this figure a file, in this case, `ISRs.c` can be viewed and edited by double clicking on the file name in the directory structure.

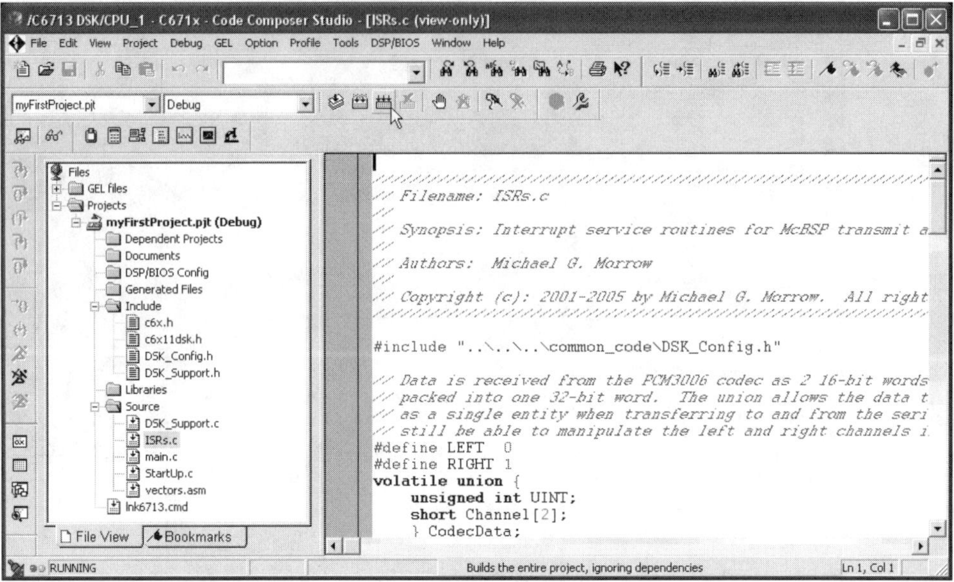

27. When you click on the "Rebuild All" button, an additional panel opens below the previously displayed CCS window; this new panel tells you the status of the compiling and linking process.

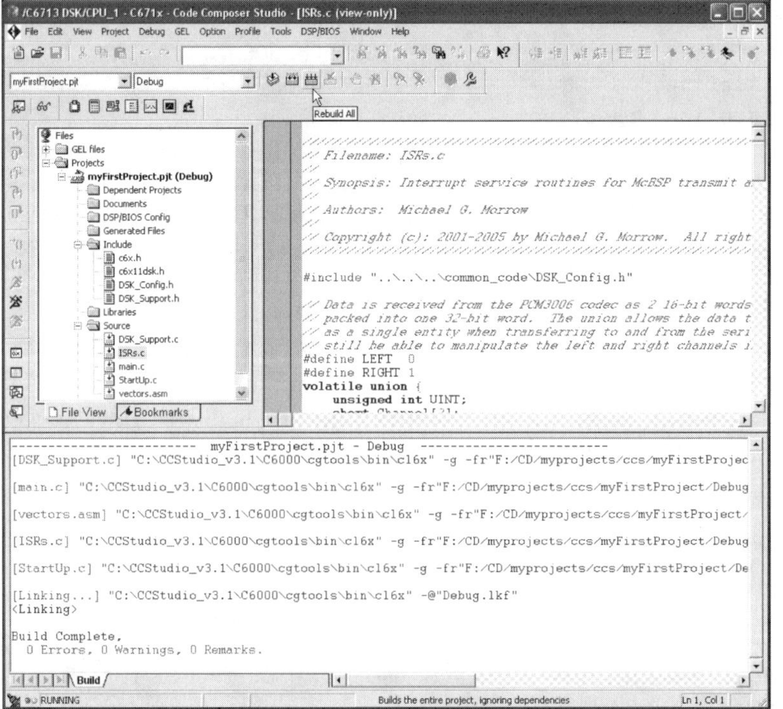

To the left of the "Rebuild All" button is a pull down window that is currently set to "Debug." The other option for this pull down menu is "Release." As their names imply, "Debug" and "Release" are the two versions of your code that the compiler can create. These two versions of the code are stored in either the "Debug" or "Release" directories, respectively. The panel, buttons, and pull down window can be seen in the figure shown above. Leave the setting on "Debug."

A.9 Loading the Program on the DSK

28. Once the build is complete, the output file must be loaded into the DSP. This is accomplished by clicking on "File," and "Load Program" as shown in the figure below.

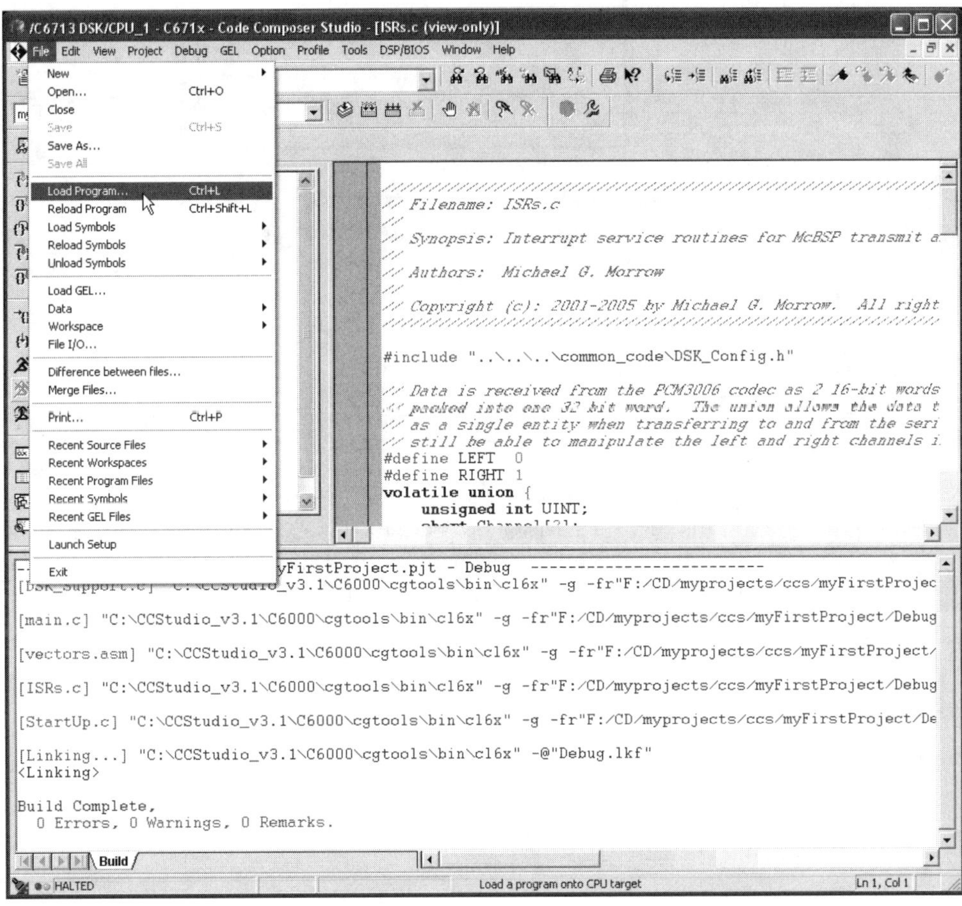

29. This results in the "Load Program" dialog box opening as shown in the figure below. You need to navigate to either the "Debug" or "Release" directory/folder depending on which version you selected prior to the build process.

Notice that CCS is by default looking in the last directory or folder from which it loaded a program to find this new program to load. This may *not* be the directory in which you are currently working! Not noticing this seemingly minor point can cause great frustration.

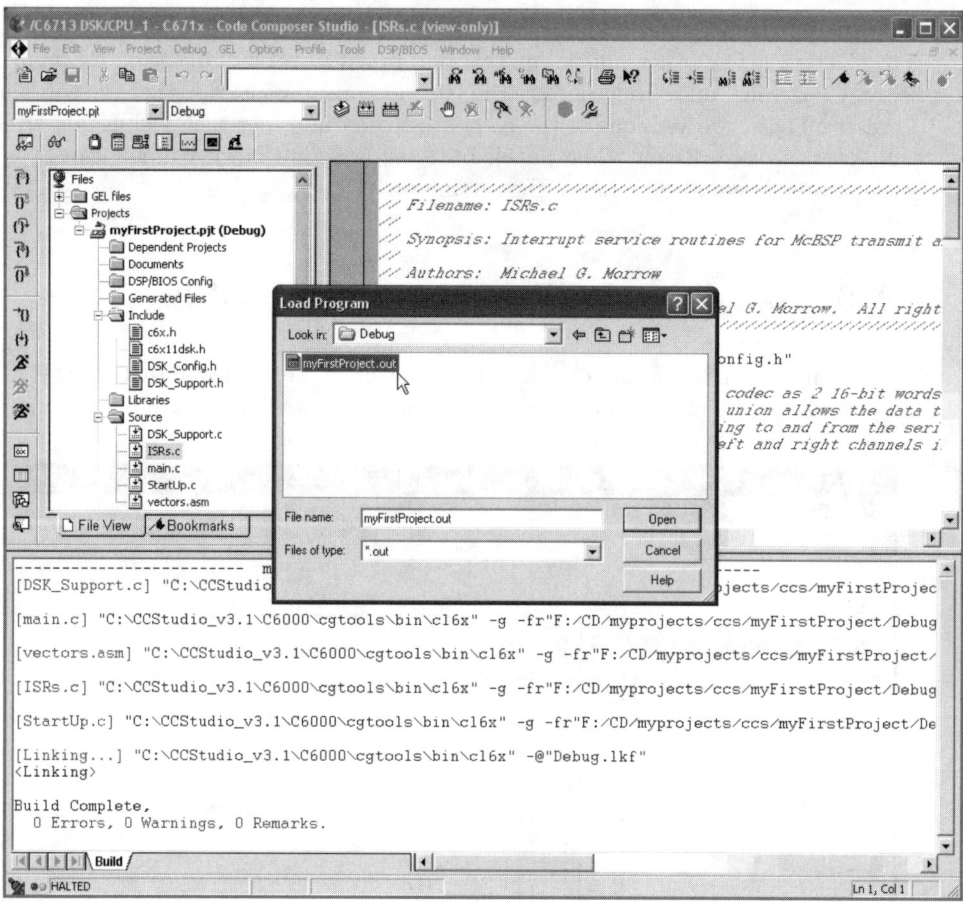

30. You will need to navigate through the directory/folder structure to find the desired
    output program file in your current project. In this example the output file is named
    `myFirstProject.out`. Selecting and opening this file will result in a program load
    via the attached USB cable. This process should only take a few seconds.

31. After the program is loaded onto the DSK, the CCS window should look similar to the figure shown below. The right panel is displaying a disassembled version of the program code.

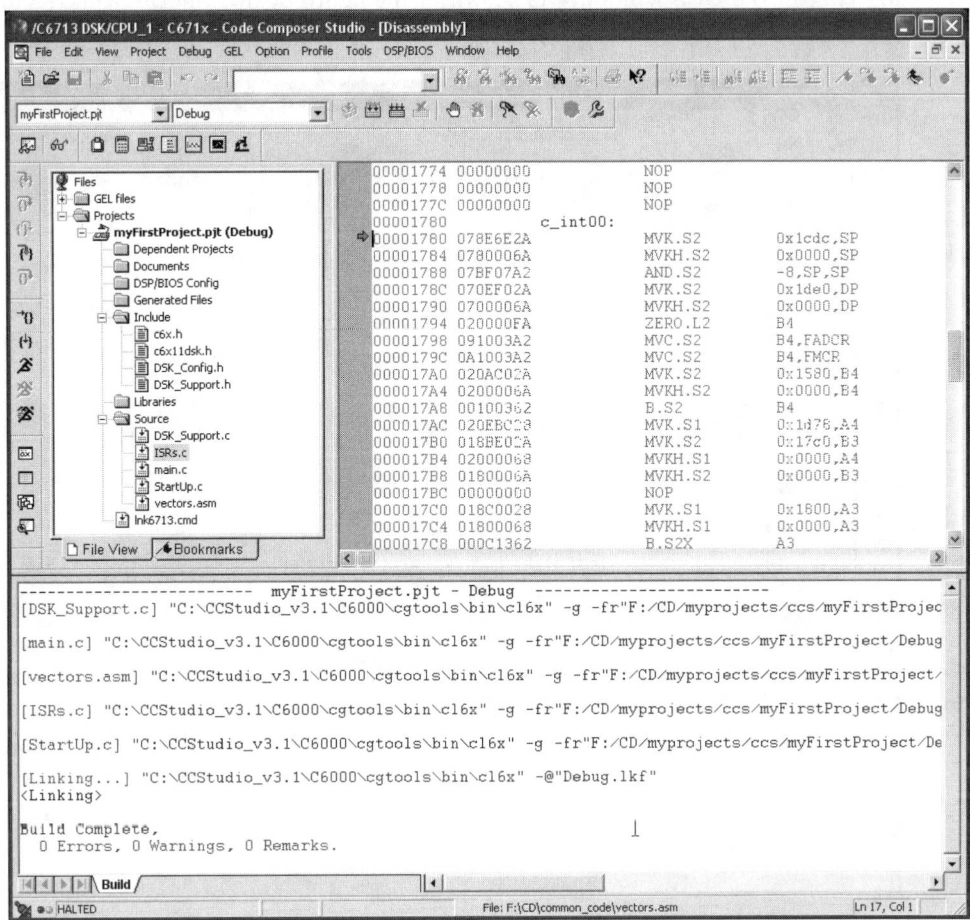

A.10   Running the Program on the DSK

32. To execute or run the program, click on the "Run Target" icon that is located on the left edge of the CCS window. As shown in the figure below, the run icon resembles a profile view of someone running.

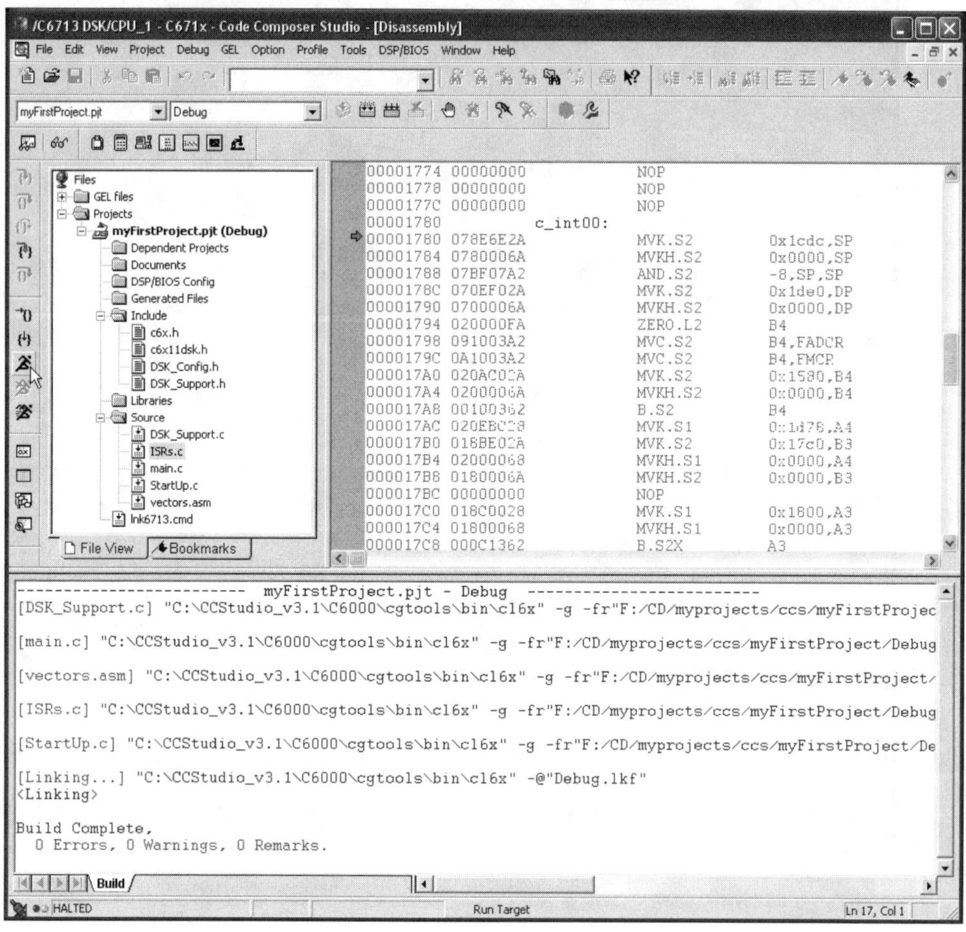

33. To stop or halt the program, click on the halt icon that is directly below the run icon.

34. You may modify and rebuild your code even while the DSK is running. Using older versions of CCS, reloading the program will automatically halt the DSK before the program load is initiated. Using CCS version 3.1, you must manually halt the DSP Target *before* you can reload the program. You must reload a program after *any* changes are made in the code if you wish to see the results of those changes!

35. When you are finished with CCS and the DSK, halt the DSK and close the project. This is accomplished by clicking on "Project" and "Close" as shown in the figure below.

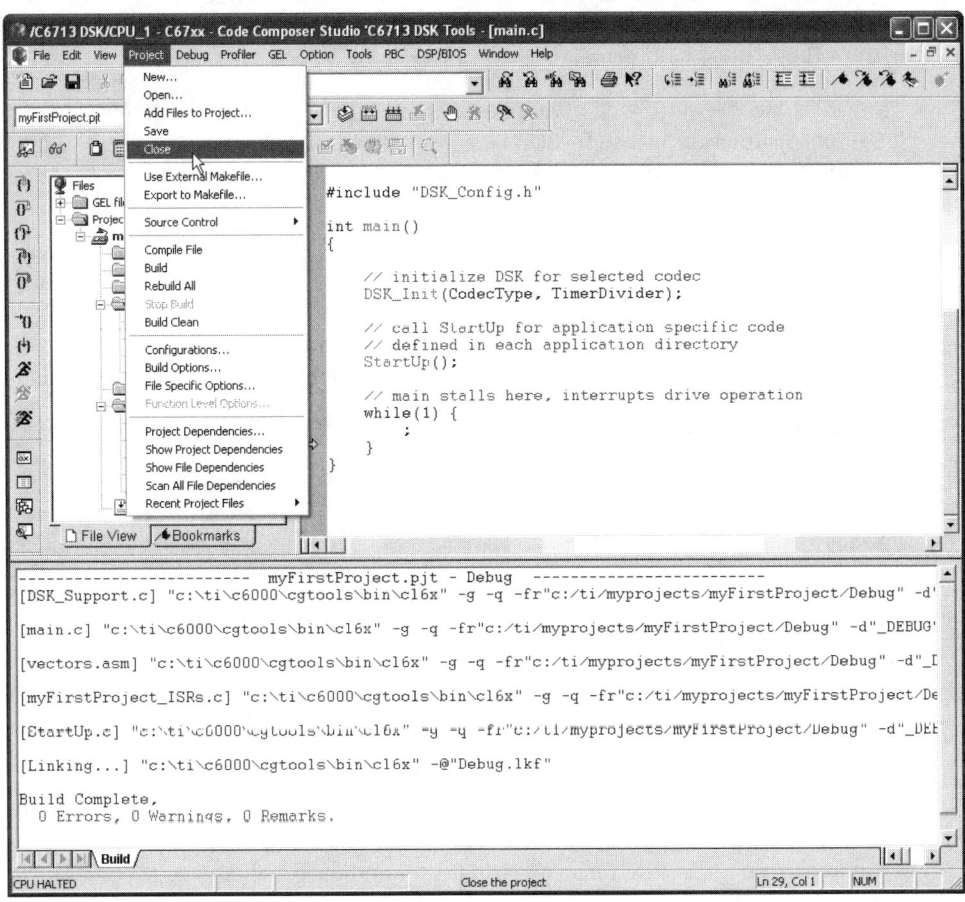

36. With the project closed, you may exit CCS and power down your DSK.

A.11 Get to Know CCS

As we mentioned at the start of this appendix, we *highly* recommend that you invest some time to become familiar with the CCS interface and its operation. It is the cornerstone of real-time DSP using the C671x DSKs.

A.12 Typical Files for Our CCS Projects

Throughout this text we have discussed examples of real-time code written mostly in C and presented as Code Composer Studio (CCS) projects. To make these examples easier for the reader to follow, and to demonstrate some basic principles of software engineering, we used a consistent and standardized format when we created all the CCS projects. The discussion below presents a typical collection of files that you will find in our CCS projects, and briefly explains their purpose.

We will assume that this example project, called `myProject`, has the following files associated with it. In the `common_code` directory: `main.c`, `DSK_Support.c`, `DSK_Config.h`, `DSK_Support.h`, `c6x11dsk.h`, `vectors.asm`, and `lnk7.cmd`. In the separate directory under `ccs` for the particular chapter: `myProject.pjt`, `StartUp.c`, and `ISRs.c`.

The files that are stored in the `common_code` directory are those that are used by multiple projects and don't need to be changed from project to project.

- `main.c` is the top level file that contains the initial starting code for the program. The main() function in the file `main.c` intentionally stalls in an endless "while" loop so that the program can respond as needed to real-time interrupts.

- `DSK_Support.c` contains a variety of functions used to configure and initialize the DSP peripheral devices. This code is quite involved and detail oriented; thankfully there will be no cause to change any of it.

- `DSK_Config.h` contains a series of `#define` statements. By uncommenting only the desired line, the correct DSP and codec initialization software are automatically run.

- `DSK_Support.h` contains the function prototypes for `DSK_Support.c`, and several enumerations for the supported codec types.

- `c6x11dsk.h` contains declarations for many of the peripheral registers found in the DSP.

- `vectors.asm` is used to fill the DSP's interrupt vector table. When the DSP receives an enabled interrupt, the DSP will jump to the location in the vector table corresponding to the interrupt number and begin execution. In this file, branch instructions have been placed at the interrupts that are configured for the serial port servicing the codec. This causes the functions McBSP_Rx_ISR() and McBSP_Tx_ISR() to be run in response to multi-channel buffered serial port (McBSP) interrupts.

- `lnk7.cmd` controls the operation of the linker. The linker takes all of the object files resulting from the *.c and *.asm files, and joins them together into a single executable program. To do this, the linker command file indicates where the physical memory is located, and how code sections are to be placed into the possible physical locations.

The files that are stored in individual CCS project directories under their associated chapters are specific to that project.

- `myProject.pjt` is a file used by Code Composer Studio to store all the information related to the project. This includes a list of the files that are part of the project and the many possible project options. These options include directories to search, optimization settings, file generation, and many more. In production software, the management of these details becomes nearly as involved as the software itself. In general, this file should not be edited directly.

- `StartUp.c` provides a place for code that must be executed for the project before the main() function intentionally stalls in its "while" loop. This could include coefficient or buffer initialization. The StartUp() function is called by the main() function in the file `main.c`. In many projects, this function is left empty.

- `ISRs.c` contains the interrupt service routines (ISRs) for receiving input and transmitting output (e.g., ISR functions McBSP_Rx_ISR() and McBSP_Tx_ISR()). The ISRs are executed in response to McBSP interrupts. This is the actual *real-time* code that must complete execution before the next interrupt arrives, or data will be lost. The vast majority of code that will be developed for a DSP algorithm is placed in these two functions.

Given this brief discussion, we hope the format of all our CCS projects will make some sense to you.

Appendix B

DSP/BIOS

B.1 Introduction

DSP/BIOS is a real-time operating system specific to the Texas Instruments DSPs [68, 69]. This appendix provides a short description of DSP/BIOS, and three projects intended to get you started in the DSP/BIOS environment.

B.1.1 DSP/BIOS Major Features

The major features of DSP/BIOS include:

- The scheduler and its associated thread classes provide a mechanism for arranging and controlling the software's execution. The scheduler is preemptive, meaning that it will periodically interrupt the currently executing thread, determine what the highest priority thread is that is ready to execute, and start that thread running. One of the DSP's hardware timers is used to implement this preemptive behavior. The available thread types will be discussed in more detail later in the chapter.

- Memory manager to control the operation of the memory/cache architecture and control allocation of memory resources.

- Instrumentation that provides deterministic, minimally-invasive analysis, profiling, and statistical functions.

- Communications resources including queues, pipes and streams, and a device driver mechanism.

- Support libraries providing standardization of access and hardware abstraction across multiple DSPs. These include the board support library (BSL) providing board level functional support and the chip support library (CSL) providing DSP device level support.

B.1.2 DSP/BIOS Threads

DSP/BIOS provides several classes of threads that can be scheduled for execution:

- Hardware Interrupt (HWI): Executed in response to a hardware interrupt, so should be very short and fast. Typically, these threads simply transfer data, and schedule a software interrupt for any further processing. The DSP/BIOS interrupt dispatcher can be used to permit ordinary C functions to serve as interrupt service routines.

- Software Interrupt (SWI): Software interrupts are typically posted (scheduled) by a HWI, and handle more involved interrupt processing while allowing hardware interrupts to be processed without delay. The posting mechanism has a mailbox variable that can be used to condition posting of ISR (i.e., can countdown to posting, or use bits as flags). SWIs are preempted by HWIs and higher priority SWIs.

- Periodic Function (PRD): Periodic functions are a class of SWI that are scheduled at regular intervals. DSP/BIOS automatically implements a hardware timer HWI and SWI for PRD scheduling. PRDs are preempted by HWIs and higher priority SWIs.

- Tasks (TSK): Tasks are functions that run to completion once scheduled, and are intended for longer duration, more complex processing that must be done when required. TSKs are preempted by HWIs, SWIs, PRDs, and higher priority TSKs.

- Idle Functions (IDL): Idle functions execute when DSP/BIOS has no other pending threads. They are most useful for true background tasks, like system maintenance and self-test. If multiple IDLs are present, they are run to completion in round-robin order.

B.2   Using the DSP/BIOS Scheduler

This section will give a hands-on demonstration of how to use the different DSP/BIOS thread classes to create a complete application. It uses the DSK's LEDs to signal which thread is currently running, and explores some of the DSP/BIOS analysis tools available in CCS. The complete source code is available in the **threads** directory of Appendix B.

In CCS, select File→New, and a *New Project* dialog box will appear. Select the desired directory, name the project *threads*, set the project type to *Executable (*.out)*, and set the target to be *TMS320C67XX*. Click Finish, and your project will appear in the left (project) window of CCS.

To use DSP/BIOS, we need to add a DSP/BIOS configuration file to our project. Create the configuration file by selecting File→New→DSP/BIOS Configuration..., which will start the DSP/BIOS configuration wizard. Select the **dsk6711.cdb** or **dsk6713.cdb** configuration template as appropriate, and click OK. The new DSP/BIOS configuration file will open in CCS. Save this file as **threads.cdb** in the project directory. For now, just leave the configuration file as it is.

Now, add the DSP/BIOS configuration file to the project. Right-click on **threads.pjt** in the project window (or select Project→Add files to project... from the menu), and add **threads.cdb**. We also need to add the linker command file that DSP/BIOS generates to the project; it is **threadscfg.cmd**.

Select Project→Scan all dependencies, and note that a number of include files now appear in the project tree. The DSP/BIOS configuration generated these references automatically.

Now, create a C source file for the main function. Other functions will be added to this file during the exercise. Select File→New→Source File, and then save the file as **main.c**. Add the following code to the file:

```c
#include <std.h>
#include "threadscfg.h"

unsigned int LED_Mask = 0;
#define LED1_BIT 1
#define LED2_BIT 2
#define LED3_BIT 4
```

```c
int main()
{
    return 0; // return to DSP/BIOS
}


// drive LEDs on the 6711/6713 DSKs
void UpdateLEDs(unsigned int state) {
        // comment out the line below if using a 6711 DSK
        *(volatile unsigned char *)(0x90080000) = state << 1;
        // comment out the line below if using a 6713 DSK
        *(volatile unsigned int *)(0x90080000) = (~state) << 24;
}
```

Comment out the appropriate line in the `UpdateLEDs` function. This function will handle the actual writes to the LED port. Add the `main.c` file to your project. At this point, the project could be built and run, although it wouldn't do very much.

B.2.1 Adding a PRD Function

Add a new function `PRD_function` (as shown below) to the file `main.c`. This code will toggle LED#1 on the DSK every time it is run.

```c
void PRD_function()
{
    LED_Mask ^= LED1_BIT; // toggle LED #1
    UpdateLEDs(LED_Mask);
}
```

Now, the function needs to be hooked to the PRD scheduling mechanism. To do this, open the `threads.cdb` configuration file, and expand the *Scheduling* entry. Under Scheduling, select *PRD Periodic Function Manager*, and note that there is nothing associated with it yet. Right-click *PRD Periodic Function Manager*, and select *Insert PRD*. This creates the entry `PRD0`. To configure it, right-click `PRD0`, and select *Properties*. Set the *period* to 500 ticks (the default tick setting is 1ms), the *function* to `_PRD_function` (note the leading underscore), *continuous mode*, and accept the remaining defaults.

Now, build the program, download to the DSK, and run. The periodic function should be scheduled to run every 500ms, with the result being LED #1 flashing at a 1 Hz rate.

B.2.2 Adding a HWI Function

To add an HWI function, you again need to create the function itself and then configure DSP/BIOS to schedule it appropriately. To use a hardware interrupt, there are two basic choices: to use the DSP/BIOS dispatcher or not. There is some overhead associated with the dispatcher, but then any standard C function can be used as the interrupt service routine. If the dispatcher is not used, the DSP/BIOS entry and exit macros must be added to the function.

The dispatcher will be used in this example. Create a new function in `main.c` named `HWI_function` that will toggle LED #2, as shown below.

```c
void HWI_function()
{
    LED_Mask ^= LED2_BIT; // toggle LED #2
```

```
      UpdateLEDs(LED_Mask);
5 }
```

Now, the function must be linked to the desired interrupt source and priority, the interrupt source configured, and interrupts enabled. Open **threads.cdb**, and expand Scheduling→HWI-Hardware Interrupt Service Routine Manager. Note the various interrupt priorities, including the two reserved ones (these are actually used as part of RTDX). Now, select **INT4** (the highest priority available) and get its *Properties* page. Set **Timer_1** as the interrupt source, make the function name **_HWI_function**, and select *Use Dispatcher*. The remaining defaults are fine. This sets up the interrupt structure such that INT4 will triggered by TIMER1, and the dispatcher will be used to invoke our HWI function.

Now, to configure Timer1, expand *CSL→TIMER*. This shows two sub-categories, *Timer Configuration Manager* and *Timer Resource Manager*. To use this, first set up a timer configuration, and then assign it to be used for Timer1. Select *Timer Configuration Manager→Insert timerCfg*, then edit the properties of **timerCfg0** to a *Period Value* of **3,750,000**, set *Timer Operation* to *Start with reset*, and set *Input Clock Source* to *(CPU clock)/4* clock source. The remaining defaults are fine. Then, to tell DSP/BIOS to use this to configure Timer1, select *Timer Resource Manager→Timer_Device1*, check *Open Timer Device* and *Enable Pre-Initialization*, and pre-initialize with **timerCfg0**. This will initialize Timer1 to generate interrupts at a 10 Hz rate, which will cause the HWI to be invoked at a 10 Hz rate.

The only remaining item is to enable the Timer1 interrupt using a DSP/BIOS function that is designed to do just that as part of a larger set of APIs for manipulating interrupts. Add the code below to the **main** function before the **return** statement.

```
1 C62_enableIER(0x0010); // enable INT4
```

Build and run the program. LED #1 should be flashing at a 1Hz rate due to the PRD function, and LED #2 should be flashing at a 5Hz rate due to the HWI function.

B.2.3 Adding a SWI Function

To add a SWI, follow the same sort of steps as adding a PRD or HWI. First, create a new function in **main.c** named **SWI_function** that toggles LED #3, as shown below.

```
1 void SWI_function() {
      LED_Mask ^= LED3_BIT; // toggle LED #3
3     UpdateLEDs(LED_Mask);
  }
```

Expand *Scheduling→SWI - Software Interrupt Manager* in the DSP/BIOS configuration file. Right-click *SWI - Software Interrupt Manager* and select *Insert SWI*. This will create a new entry SWI0. Edit SWI0's properties to set *function* to *_SWI_function*, *priority* to 2, and accept the remaining defaults. This establishes a new SWI that can be scheduled.

To get DSP/BIOS to actually schedule SWI0, we will post the SWI from within our HWI. We'll use the function **SWI_post**, which takes a pointer to a **SWI_Obj**. The **SWI_Obj** is actually created by DSP/BIOS in the file **threadscfg.s62**, an assembly language file. To reference it in our code, it must be declared as an **extern**, which is conveniently done by DSP/BIOS in **threadscfg.h**. Locating both of these references is suggested.

To actually post the SWI, add the following code to your HWI.

```
  SWI_post(&SWI0); // post SWI
```

Build and run the code. Each time the HWI is run, it posts our SWI and the SWI runs, so LED #3 flashes at the same rate as LED #2.

Typically though, we might want to post the SWI every **n** invocations of the HWI. We could add code to do this in the HWI, but DSP/BIOS provides a number of handy mechanisms to control posting based on the mailbox associated with the SWI. Edit the SWI0 configuration to give it a *mailbox* value of 8, then use the `SWI_dec` function instead of `SWI_post`, as shown below.

```
SWI_dec(&SWI0); // post SWI
```

`SWI_dec` decrements the value in the mailbox each time it is called, and only posts the SWI when the mailbox goes to 0, whereupon it conveniently reloads the mailbox with the value 8. Build and run the code again. Note that LED #3 now flashes at a slower rate than LED #2 since the SWI is being run only every eighth HWI invocation.

B.2.4 Adding a TSK Function

Once invoked, tasks run to completion, although they will be preempted by HWIs, SWIs, PRDs, and higher priority tasks. They can also yield, which will allow the scheduler to run another task. Once a task has run to completion, it will not be run again until it is explicitly scheduled again.

Create a new function in `main.c` named `TSK_function` that turns off all LEDs, as shown below.

```
void TSK_function() {
    int i;

    for(i = 0;i < 40000000;i++) {
        LED_Mask = LED1_BIT | LED2_BIT | LED3_BIT; // all LEDs on
        UpdateLEDs(LED_Mask);
    }
}
```

Expand *Scheduling→TSK - Task Manager* in the DSP/BIOS configuration file. Right-click *TSK - Task Manager* and select *Insert TSK*. This will create a new entry TSK0. Edit TSK0's properties to set *Task function* to `_TSK_function` and accept the remaining defaults. This establishes a new TSK that will be scheduled to run on start-up. In this case, the task will continually turn on all of the LEDs for a period of several seconds. During this time, the PRD, HWI and SWI will be preempting the task and changing the LED state, but as soon as those functions return, the task will be running again and set the LEDs back on, so they will appear to remain on.

Build and run the code. Note that all LEDs are in fact on for several seconds, then begin flashing as before. This use of a task to run once at the beginning of the program execution is a good example of how start-up code could be run under DSP/BIOS. Tasks can be used for much more sophisticated functionality as well.

B.2.5 Real-Time Analysis Tools

From the CCS menu, select *DSP/BIOS→CPU Load Graph*. This allows to us to see what the CPU loading is (it actually measures idle time, which is indicative of the degree of CPU loading). If the program from above is still running, the load should be extremely small, since the program is idle most of the time. If the program is reloaded and started again, note that the CPU load is initially almost 100%, then decreases dramatically when the task finishes.

While this is useful information, it doesn't show how the thread scheduling is actually working out. For that, we can use *DSP/BIOS→Execution Graph*. This shows when threads are being executed. [Note that the time scale is not linear; i.e., look at the spacing of the PRD ticks.] To make this display easier to see, halt the program and make the following changes to the the DSP/BIOS configuration file:

- Set PRD0 to a 2 tick interval.

- Change Timer1 to 1KHz interrupt rate by setting the Period Value to 37500.

- Under *Instrumentation→LOG - Event Log manager→LOG system*, set the `buflen` parameter to 512. This will give a longer continuous event display.

Rebuild the program, load, and run. Now you should see the different threads being posted and executed, although the HWI is not displayed since it is not scheduled by the DSP/BIOS scheduler.

These and many other tools are available in CCS; you will want to spend some more time familiarizing yourself with CCS and DSP/BIOS. As always, the tools only increase our vision of the system. The important part is the interpretation of what we are seeing.

B.3 A DSP/BIOS Talk-Through Application for the 6711 DSK

This application will implement audio talk-through using the C6711 DSK's onboard codec. The larger sections of code required to complete this section are available for cut-and-paste from the file `6711\B_3_6711_text_to_paste.txt` under Appendix B.

To begin, create a new CCS project in the directory of your choice. In CCS, select File→New, and a *New Project* dialog box will appear. Select the desired directory, name the project *talkthru*, set the project type to *Executable (*.out)*, and set the target to be *TMS320C67XX*. Click Finish, and the project will appear in the left (project) window of CCS.

Then, since this will be a DSP/BIOS-based application, we need to create a DSP/BIOS configuration file. To do this, select File→New→DSP/BIOS Configuration... from the menu and choose `dsk6711.cdb` on the window that appears. This will create a DSP/BIOS configuration file that is tailored to the 6711 DSK. Save the created file as `talkthru.cdb` in the project directory. We will come back to it and make modifications later. For now, we will just add it to our CCS project. Select Project→Add Files to Project... from the menu, locate the `talkthru.cdb` file, and select Open. Note that the `talkthru.cdb` file is now part of the project, and that two additional files were also added under the *Generated Files* section. These files are regenerated every time you make a change to the DSP/BIOS config-uration, so you should not edit them. The DSP/BIOS configuration also generates a linker command file, but it is not added automatically. Select Project→Add Files to Project... from the menu, locate the `talkthrucfg.cmd` file, and select Open. Now, we need to create our source code files and add them to the project. In this case, we will create two source files, one for the main function and another for the interrupt handlers.

B.3.1 main.c

Select File→New→Source File from the menu, then save the file as `main.c` and add it to the project. Then, add the following code to the file.

Listing B.1: main.c

```
#include "talkthrucfg.h"
2
void main()
4 {
    return;
6 }
```

For now, leave main as an empty function. The header file `talkthrucfg.h` is generated automatically from the DSP/BIOS configuration file. Remember that DSP/BIOS programs require that the main function return so that the DSP/BIOS scheduler can take control of program execution; this admittedly seems quite odd at first to experienced C programmers. In this program, we will need to add code to main to start up the serial port and enable interrupts, but we will come back to that later.

B.3.2 hwi.c

Create another new source code file, save it as `hwi.c`, and add it to the project. In this file, we will add the functions that are used to service the serial port interrupts. We will be using the DSP/BIOS interrupt dispatcher, which will require us to use standard functions as opposed to the functions declared with the **interrupt** keyword. Add the following code to the file and save it.

Listing B.2: hwi.c

```
#include "talkthrucfg.h"
2
volatile short audio_data = 0; // global audio data variable
4
void HWI_McBSP_Tx(void)
6 {
    return;
8 }

10 void HWI_McBSP_Rx(void)
{
12    return;
}
```

The convention of prefixing functions with the scheduler class that they will use is a good idea to help prevent errors. We will need to add code to these functions to read and write the serial port data, but first, we will go back to the DSP/BIOS configuration file and make some changes.

B.3.3 DSP/BIOS HWI Configuration

Open the configuration file (`talkthru.cdb`) in the CCS editor. The first thing we will do is configure DSP/BIOS to use the functions in `hwi.c` to handle the serial port interrupts. Expand the Scheduling→HWI - Hardware Interrupt Service Routine Manager tree. We could use any available hardware interrupt, but for consistency with previous work we will use `INT11` for the transmit interrupt and `INT12` for the receive interrupt. Right-click on the entry `HWI_INT11`, and select *Properties*. For *interrupt source:*, select `MCSP_0_Transmit`. For *function:*, enter `_HWI_McBSP_Tx`. (Note that a leading underscore must be appended to the function name.) On the *Dispatcher* tab, check the *Use Dispatcher* box, and click OK. What we have just done is instructed DSP/BIOS to use its interrupt dispatcher to respond to the McBSP0 transmit interrupt which it will map to INT 11, and to call our function HWI_McBSP_Tx to perform the interrupt processing. Similarly, right-click on HWI_INT12, and select *Properties*. Configure as before, except the interrupt source will be MCSP_0_Receive and the function will be _HWI_McBSP_Rx.

B.3.4   DSP/BIOS Serial Port Configuration

The other issue to be completed in the configuration file is the serial port initialization. In the configuration tool, expand the CSL - Chip Support Library→MCBSP - Multichannel Buffered Serial Port tree. First, we will create a McBSP configuration by right-clicking *MCBSP Configuration Manager* and selecting *Insert mcbspCfg*. This will create a new entry under the *MCBSP Configuration Manager* named mcbspCfg0. Right click mcbspCfg0, and select *Properties*. The various tabs permit you to set the configuration of the many parameters and operational modes of the McBSP. For now, we will use the *Advanced* tab to allows us to directly enter the register values. Set RCR and XCR to 0x00010040, PCR, MCR, RCER and XCER to 0x00000000, and SPCR to 0x00002000. Click OK to save your updates. Now, we will instruct DSP/BIOS to use this configuration and to automatically apply it to McBSP0. Under *MCBSP Resource Manager*, right-click Mcbsp_Port0 and select *Properties*. Check the *Open MCBSP Port* box and the *Enable Pre-Initialization* box. Select the configuration you created (mcbspCfg0) and click OK. Save the changes you made to the configuration file.

B.3.5   Source Code

Now, we need to add some code to our source files. Open main.c in the CCS editor. Add the following code to the main function.

Listing B.3: main() function code.

```
1  MCBSP_enableXmt(hMcbsp0);
   MCBSP_enableRcv(hMcbsp0);

3
   C62_enableIER(IRQ_MASK_11 | IRQ_MASK_12);
```

The first two lines use the Chip-Support Library functions to enable the McBSP0 transmitter and receiver. Note that the argument is the handle name that was specified in the DSP/BIOS configuration file. The next line enables the two hardware interrupts we selected in the HWI configuration. You must always explicitly enable hardware interrupts. That's all for the main function, so save the changes.

Open hwi.c in the CCS editor. Add the following code to the HWI_McBSP_Rx function. This simply reads the serial port data register and stores it in the global variable data. We do not have to test to see if data is available since this is an interrupt service routine that is only invoked when data is available.

Listing B.4: HWI_McBSP_Rx function code.

```
audio_data = MCBSP_read(hMcbsp0);
```

Add the following code to the HWI_McBSP_Tx function. This gets the value from the global variable data and writes it to the serial port transmitter. The AND operation with 0xFFFE is necessary to ensure that the least significant bit is always 0, to avoid reprogramming the codec. (The onboard codec of the C6711 uses the least significant bit to signal that a programming word will be transmitted next; we don't want that to happen unintentionally.)

Listing B.5: HWI_McBSP_Tx function code.

```
1    MCBSP_write(hMcbsp0, audio_data & 0xFFFE);
```

Now, build the project, download to your DSK and run. An audio signal on the codec input will now be transferred to the codec output. This project can serve as a building block for future work by adding code to process the data in different ways.

B.4 Modification Needed for the Stereo Codecs

The program in the preceding section used the onboard codec of the C6711 DSK. This codec is quite limited in capability, so there are a number of codec daughtercards available to provide enhanced capability. This section shows how to modify the program in the previous section to work with the Texas Instruments Analog Daughtercard and the Educational DSP PCM3006 daughtercard. The primary changes required when modifying the audio talk-through application to run with the the various stereo codecs are differences in the serial port usage and configuration, and reading/writing the data from serial port. These two daughtercards use McBSP1, and the port is configured to transfer data in 32-bit words containing both the left and right channel information. The required changes are summarized below. The larger sections of code required to complete this section are available for cut-and-paste from the file 6711\B_4_6711_text_to_paste.txt under Appendix B.

B.4.1 main.c

The code in the `main()` function must be replaced with that shown below.

Listing B.6: main () function code.

```
1  // comment out these two lines if using the TI daughtercard
   MCBSP_enableSrgr(hMcbsp1);
3  MCBSP_enableFsync(hMcbsp1);

5  // these two lines are required for both codecs
   MCBSP_enableXmt(hMcbsp1);
7  MCBSP_enableRcv(hMcbsp1);
   C62_enableIER(IRQ_MASK_11 | IRQ_MASK_12);
```

These first two lines are only required with the Educational DSP daughtercard; comment them out if using the TI daughtercard. The first line of code starts the McBSP's sample rate generator, then the next line starts the frame sync generator. Note that all of the arguments now reference McBSP1 instead of McBSP0.

B.4.2 hwi.c

The global variable declaration must be changed to reflect that the serial port data will be two 16-bit values packed into a single 32-bit word. The use of the union as shown below permits the data to be treated as a single 32-bit value (as in this case), but permits easy modification to the 16-bit data as well. The code in the serial port functions is also updated to reflect the data type.

Listing B.7: hwi function code.

```
   #include "talkthrucfg.h"
2
```

```c
#define LEFT   0
#define RIGHT  1
volatile union {
    unsigned int uint;
    short channel[2];
} PcmData = {(unsigned int)0}; // global audio data

void HWI_McBSP_Rx(void)
{
    PcmData.uint = MCBSP_read(hMcbsp1);

    // to modify individual channel data
    // use PcmData.channel[LEFT] or
    // use PcmData.channel[RIGHT], i.e
    // PcmData.channel[LEFT] *= Gain;
}

void HWI_McBSP_Tx(void)
{
    // to modify individual channel data
    // use PcmData.channel[LEFT] or
    // use PcmData.channel[RIGHT], i.e
    // PcmData.channel[LEFT] *= Gain;

    MCBSP_write(hMcbsp1, PcmData.uint);
}
```

B.4.3   DSP/BIOS HWI Configuration

Both the transmit and receive interrupt configurations under the *HWI - Hardware Interrupt Service Routine Manager* must now use McBSP1 instead of McBSP0.

B.4.4   DSP/BIOS Serial Port Configuration

The serial port configuration parameters under the `MCBSP Configuration Manager` should be changed to the values shown in the table below to support the required protocol. Then, the configuration must be applied to `Mcbsp_Port1` under the *MCBSP Resource Manager* instead of `Mcbsp_Port0`.

	eDSP PCM3006	TI Analog Daughtercard
SPCR	0x00000000	0x00000000
RCR	0x000400A0	0x000400A0
XCR	0x000400A0	0x000400A0
SRGR	0x101F0F07	0x00000000
MCR	0x00000000	0x00000000
RCER	0x00000000	0x00000000
XCER	0x00000000	0x00000000
PCR	0x00000A03	0x00000000

Now, build the project, download to your DSK and run. An audio signal on the codec input will now be transferred to the codec output.

B.5 A DSP/BIOS Talk-Through Application for the 6713 DSK

This application will implement audio talk-through using the C6713 DSK's onboard stereo codec. The larger sections of code required to complete this section are available for cut-and-paste from the file 6713\B_5_6713_text_to_paste.txt under Appendix B.

To begin, create a new CCS project in a directory of your choice. In CCS, select File→New, and a *New Project* dialog box will appear. Select the desired directory, name the project *talkthru*, set the project type to *Executable (*.out)*, and set the target to be *TMS320C67XX*. Click Finish, and the project will appear in the left (project) window of CCS.

Then, since this will be a DSP/BIOS-based application, we need to create a DSP/BIOS configuration file. To do this, select File→New→DSP/BIOS Configuration... from the menu and choose the dsk6713.cdb seed on the window that appears. This will create a DSP/BIOS configuration file that is tailored to the 6713 DSK. Save the created file as talkthru.cdb in the project directory. We will come back to it and make modifications later. For now, we will just add it to our CCS project. Select Project→Add Files to Project... from the menu, locate the talkthru.cdb file, and select Open. Note that the talkthru.cdb file is now part of the project, and that two additional files were also added under the Generated Files section. These files are regenerated every time you make a change to the DSP/BIOS configuration, so you should not edit them. The DSP/BIOS configuration also generates a linker command file, but it is not added automatically. Select Project→Add Files to Project... from the menu, locate the talkthrucfg.cmd file, and select Open. Now, we need to create our source code files and add them to the project. In this case, we will create two source files, one for the main function and another for the interrupt handlers.

B.5.1 main.c

Select File→New→Source File from the menu, then save the file as main.c and add it to the project. Then, add the following code to the file.

Listing B.8: main.c

```
#include "talkthrucfg.h"

void main() {
    return;
}
```

Main will remain an empty function. Remember that DSP/BIOS programs require that the main function return so that the DSP/BIOS scheduler can take control of program execution — this admittedly seems quite odd at first to experienced C programmers. The header file talkthrucfg.h is generated automatically from the DSP/BIOS configuration file. In this program, we will need to add code to main.c to start up the serial port and enable interrupts, but we will come back to that later.

B.5.2   hwi.c

Create another new source code file, save it as `hwi.c`, and add it to the project. In this file, we will add the functions that are used to service the serial port interrupts. We will be using the DSP/BIOS interrupt dispatcher, which will require us to use standard functions as opposed to the functions declared with the `interrupt` keyword. Add the following code to the file and save it.

Listing B.9: hwi function code.

```c
#include "talkthrucfg.h"

#define LEFT  0
#define RIGHT 1
volatile union {
    unsigned int uint;
    short channel[2];
} PcmData = {(unsigned int)0}; // global audio data


void HWI_McBSP_Rx(void)
{
    return;
}

void HWI_McBSP_Tx(void)
{
    return;
}
```

The global variable declaration is used to reflect that the serial port data will be two 16-bit values packed into a single 32-bit word. The use of the union as shown above permits the data to be treated as a single 32-bit value, but permits easy modification to the 16-bit data as well.

The convention of prefixing functions with the scheduler class that they will use is a good idea to help prevent errors. We will need to add code to these functions to read and write the serial port data, but first, we will go back to the DSP/BIOS configuration file and make some changes.

B.5.3   DSP/BIOS HWI Configuration

Open the configuration file (`talkthru.cdb`) in the CCS editor. The first thing we will do is configure DSP/BIOS to use the functions in `hwi.c` to handle the serial port interrupts. Expand the Scheduling→HWI - Hardware Interrupt Service Routine Manager tree. We could use any available hardware interrupt, but for consistency with previous work we will use `INT11` for the transmit interrupt and `INT12` for the receive interrupt. Right-click on the entry HWI_INT11, and select *Properties*. For *interrupt source:*, select MCSP_1_Transmit. For *function:*, enter _HWI_McBSP_Tx. (Note that a leading underscore must be appended to the function name.) On the *Dispatcher* tab, check the *Use Dispatcher* box, and click OK. What we have just done is instructed DSP/BIOS to use its interrupt dispatcher to respond to the McBSP1 transmit interrupt which it will map to INT 11, and to call our function HWI_McBSP_Tx to perform the interrupt processing. Similarly, right-click on HWI_INT12, and select *Properties*. Configure as before, except the interrupt source will be MCSP_1_Receive and the function will be _HWI_McBSP_Rx.

B.5.4 DSP/BIOS Serial Port Configuration

The other issue to be completed in the configuration file is the serial port initialization. The codec on the C6713 DSK uses two serial connections, one to configure the codec (McBSP0) and one to transfer the audio data (McBSP1) In the configuration tool, expand the CSL - Chip Support Library→MCBSP - Multichannel Buffered Serial Port tree. First, we will create a McBSP configuration by right-clicking *MCBSP Configuration Manager* and selecting *Insert mcbspCfg*. This will create a new entry under the *MCBSP Configuration Manager* named `mcbspCfg0`. Right click `mcbspCfg0`, and select *Properties*. The various tabs permit you to set the configuration of the many parameters and operational modes of the McBSP. For now, use the *Advanced* tab to directly enter the register values as shown below for McBSP0, and click OK to save your updates. Now, we will instruct DSP/BIOS to use this configuration and to automatically apply it to McBSP0. Under *MCBSP Resource Manager*, right-click `Mcbsp_Port0` and select *Properties*. Check the *Open MCBSP Port* box and the *Enable Pre-Initialization* box. Select the configuration you just created (`mcbspCfg0`) and click OK.

Now, repeat this process to create another configuration `mcbspCfg1`, set it to the parameters listed below for McBSP1, and assign it to initialize McBSP1. Save the changes you made to the configuration file.

	McBSP0	McBSP1
SPCR	0x00001000	0x00000000
RCR	0x00000000	0x000400A0
XCR	0x00000040	0x000400A0
SRGR	0x20100F04	0x00000000
MCR	0x00000000	0x00000000
RCER	0x00000000	0x00000000
XCER	0x00000000	0x00000000
PCR	0x00000A0A	0x0000000F

B.5.5 Source Code

Now, we need to add some code to our source files. Open `hwi.c` in the CCS editor. Add the following code to the `HWI_McBSP_Rx` function. This simply reads the serial port data register and stores it in the global variable **data**. We do not have to test to see if data is available since this is an interrupt service routine that is only invoked when data is available.

Listing B.10: HWI_McBSP_Rx function code.

```
PcmData.uint = MCBSP_read(hMcbsp1);

// to modify individual channel data
// use PcmData.channel[LEFT] or
// use PcmData.channel[RIGHT], i.e.
// PcmData.channel[LEFT] *= Gain;
```

Add the following code to the `HWI_McBSP_Tx` function. This gets the value from the global variable data and writes it to the serial port transmitter.

Listing B.11: HWI_McBSP_Tx function code.

```
// to modify individual channel data
// use PcmData.channel[LEFT] or
// use PcmData.channel[RIGHT], i.e.
// PcmData.channel[LEFT] *= Gain;

MCBSP_write(hMcbsp1, PcmData.uint);
```

The last piece of source code we have to add is a task that will configure the codec as soon as DSP/BIOS has finished its initialization. Open `main.c` in the CCS editor. Add the following new function to the file below the `main()` function.

Listing B.12: TSK_CodecInit() function code.

```
void TSK_CodecInit()
{
    #define NUM_AIC23_REGISTERS 10
    int i;
    short AIC23_data[NUM_AIC23_REGISTERS] = {
        (0<<9) | 0x017, // reg 0-left line volume (0 db)
        (1<<9) | 0x017, // reg 1-right line volume (0 db)
        (2<<9) | 0x0f9, // reg 2-left headphone volume (0 db)
        (3<<9) | 0x0f9, // reg 3-right headphone volume (0 db)
        (4<<9) | 0x012, // reg 4-analog path (line input)
//      (4<<9) | 0x014, // reg 4-analog path (mic input, 0db)
//      (4<<9) | 0x015, // reg 4-analog path (mic input, +20db)
        (5<<9) | 0x000, // reg 5-digital path (no deemphasis)
        (6<<9) | 0x000, // reg 6-power down (all on)
        (7<<9) | 0x053, // reg 7-digital interface (DSP mode)
        (8<<9) | 0x001, // reg 8-sample rate (48kHz)
        (9<<9) | 0x001  // reg 9-digital interface (activate)
    };
    MCBSP_enableSrgr(hMcbsp0);
    MCBSP_enableFsync(hMcbsp0);
    MCBSP_enableXmt(hMcbsp0);

    for(i = 0;i < NUM_AIC23_REGISTERS;i++) {
        while(!MCBSP_xrdy(hMcbsp0)) // wait for codec ready
            ;
        MCBSP_write(hMcbsp0, AIC23_data[i]); // write codec
            [+]registers
    }

    MCBSP_enableXmt(hMcbsp1); // enable McBSP1
    MCBSP_enableRcv(hMcbsp1);

    C62_enableIER(IRQ_MASK_11|IRQ_MASK_12); // enable interrupts
}
```

This code uses McBSP0 to initialize all of the codec's registers for operation in the desired mode. Note that for *register 4* of `AIC23_data`, there are three possible settings to determine the codec's input source; corresponding to line input, microphone input without amplification, and microphone input with 20db of amplification. Only one of these should be uncommented in your code at any time. In the code above, the codec will use the line input. The final three lines of code enables the McBSP1 transmitter and receiver, and then enables the two hardware interrupts we selected in the HWI configuration. You must always explicitly enable hardware interrupts. That's all for the `TSK_CodecInit` function, so save the changes.

Now, the `TSK_CodecInit` function needs to be scheduled to run in DSP/BIOS. Expand *Scheduling→TSK - Task Manager* in the DSP/BIOS configuration file. Right-click *TSK - Task Manager* and select *Insert TSK*. This will create a new entry TSK0. Edit TSK0's properties to set *Task function* to `_TSK_CodecInit` (note the leading underscore) and accept the remaining defaults. This establishes a new task that will be scheduled to run on start-up, which will initialize the codec and start the interrupts.

Now, build the project, download to your DSK and run. An audio signal on the selected codec input will now be transferred to the codec output. This project can serve as a building block for future work by adding code to process the data in different ways.

Appendix C

Numeric Representations

IN the digital domain, a given number may be stored and used in a number of different representations. A number may be exactly representable in one form, but not in others. How the number is represented will affect the accuracy of calculations, the memory and bus bandwidth requirements, and the speed of calculations that can be attained.

C.1   Endianness

Typically, computer memories are addressable in bytes. If a data element is larger than one byte, then a decision must be made as to how the individual bytes of the data element are to be stored in memory. Given a 32-bit (4 byte) data element with hexadecimal value 12345678h, the data can be ordered in memory in two distinctly different ways. Assuming the data is stored at address 00001000h, it could represented in either of the orderings shown below.

Address	00001000h	00001001h	00001002h	00001003h
Big-endian	12h	34h	56h	78h
Little-endian	78h	56h	34h	12h

The choice of storage method determines what is commonly referred to as *endianness*, so named since it is based on which *end* of a number is stored in the first byte of the memory space that the number occupies. *Big-endian* organization placed the most significant byte of the data at the first address, while *little-endian* places the least significant byte of the data at the first address. There is no advantage of one over the other, and both are used in practice. In fact, the TMS320C6x DSPs are capable of operating with either endianness based on the setting of a processor pin at reset. By default, the DSK is configured to operate in little-endian mode.

To verify the DSK's endianness, open Code Composer Studio, and use View→Memory to open two memory windows at address 00000000h. Set the properties on one memory window to `32-Bit Hex - TI Style` and the other to `8-Bit Hex - TI Style`. Then, in the 32-bit window, set the value at address 00000000h to 12345678h. After making the change, observe how the bytes of the value are stored in the 8-bit window in order to determine the endianness in use.

While endianness will likely not affect typical single-processor DSP code, you should be aware of it, particularly if transferring information between processors, working in a multiprocessor environment with shared memory, or manipulating individual bytes of multi-byte data elements directly.

C.2   Integer Representations

Integer representations use a number of bits to represent just the integer part of a number; that is, they cannot represent any fractional part of a number. Integer representations can be broadly divided into those that represent signed numbers and those that represent unsigned (non-negative) numbers.

In general, an unsigned integer can represent any value in the range from 0 to $2^n - 1$, where n is the width of the unsigned integer in bits. For example, with an 8-bit unsigned integer value, the minimum value will be $00000000_2 = 0_{10}$. The maximum value will be $11111111_2 = 255_{10} = 2^8 - 1$.

Signed representations can represent both positive and negative values, using one of several conventions. The most commonly used is the 2's-complement representation. In this representation, the most significant bit determines the sign of the number, with a 1 signifying a negative number. The determination of a positive value is straight forward in that it can simply be interpreted as an unsigned number. However, if a number is negative, then it is easiest to determine its value by negating it and interpreting the resulting positive value as the magnitude of the negative value. To negate a 2's-complement number, simply complement the number (flip the value of each bit) and add 1. If the addition carries beyond the number of bits in the number, the carry is discarded.[1] The number that remains is interpreted as though it is unsigned, and a negative sign is added. This is illustrated for six different 8-bit 2's-complement numbers below, with only the first of the numbers being positive.

binary value	complement	add 1	decimal value
01100011	(not needed)	NA	99
11100011	00011100	00011101	-29
11111111	00000000	00000001	-1
11111110	00000001	00000010	-2
10000001	01111110	01111111	-127
10000000	01111111	10000000	-128

While the 2's-complement representation may seem strange, it has a great advantage over other representations in the design of computer arithmetic hardware. In general, 2's-complement integers can represent any value in the range from (-2^{n-1}) to $(+2^{n-1} - 1)$, where n is the width of the 2's-complement integer in bits. For example, with an 8-bit 2's-complement integer value, the minimum value will be $10000000_2 = -128_{10}$. The maximum value will be $01111111_2 = +127_{10}$.

Another relatively common method for representing signed binary integers is to use the sign-magnitude convention. In this case, the most significant bit again determines the sign with a 1 signifying a negative number, but the remaining bits are interpreted as an unsigned magnitude. This has an advantage in that the representation is symmetric about 0, as compared to the 2's-complement representation where it can represent one more negative value than positive value. For example, with an 8-bit sign-magnitude integer value, the lowest value will be $11111111_2 = -127_{10}$. The highest value will be $01111111_2 = +127_{10}$. The values 00000000_2 and 10000000_2 both represent 0.

Note that for an integer representation, every possible bit combination is in fact a valid value—this makes it impossible to represent or detect erroneous values in hardware without modifying the representation.

---

[1] An alternate method of negating a 2's-complement number: begin at the least significant bit, and move to the left keeping digits unchanged up to and including the first "1" that is encountered. Then complement all the remaining bits.

C.3 Integer Division and Rounding

In general, if we were asked to round the result of a division, the common solution is to see if the fraction part is 0.5 or greater; if so, add 1 to the result, then simply truncate to the integer value. However, integer division hardware produces the integer quotient and a remainder, not a fractional value. So, to do rounding, one method would be to add one to the quotient if the remainder is greater than the divisor divided by two. However, this requires the following additional steps:

1. obtain the divisor divided by two

2. compare to the remainder

3. conditionally add one to the quotient

A more efficient approach is to recognize that if we add one half of the divisor to the dividend before division, the result will be rounded. This requires the following additional steps:

1. obtain divisor divided by two (with any remainder truncated)

2. add to the dividend before the division

The truncated result will in fact be rounded, as shown below for an 8-bit unsigned integer.

Operation	Dividend	Divisor	Quotient
$47 \div 3$	00101111_2 (47)	00000011_2 (3)	00001111_2 (15)
$(47 + (3 \div 2)) \div 3$	00110000_2 (48)	00000011_2 (3)	00010000_2 (16)

When programming computer hardware that does not have a hardware integer divider (i.e., the TMS320C6x family), great pains are taken in algorithm implementation to avoid division by anything other than a power of two. This allows the division operations to be accomplished using bit-wise shifts, since every right-shift is equivalent to division by two, as shown below for an 8-bit unsigned integer. (Conversely, a left shift is equivalent to a multiplication by two.) Even in machines having a hardware integer divider, it is much faster to perform division by powers of 2 as shifts. In the above example, the division by two would best be accomplished as a 1-bit right shift.

Initial Value	Shift	Equivalent to	Result
00101111_2 (47_{10})	right 1	$\div 2$	00010111_2 (23_{10})
00101111_2 (47_{10})	right 2	$\div 4$	00001011_2 (11_{10})
00101111_2 (47_{10})	right 3	$\div 8$	00000101_2 (5_{10})
00101111_2 (47_{10})	right 4	$\div 16$	00000010_2 (2_{10})
00101111_2 (47_{10})	right 6	$\div 64$	00000000_2 (0_{10})
00101111_2 (47_{10})	left 1	$\times 2$	01011110_2 (94_{10})
00101111_2 (47_{10})	left 2	$\times 4$	10111100_2 (188_{10})
00101111_2 (47_{10})	left 3	$\times 8$	01111000_2 (120_{10})

Note that in the last result, overflow occurred and the result is erroneous. This highlights a significant limitation in the dynamic range available when using integer representations. Floating point representations help to mitigate this problem, although they are subject to similar issues as well.

C.4 Floating-Point Representations

In floating point representations, numbers are stored as a mantissa and exponent value, similar to the scientific notation we commonly use. However, instead of being stored in the form ($mantissa_{10} \times 10^{exponent_{10}}$), a floating point representation is stored in a binary format as ($mantissa_2 \times 2^{exponent_2}$). Commonly used floating point representations include the IEEE single-precision and double-precision formats [70], as illustrated in Figure C.1.

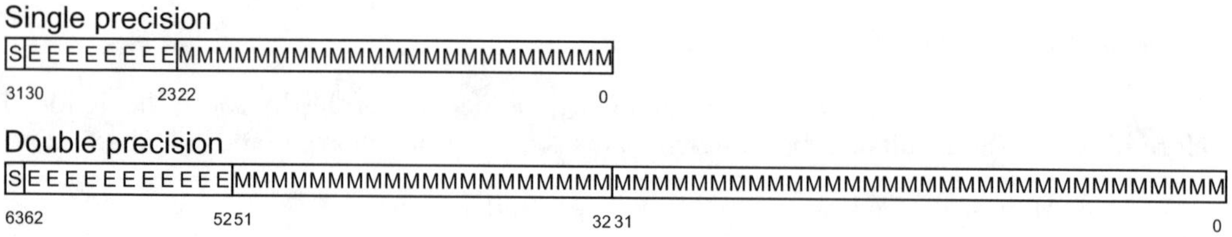

Single precision

S	E E E E E E E E	M M M M M M M M M M M M M M M M M M M M M M M

3130   2322   0

Double precision

S	E E E E E E E E E E E	M M M M M M M M M M M M M M M M M M M M	M M M M M M M M M M M M M M M M M M M M M M M M M M M M M M M M

6362   5251   3231   0

Figure C.1: IEEE-754 floating-point representations.

The following discussion focuses on the single-precision representation in detail; the double precision representation is conceptually similar. In the single precision representation, the S bit determines the sign of the number, where a 0 value indicates a positive number and a 1 value indicates a negative value. This can be expressed as -1^S. The 8-bit exponent portion of the number is biased by 127 (called "excess 127"), such that an exponent value of 00000000_2 is interpreted as a value of -127, a value of 11111111_2 (255_{10}) is interpreted as $+128$, and a value of 01111111_2 (127_{10}) is interpreted as an exponent of 0. This results in an potential exponent term multiplier of 2^{-127} to 2^{+128}, but as we shall see later some exponent values are reserved for special cases. The 23-bit mantissa portion is interpreted as a binary value 1.MMM$\cdots$MMM, where the radix point is a binary point, not a decimal point. The value of each mantissa bit therefore takes on the values shown below.

$$
\begin{array}{ccccccccc}
1 & . & M & M & M & \cdots & M & M & M \\
 & & 2^{-1} & 2^{-2} & 2^{-3} & \cdots & 2^{-21} & 2^{-22} & 2^{-23}
\end{array}
$$

By assuming the leading 1 bit (the normalized condition), the range of the mantissa can be seen to be from (1.0) where all M bits are 0 to ($2.0 - 2^{-23}$) where all M bits are 1. Sample floating point values are shown below.

Hexadecimal	SEEE EEEE EMMM MMMM MMMM MMMM MMMM MMMM
	$= -1^S \times (1.\text{MMMMMMMMMMMMMMMMMMMMMMM}_2) \times 2^{\text{EEEEEEEE}_2 - 127}$
0x3F800000	0011 1111 1000 0000 0000 0000 0000 0000
	$= -1^0 \times (1.00000000000000000000000_2) \times 2^{127-127}$
	$= 1 \times (1) \times 2^0 = 1.0$
0xBF800000	1011 1111 1000 0000 0000 0000 0000 0000
	$= -1^1 \times (1.00000000000000000000000_2) \times 2^{127-127}$
	$= -1 \times (1) \times 2^0 = -1.0$
0xC2820000	1100 0010 1001 0011 0000 0000 0000 0000
	$= -1^1 \times (1.00100110000000000000000_2) \times 2^{133-127}$
	$= -1 \times (1 + 2^{-3} + 2^{-6} + 2^{-7}) \times 2^6 = -73.5$
0x7F00000	0111 1111 0000 0000 0000 0000 0000 0000
	$= -1^0 \times (1.00000000000000000000000_2) \times 2^{254-127}$
	$= 1 \times (1) \times 2^{127} = 1.7014118 \times 10^{38}$
0x3DCCCCCD	0011 1101 1100 1100 1100 1100 1100 1101
	$= -1^0 \times (1.10011001100110011001101_2) \times 2^{123-127}$
	$= 1 \times (1 + 2^{-1} + 2^{-4} + 2^{-5} + 2^{-8} + 2^{-9} + 2^{-12} + 2^{-13}$
	$\quad + 2^{-16} + 2^{-17} + 2^{-20} + 2^{-21} + 2^{-23}) \times 2^{-4}$
	$= 0.10000000149012 \approx 0.1$

As the last entry above shows, even though the floating point number has a seemingly benign value such as 0.1, it may well not be possible to represent it exactly in a given format. In fact, 0.1 can be exactly represented only by the *infinite* binary series

$$\sum_{i=1}^{\infty} \frac{1}{2^{4i}} + \frac{1}{2^{4i+1}}$$

and so any use of it in calculations is inexact. Although the difference may seem trivial, using the value in a repeated calculation leads to larger accumulated errors, and a form of quantization noise will be found in the results.

If a leading 1 in the mantissa is assumed, then it is not possible to exactly represent 0.0. To overcome this, a special case is designated to represent 0.0 exactly; if the exponent and the mantissa fields are both all 0's, then the value of the number is defined by the IEEE standard to be exactly 0.0.

Although the floating point representation permits the representation of a much wider range of values, there exists the potential for other inaccuracies in mathematical computation. One such effect that may occur happens when adding a small number to a much larger number. In order to add or subtract two floating point numbers, they must be converted to the same exponent value, then the mantissas can be added. After adding the mantissas, the resulting number is normalized by determining the most significant 1 bit, then adjusting the exponent so that bit becomes the assumed 1 bit in the representation. If two numbers are added together where one number is substantially larger than the other, the result may not be accurate. By way of illustration assume that there are two operands with values as given below.

	SEEE EEEE EMMM MMMM MMMM MMMM MMMM MMMM	
Operand1	0100 1011 1000 0000 0000 0000 0000 0000	(16,777,216.0)
Operand2	0011 1111 1000 0000 0000 0000 0000 0000	(1.0)

After converting both numbers to the same exponent by converting Operand2 to an exponent of 151 (10010111_2), the mantissa addition would take place as shown below (exponents not shown).

```
              1.MMM MMMM MMMM MMMM MMMM MMMM MMMM MMMM MMMM MMMM MMMM
  Mantissa1   1.000 0000 0000 0000 0000 0000 ---- ---- ---- ---- ----
  Mantissa2   0.000 0000 0000 0000 0000 0000 1000 0000 0000 0000 0000
        Sum   1.000 0000 0000 0000 0000 0000 1000 0000 0000 0000 0000
Sum mantissa    000 0000 0000 0000 0000 0000
```

In order to store the resultant mantissa, it must be truncated to 23 bits. When this is done, it can be seen that the original Operand1 mantissa value has not changed, and therefore the addition operation had no effect. A more complete discussion of numerical accuracy in computing can be found in [71, 72].

In addition to storing normalized numbers, special representations are used in the IEEE floating-point standard to express non-normalized numbers as well as various invalid numbers and error conditions, such as `NaN` (not a number). A complete discussion of these can be found in [70].

C.5   Fixed-Point Representations

Although floating-point data representation and manipulation are more intuitive to most engineers, the use of floating-point is not without penalty. First, the floating-point representations may be wider than necessary and therefore increase system cost and power consumption with no benefit. Also, integer hardware is significantly simpler than floating point hardware, so it can be designed to operate faster and consume less power. These attributes are the reason that floating-point DSPs are almost never used in high volume portable applications, notably in most cellular telephones. In these commodity market environments, cost and low power consumption are dominating factors, and the additional programming complexity required to implement algorithms on fixed-point hardware can be amortized over a large number of units. Also for a given width (i.e., 32 bits), the fixed-point number will have better resolution (and hence a lower noise floor) than a floating-point representation (assuming the proper scaling is performed). This is due to the fewer number of bits in the floating-point mantissa as compared to the fixed-point number.

Fixed-point representations are used to allow the implementation of fractional arithmetic using only integer arithmetic hardware. In fixed-point, a binary point is assumed to exist at a fixed location in a 2's-complement number, typically in a 16-bit value. The location of the binary point is indicated by the `Q-number` notation, such that a `Qn` number is assumed to have `n` bits to the right of the binary point. The most commonly used are the `Q15` and `Q12` formats, as illustrated below.

```
Q15   S.XXX XXXX XXXX XXXX    1000000000000000₂ = −1.0000000
                              0111111111111111₂ = +0.9999695
Q12   SXXX. XXXX XXXX XXXX    1000000000000000₂ = −8.0000000
                              0111111111111111₂ = +7.9997559
```

To determine the decimal value of a fixed-point number, first determine if it is positive or negative. The sign bit `S` is 1 for negative numbers and 0 for positive numbers. For a negative number, convert it to the corresponding positive value as was done for 2's-complement integers. Then determine the value based on sum of the individual bits multiplied by their weight.

	binary value	complement	add 1	decimal value
Q15	0.110100000000000	not needed	NA	$\frac{1}{2} + \frac{1}{4} + \frac{1}{16} = 0.8125$
Q12	0110.100000000000	not needed	NA	$4 + 2 + \frac{1}{2} = 6.5$
Q15	1.110100000000000	0.001011111111111	0.001100000000000	$-(\frac{1}{8} + \frac{1}{16}) = -0.1875$
Q12	1110.100000000000	0001.011111111111	0001.100000000000	$-(1 + \frac{1}{2}) = -1.5$

It is important to note that fixed-point arithmetic is intended to be accomplished on standard 2's-complement integer arithmetic hardware, so it must yield correct results without modification. Using the Q15 values of 0.375 $(\frac{3}{8})$ and -0.25 $(-\frac{1}{4})$, addition using integer logic is illustrated below.

$$
\begin{array}{rl}
0.375 = & 0.011000000000000 \\
+ \quad \underline{-0.250} = & \underline{1.110000000000000} \\
& 10.001000000000000 \qquad \textit{but carry out is lost} \\
& 0.001000000000000 \qquad\qquad = 0.125 \ \checkmark
\end{array}
$$

Binary multiplication of two n-bit numbers results in a $2n$-bit result. The result of multiplying two Q15 numbers is a 32-bit number in Q30 representation with a redundant sign bit. To return the result to Q15 format, it is shifted right 15 bits and truncated to 16 bits. (Note that multiplying negative numbers requires a different algorithm, not illustrated here.)

$$
\begin{array}{rl}
0.375 = & 0011\ 0000\ 0000\ 0000 \\
\times \quad \underline{0.750} = & \underline{0110\ 0000\ 0000\ 0000} \\
& 0\ 0110\ 0000\ 0000\ 000 \\
& \underline{00\ 1100\ 0000\ 0000\ 00} \\
& 0001\ 0010\ 0000\ 0000\ 0000\ 0000\ 0000\ 0000 \\
\text{(shift right 15 bits)} \rightarrow & 0\ 0010\ 0100\ 0000\ 0000 \\
\text{(truncate to 16 bits)} \rightarrow & 0010\ 0100\ 0000\ 0000 = 0.28125 \ \checkmark
\end{array}
$$

It is useful to note that the exponent in floating point provides automatic scaling, whereas in fixed-point the scaling must be performed in the algorithm as required to prevent overflow. Q15 is commonly used in signal processing since multiplication of two Q15 numbers will never overflow beyond the range of the Q15 number with the exception of the (-1×-1) case.

C.6 Summary of Numeric Representations

A summary of the numeric range, precision, and dynamic range of various number representations is provided below. Signed integer calculations are based on the 2's-complement representation. Numeric range shows the most negative and most positive values that can be represented. Precision is defined as the smallest increment that the given representation can represent in normalized form. For floating point representations, the precision figure is based on the smallest mantissa increment that can be used in an addition/subtraction with 1.0 and still affect the resultant mantissa value. Dynamic range for all representations is then calculated as $20\log(\frac{\text{numeric range}}{\text{precision}})$. Note that the floating point representations can represent a much larger range of values than the precision represented by the mantissa (for example, a range in excess of 1600 dB for single-precision values), however, for any given value being represented there is a related range of values that can be added to the current value without the computation's effect being lost. In a sense, the exponent of the floating-point number could be thought of as determining where the current dynamic range of the floating-point number exists in the much larger value space of the numeric representation.

Table C.1: Summary of numeric representations.

Bits		Numeric Range	Precision	Dynamic Range
8	Unsigned integer	$0 \rightarrow +255$	1	≈ 48 dB
8	Signed integer	$-128 \rightarrow +127$	1	≈ 48 dB
16	Unsigned integer	$0 \rightarrow +65,536$	1	≈ 96 dB
16	Signed integer	$-32,768 \rightarrow +32,767$	1	≈ 96 dB
16	Fixed-point (Q12)	$-8.0 \rightarrow \approx +7.999756$	≈ 0.000244	≈ 96 dB
16	Fixed-point (Q15)	$-1.0 \rightarrow \approx +0.9999695$	≈ 0.0000305	≈ 96 dB
32	Unsigned integer	$0 \rightarrow +4,294,967,296$	1	≈ 193 dB
32	Signed integer	$-2,147,483,648 \rightarrow +2,147,483,647$	1	≈ 193 dB
32	Single-precision	$\approx \pm 3.402823 \times 10^{38}$	$\approx 1.19 \times 10^{-7}$	≈ 138 dB
64	Double-precision	$\approx \pm 1.797693 \times 10^{308}$	$\approx 2.22 \times 10^{-16}$	≈ 314 dB

Appendix D

TMS320C6x Architecture

T HE first section of this appendix is intended to serve as a basic primer on computer architecture to the extent necessary to have a better understanding of the TMS320C6x architecture. It is assumed that the reader has a basic understanding of microprocessor operations. The second section section then discusses the TMS320C6x architecture in some detail. Readers familiar with computer architecture may find it unnecessary to read the first section in order to understand the second section. For more detailed information on the TMS320C6x processors, the reader is referred to the technical documentation from Texas Instruments (for example [53, 73]).

D.1   Computer Architecture Basics

One definition of computer architecture is "... an abstract interface between the hardware and the lowest level software of a machine that encompasses all the information necessary to write a machine language program that will run correctly, including instructions, registers, memory size, and so on" [74]. When selecting and using a processor to perform a given task, the underlying construction of the processor will determine what it does well and what it does not, and ultimately, whether it meets the design requirements.

A basic microprocessor system is illustrated in Figure D.1. The central processing unit (CPU) contains the register file and the functional units, in this case just an arithmetic-logic unit (ALU), all interconnected by signalling buses and sequenced by the control unit. The memory system and input/output (I/O) subsystems are connected to the CPU by the system bus. The clock generator provides the timing for the logic operation and the reset circuit ensures the processor starts from a known state. A number of peripheral devices are commonly found in the I/O subsystem of a microprocessor system.

- Parallel I/O ports interface to external devices for control or sensing.

- Serial ports facilitate communications with local or distant devices using various serial protocols and hardware.

- Counter/timers are used to established precise timing intervals, generate rectangular waveforms, and to count external events.

Although the I/O and memory are shown in Figure D.1 as external to the CPU, in practice they may be integrated onto the actual processor device.

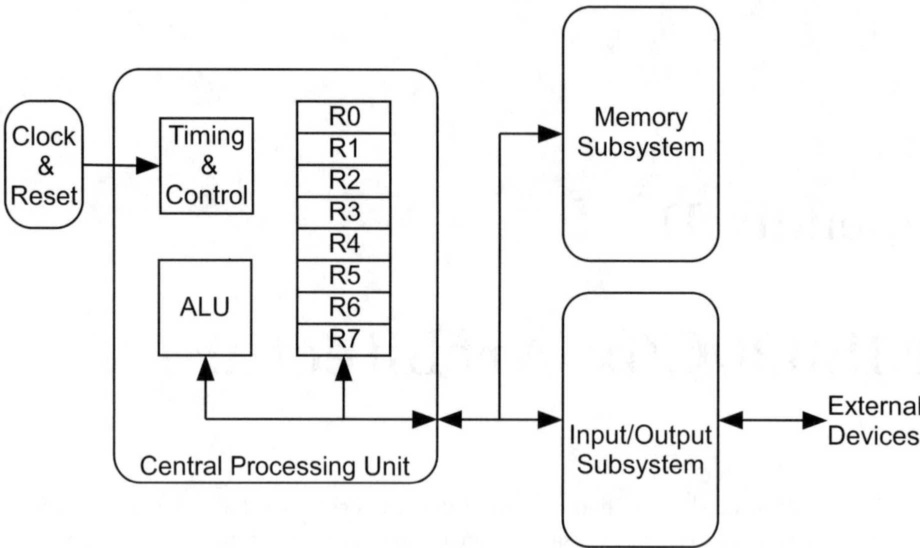

Figure D.1: Basic microprocessor system.

D.1.1   Instruction Set Architecture

One way to differentiate processors is in the make-up of their *instruction set architecture* (ISA), the set of commands that the processor can execute and the hardware required to execute them. On one end, the *complex instruction set computer* (CISC) supports instructions that can perform complicated tasks. For example, a single instruction may execute a complete FIR filter routine, or search for a given value in an array. At the other extreme, *reduced instruction set computers* (RISC) have only a limited number of low-level instructions. So, more complex tasks must be accomplished as a series of simple instructions. In particular, a RISC instruction set usually only permits operations on data stored in the processor's registers, and has a very limited number of instructions that move data between memory and the registers. Since the CISC machine effectively executes complex tasks in hardware, it should have (and often does have) a performance advantage on a limited set of very specific task(s). However, complex instruction sets make it extremely difficult to optimize the performance of a processor in a general sense (and to develop efficient compilers for high-level languages), so nearly all general-purpose microprocessors are now RISC. (Although the ubiquitous Intel 80x86 architecture is programmed using a CISC instruction set, Pentium Pro and later implementations perform real-time translation of those instructions into RISC-like micro-operations that are then executed by the processor core.)

D.1.2   Register Architectures

Another broad classification of processor architectures is based on the possible locations of an instruction's source and destination operands. There are two common architectures classified in this way.

Register-memory: architectures allow one or more instruction operands to be located in memory. This architecture is commonly used in CISC machines.

Load-store: architectures require that all instruction operands be in registers. Only a very few instructions have access to memory, and these typically perform only simple

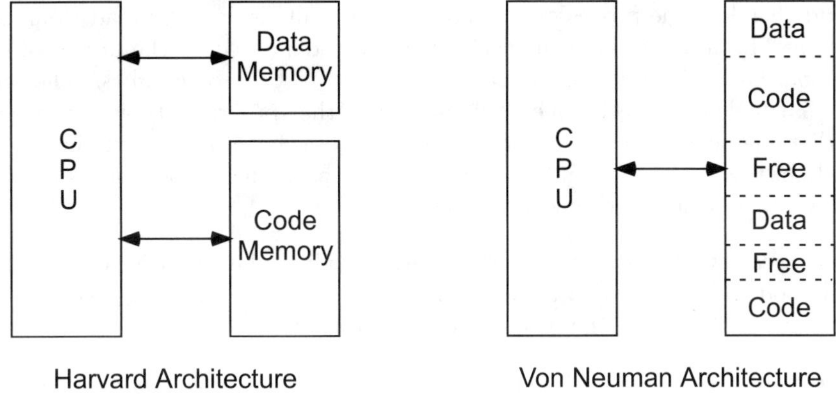

Figure D.2: Memory architectures.

transfers from memory to a register (load) or from a register to memory (store). The TMS320C6x DSPs use a load-store architecture.

D.1.3 Memory Architectures

The organization of processor memory space is a key element of the overall operation of the system. The information in memory consists of two distinct types:

1. instructions that the processor will execute exist in *code space,* and

2. information that the processor accesses as part of its program execution is stored in *data space.*

In the design of the processor, these two spaces can be placed into physically independent memories, or can exist in the same physical memories, as shown in Figure D.2. Architectures with separate code and data spaces are referred to as *Harvard* architectures. The primary advantage of the Harvard architecture is that there can be simultaneous operations in code and data memory, increasing memory bandwidth. Also, since the code and data spaces have different buses to the processor, their widths can be different in order to optimize each, and the code can be absolutely protected from inadvertent corruption by data operations. The primary disadvantage is that code and data can only be placed in their respective spaces, so any free memory in code space cannot be used by for data, or vice versa. The Harvard architecture is commonly used in microcontrollers and special purpose processors, including the Texas Instruments TMS320C2x and TMS320C5x families of DSPs.

Processors where the code and data exist in the same memory space are referred to as *Von Neumann* (or *Princeton*) architectures. The primary advantage is that all of memory is available for code or data in any proportion, giving it much greater memory allocation flexibility. The disadvantages include the fact that code and data cannot be simultaneously accessed, and that the possibility exists for corruption between the code and data spaces. This architecture is used by nearly all general purpose microprocessors and numerous specialized processors as well, due primarily to its flexibility.

In addition to their main memories, modern processor systems often have *cache memories.* These are very fast local memories of a limited size that are often internal to the processor itself, and typically operate without any explicit control by the processor. The purpose of the cache memory is to retain a subset of all instructions in the hope that a significant percentage of the overall instruction stream will be present in the cache when needed and therefore the processor will not have to wait for it to be read from the slower main memory. If an object is in the cache when it is requested by the processor, it is said to be a *cache hit*. If it is not in the cache, it is said to be a *cache miss*. The underlying principle that makes this technique very effective is that programs typically display somewhat predictable behavior that can be exploited. (Note that if the instruction stream were truly random, then a cache would be of no use.) In particular, most caches are optimized to take advantage of the following properties.

Temporal locality: this is the property whereby code or data that is accessed has a high likelihood of being needed again in the near future. By keeping recently used code and data in the cache, the possibility of reuse is high.

Spatial locality: this is the property whereby code or data that is close to a memory location that was accessed is more likely to be needed than other memory locations that are more distant. To exploit spatial locality, most processor caches automatically fetch a larger block of data from the main memory when any location in the block is accessed.

Since cache memories have limited size, various algorithms are used to try to maintain only the information that is most likely to be needed in the cache. Caches can provide a substantial performance improvement, in fact, modern general-purpose processors rely heavily on multi-level cache hierarchies to achieve good performance. However, caches make it difficult to predict program execution time since it becomes probabilistic whether the needed instructions or data are in the cache or must be fetched from slower main memory. This is a common theme in modern computer architecture, where many of the techniques used to speed up the execution of general-purpose processors also make it difficult or impossible to predict the program execution time.

D.1.4   Fetch-Execute Model

In its simplest form, a processor operates by reading an instruction from memory (referred to as a fetch), then it executes the instruction. In order to execute an instruction, the processor must first decode the instruction to determine what it is to do, read in any required operands, perform the required action, and then write the result to the proper location. While this sequential behavior results in a very simple design, performance suffers. For example, once the instruction is fetched, the bus to code memory will be idle until the instruction is completely executed. Similarly, the functional units that actually perform the operation (i.e., an adder or multiplier) are idle during the instruction fetch, the instruction decode, and while the results are being written out. Obviously, if all of the parts of the system could be kept busy simultaneously, performance would be improved.

D.1.5   Pipelining

In order to improve the utilization of the processor hardware, it is divided into stages, each of which handles a distinct portion of the total processing of an instruction. As an instruction is processed through one stage of the pipeline, it is passed to the next stage. Since the pipeline can only go as fast as the slowest stage, it is designed to be balanced where each stage has a similar delay. A representative pipeline with four stages is shown below in Figure D.3 (note that modern processors may use deeper pipelines with more than 20 stages). The pipeline stages are defined as;

F (fetch) is responsible for reading the next instruction from memory.

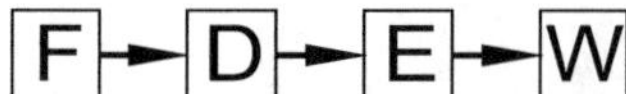

Figure D.3: Pipeline stages.

D (decode) decodes the instruction to determine what action and operands are required.

E (execute) loads operands (if any) and performs the required operation.

W (write-back) stores the results of the operation into the destination.

If we assume that each stage of the pipeline has a 10ns delay, then if each instruction were required to complete before the next could be fetched it would take 40ns for each instruction, and the processor could execute 25 million instructions per second (MIPS). With each stage of the pipeline able to operate independently on a different instruction, it will still take 40ns for an instruction to be completely executed. This is an important observation: in general, pipelining does not improve the latency of a system. However, an instruction now finishes the journey through the pipeline every 10ns, so the processor now operates at 100 MIPS, a 400% increase in throughput. In this case, the processor would have 4 instructions *in flight* (being executed) at once. This performance increase can only be achieved if the pipeline receives a steady stream of instructions and has unfettered access to the operands required as input and output of the instructions.

Consider the example instruction stream shown in Figure D.4. The operation of the four stage pipeline with instruction stream from Figure D.4 is shown in Figure D.5. Each horizontal division represents one pipeline clock interval. This figure illustrates several common issues that limit pipeline performance.

- In the execution of `Instruction1`, the execute stage is shown as being delayed waiting for the required operands. This could be due to the delay in reading from memory after a cache miss, or waiting for a previous instruction's result to become available (in a longer pipeline). This delay is referred to as a *pipeline stall*, and causes all instructions at earlier stages to be delayed as well. Although not shown here, later pipeline stages could continue to operate and the instructions in them move forward, creating a *pipeline bubble* (empty pipeline stages).

- When `Instruction4` causes a jump or branch, execution now proceeds out of sequence. In this case, the address to fetch the next instruction from is not known until the completion of the write-back stage. The later instructions currently following this instruction must be discarded in a *pipeline flush*. Then the pipeline begins to refill as fetches occur in sequence again. Pipeline flushes are very costly in terms of time, especially in deeper pipelines. For this reason, modern high performance processors use *branch prediction* to "guess" what the next instruction will be, and execute that instruction. If the guess is correct, then the pipeline remains full and maximum performance is obtained. If the guess is wrong, then those instructions must be discarded and the correct instructions executed. Fortunately, branch prediction algorithms typically have an accuracy in excess of 95%. Another technique used to minimize pipeline flushes is *predicated execution*, where an instruction's execution is predicated on a value in a specified register. Depending on the value in the predicate register, the instruction either executes normally, or it is passed through the pipeline as a no-operation (NOP) instruction. This permits conditional execution without flushing the pipeline.

Address	Instruction	Action
00001000h	Instruction1	
00001004h	Instruction2	
00001008h	Instruction3	
0000100Ch	Instruction4	(jump to 00001040h)
00001010h	Instruction5	
00001014h	Instruction6	
00001018h	Instruction7	
⋮	⋮	
00001040h	InstructionX	
00001044h	InstructionY	
00001048h	InstructionZ	

Figure D.4: Example instruction sequence with an unconditional jump.

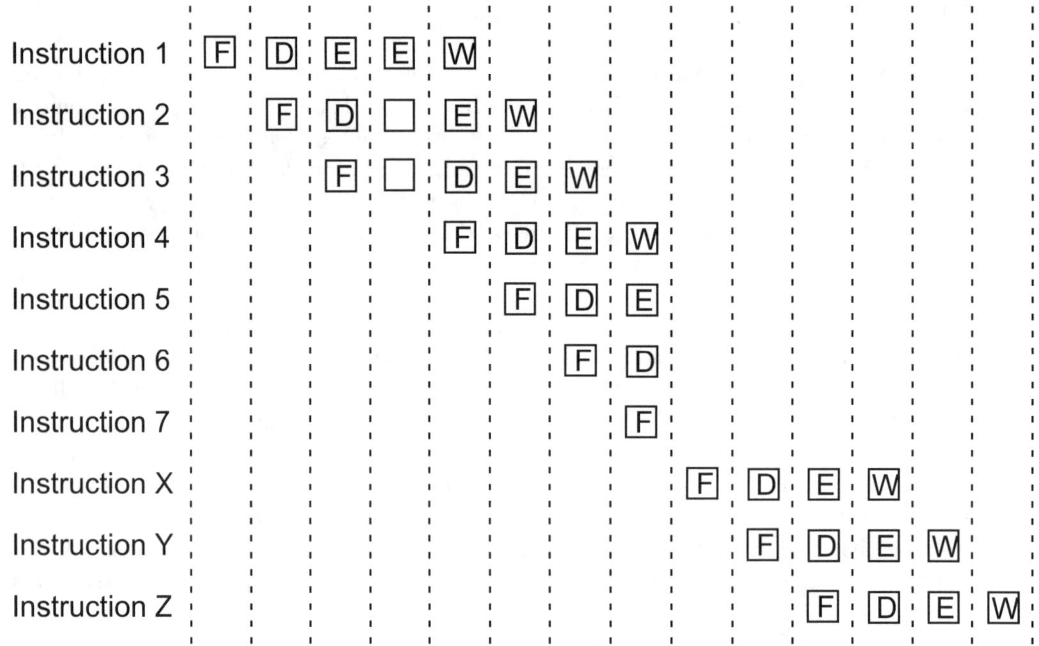

Figure D.5: Pipeline operation of instructions in Figure D.4.

D.1.6 Single- versus Multiple-Issue

The processors described to this point have been *single-issue* processors, that is, they only execute one operation at a time. Even though pipelining introduces a form of parallelism, the actual execution of the instructions is still sequential. For example, if there were a series of three ADD instructions, the actual addition operations would occur in the Execute stage of the pipeline, and so would occur in sequence. In order to speed up execution further, a *multiple-issue* processor is designed to execute more than one operation in parallel. Multiple-issue processors require the evaluation of a number of considerations.

- Clearly, to do more than one operation requires that the processor have multiple instances of the required hardware. Examples of *functional units* would typically be ALUs, specialized adders and multipliers, and load-store units.

- If the processor is going to execute instructions in parallel, then it must be able to determine which can be safely executed at any given time. This issue will be discussed in more detail in Section D.1.7.

- It is likely that the functional units will have unequal delays, so even if instructions are executed in order it is possible (and likely) that the operations will be completed out of order. This makes the state of the processor at any instant (i.e., which instructions are executed and which are not) difficult to determine. So, there must be a mechanism in place to ensure that the operation results are stored in the proper order to preserve their interdependencies, and to permit the processor to precisely suspend and restore execution in order to service interrupts.

D.1.7 Scheduling

On a multiple-issue processor, *scheduling* is necessary to determine the order in which instructions can safely be executed. The scheduling must ensure that any dependencies between instructions are preserved as they are executed and their results stored. For example, in the example below, the first and third instructions can be executed in any order since they are independent, but the second instruction may *only* be executed when the result from the first instruction is available.

```
ADD R1, R2, R3    adds R1 & R2 and places result in R3
MUL R1, R3, R4    multiplies R1 & R3 and places result in R4
ADD R1, R5, R6    adds R1 & R5 and places result in R6
```

Scheduling can either be completed in real-time on the processor hardware (dynamic scheduling) or can be done in advance by the code generation tools (static scheduling). This choice leads to fundamentally different architectures that are each well suited to specific environments.

Dynamic scheduling is primarily used with code that was not written to be intrinsically parallel. The processor attempts to discover parallelism in the code (*instruction level parallelism*, or ILP) by examining the interdependencies between instructions, managing functional unit assignments, and ensuring that results are store in the proper order. This gives dynamic scheduling a great advantage in that it can exploit parallelism in code that was not written in a parallel fashion, permitting it to run unaltered serial code on a parallel architecture (dynamic scheduling is used in nearly all processors used in computer workstations). The primary disadvantage is that scheduling is a non-trivial process, so doing it in real-time requires a significant additional amount of hardware with the attendant cost and power consumption. Due to the hardware requirements, typical dynamic scheduling hardware is only able to look at a window of only several hundred instructions, limiting its ability to find parallelism. *Superscalar* processors use multiple functional units and dynamic scheduling, with the processor enforcing all dependencies between instructions. The exact execution order is not known until run-time, but execution is guaranteed to produce the same results as would serial execution of the code.

Static scheduling removes the burden of scheduling from the processor and instead requires code that explicitly specifies what instructions may be executed in parallel. In this case, the compiler is responsible for determining which instructions may be executed in parallel, and for ensuring that the result of an operation will in fact be ready before it is used in another instruction. The primary advantages of static scheduling are that the compiler can look for parallelism across the entire program and so hopefully determine a more efficient execution order, and that it eliminates the need for the dynamic scheduling hardware — a significant power and cost savings. The main disadvantage is that the code that is generated is very machine dependent and may be less adaptive to changing system dynamics. The statically scheduled *very-long instruction word* (VLIW) architecture used in the TMS320C6x fetches eight instructions in parallel (a *fetch packet*) to simultaneously pass to its eight functional units. If a functional unit is not used, then it is passed a no-operation (NOP) instruction. The high-level language compilers do all instruction scheduling and enforce dependencies. Writing assembly language for this architecture is a challenge, but is typically only done for very time critical code to maximize the functional unit utilization and reduce execution time.

For more detailed reading on VLIW processor design and their use in embedded systems, the reader is referred to the book by Fisher, Faraboschi, and Young [75].

D.2 TMS320C671x Architecture

The heart of the TMS320C671x DSP is an 8-way VLIW core with 32 general purpose registers (A0-15, B0-15) and eight functional units split into two clusters as shown in Figure D.6. Each functional unit has a primary specialization but most are capable of multiple operations. The primary functions of each are:

Unit	Integer operations	Floating-point operations
.L	Logical	Arithmetic
	Arithmetic / compare	Integer/floating-point conversions
.S	Shifts and bit fields	Compare
	Logical	Reciprocal
	Arithmetic	Reciprocal square root
	Branches	Absolute value
	Constant generation	Single-/double-precision conversions
.M	Multiply	Multiply
.D	Load and store	Load and store
	Address calculation	
	Addition/subtraction	

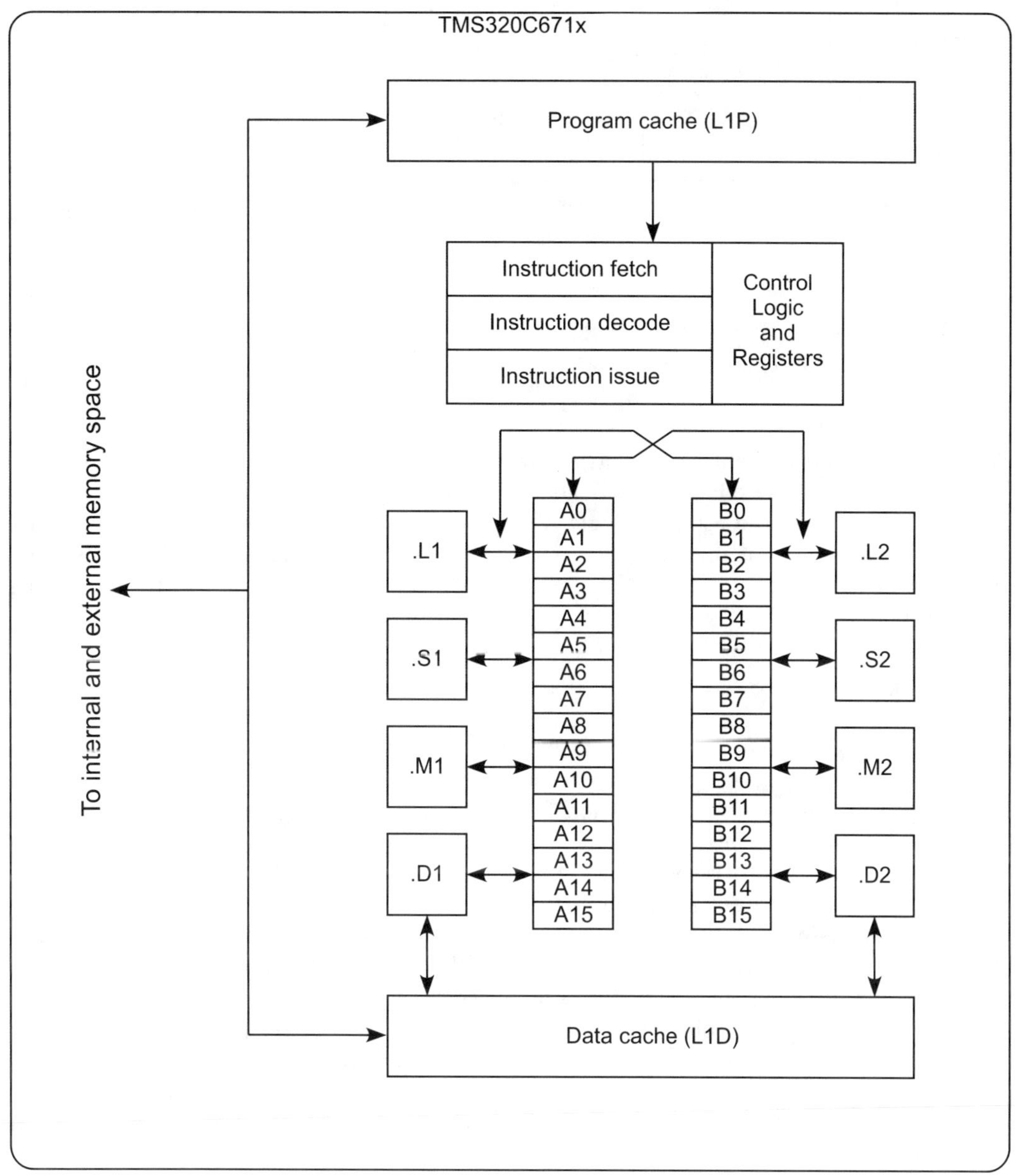

Figure D.6: TMS320C671x core.

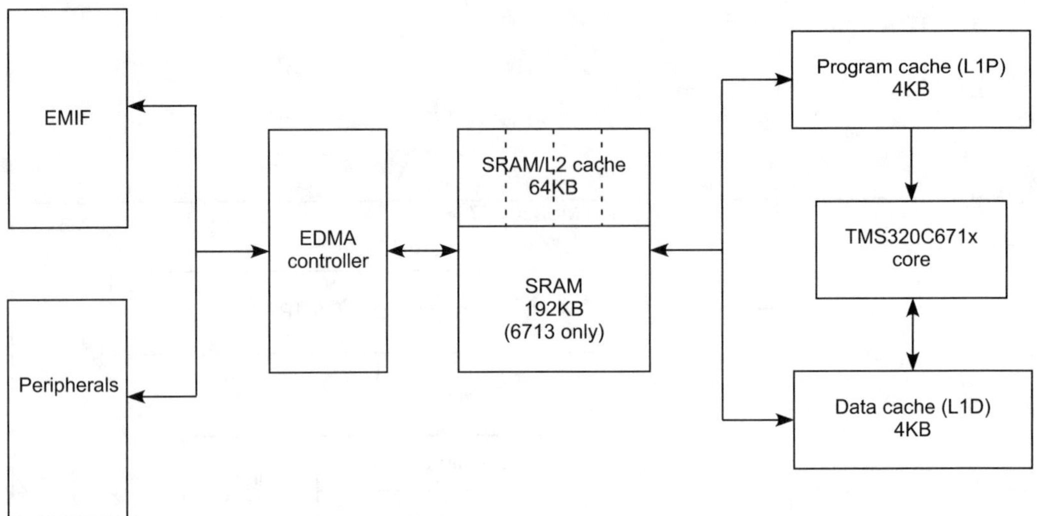

Figure D.7: TMS320C671x memory organization.

The A and B register banks both have data buses for transferring data to and from the functional units associated with them, as well as for loading and storing operands. Two cross paths to permit the use of a single A-side register with a B-side functional unit, and vice versa. As one might imagine, determining the optimum combination of registers and functional units to use and efficiently scheduling the instruction stream can be a daunting task.

D.2.1 Memory System

The overall layout of the TMS320C671x processor's memory organization is shown in Figure D.7. The processor core must access memory through one of two caches, one for data operands (L1D) and one for instruction fetches (L1P). This notation is used to indicate that these are level 1 caches, that is, the caches closest to the processor core. These caches do not appear in the processor memory map, and are not directly accessible to the programmer. If the data is not available in the L1 cache or is being written to memory, then the request is passed to the L2 cache/SRAM controller. This controller manages a 64KB block of memory that can be used either as a level 2 cache or as simply memory, in 16KB blocks, by programming the associated control registers. For programs that need 64KB or less of memory, the entire block can be left as memory. If a larger program is being run from off-chip memory, then the cache memory can significantly improve execution time. Or, if a program may be able to be made more efficient by using some portion of the memory as L2 cache and the rest used for on-chip memory for frequently accessed data or code. In the TMS320C6713, there is an additional 192KB block that can only be used as on-chip memory.

If a memory request cannot be filled by the L2 controller (the requested data is not in the L2 cache or memory), it is passed to the enhanced direct memory access (EDMA) controller, which then executes the transfer. The EDMA controller is an on-chip peripheral device that can be programmed to automatically transfer data under software control, however, those transfers requested by the L2 controller are transparent to the programmer. Since the processor may well be fetching instructions and transferring data at the same time, the EDMA controller is designed to handle simultaneous requests according to a priority scheme. Based on the address of the memory request, the data transfer will be made to on-chip memory, one of the on-chip peripherals, or to an external memory device using the external memory interface (EMIF).

The EMIF is designed to provide a *glueless* interface to a number of different external memory devices, including various dynamic memories (SDRAM, SBSRAM, etc.) as well as static memory devices (ROM, SRAM, FIFOs, etc.). The term glueless refers to the fact that the processor and the memory device can be connected directly together without requiring any additional logic devices to create a compatible interface (this interface logic is commonly referred to as *glue logic*).

D.2.2 Pipeline and Scheduling

The TMS320C671x's VLIW architecture utilizes static scheduling. The hardware handles pipeline stalls typically caused by cache misses, but does not enforce any scheduling. Branch prediction is not used, but predicated execution instructions allow for conditional code without forcing pipeline flushes. The functional units require different amounts of time to complete their operations, for example, the ADD instruction result is available in the next clock cycle, but the MPY (multiply) instruction result is not available until after an additional 1 clock delay. The branch instructions execute such that branch does not actually occur until 5 clocks after it is executed. The combination of all these constraints makes simply reading the assembly language difficult, and makes assembly language programming much more complex and tedious. Because of this, most programming is done in a high-level language such as C, and only very time-critical routines are hand-coded in assembly language.

Since the functional units are pipelined, they also can contain a number of instructions in the process of execution. For example, to perform a tight loop where a single execution packet is repeatedly executed, the branch unit would have multiple branch instructions in-flight, with a branch instruction completing at every clock cycle until the loop finished.

D.2.3 Peripherals

The TMS320C671x DSP incorporates a number of on-chip peripheral devices. Timers are typically used to provide precise periodic interrupts to the DSP, to generate rectangular waveforms for external devices (i.e., a clock for an ADC), and to count external events. Each timer consists of a counter register that is incremented by the DSP clock or an external signal, a period register that is used to determine when to reset the counter register, and a control register that is used to configure the timer. Basically, the timer operates by incrementing the counter register until it matches the period register; when they match the counter is reset to 0 and an interrupt is generated. This interrupt provides a very accurate time reference. One of the common uses of the timer interrupt is to permit the DSP/BIOS operating system to gain control of the DSP at prescribed intervals in order to perform preemptive multitasking.

The EDMA controller is used to offload data movement from the DSP. As mentioned above, the EDMA controller operates transparently in handling transfers required by the L2 cache controller. It can also be explicitly programmed to service interrupts by transferring data between the device and memory, or to move data from one region in memory to another. The EDMA controller is capable of sophisticated data movements, such as moving a two-dimensional region of a large array to a smaller array so that the extracted data is contiguous in memory.

Serial ports provide an interface mechanism to communicate with serial devices (i.e. many codecs) using a number of different formats. Once configured, they permit the DSP to simply write a value to the serial port in order to have it sent in the appropriate serial format (or to read a value that has been received). The serial ports can be serviced directly by the DSP, or the EDMA controller can be used to automatically transfer serial port data to and from memory. The EDMA controller is then programmed to interrupt the DSP when a complete frame of data has been transferred. This eliminates the need for the processor to perform basic data movements, freeing up processing power for more complex tasks.

D.2.4   Host Port Interface

The TMS320C6x DSPs all incorporate a Host Port Interface (HPI). The HPI makes it possible for an external (host) processor to access the entire memory space of the DSP. In addition to reading and writing memory locations, the host processor can also configure any of the DSPs peripherals. The DSP is also designed so that it can be forced to boot using the HPI port. In this mode, the DSP holds itself in reset until signalled by the host processor to begin execution. While the DSP is in reset, the host can load the desired program into the memory of the DSP, configure the peripherals, and then issue a *host port interrupt* to start the DSP running. While the DSP is running, the host processor can still read and write to the DSP memory space.

The Host Port Interface provides a very effective means for a central processor to control the operation of multiple DSPs. It is also used for high-speed communication between two DSPs, or a DSP and general-purpose processor. The winDSK6 software and the other software tools described in Appendix D all use the HPI.

Appendix E

Related Tools for DSKs

E.1   Introduction

THIS appendix contains information on several tools that are available for use with the TMS320C6211, TMS320C6711 and TMS320C6713 DSKs. All of these tools use the DSP's Host Port Interface (HPI) for control and communication. (Note that to use the tools with the TMS320C6713 DSK, the board must be equipped with the Educational DSP, LLC HPI daughtercard.) The Host Port Interface provides external access to the DSPs memory space, as described in Appendix D. This allows the tools to download and start programs, then read and write DSP memory locations to get data back from the DSK and to control the program.

E.2   Windows Control Applications

To control the DSK from a Windows application, programs must be created for both the host computer and the DSK. The sample host computer Windows application is written in `MicrosoftVisualC++6.0`. The interface between the host computer and the DSK can be parallel, serial, or USB. The details of this interface are hidden in a dynamic link library (DLL) file that is included with the host computer program. To transfer data to and from the DSK, the host computer must know the variable addresses where the data is stored on the DSK. To simplify this process, a predefined data structure is used, and the interface software has the ability to determine where the data structure is located in the DSK's memory space.

The host computer can perform a few basic operations:

- Reset the DSP.

- Load a program onto the DSP.

- Start the DSP program.

- Read and write DSP memory.

Control of a DSP program is implemented by writing to variables in the DSP memory space. Program status and output data is obtained by reading from variables in the DSP memory space. However, keeping track of the specific addresses of all the variables in the DSP program is tedious and error-prone, since variable locations can change each time a program is recompiled. To simplify the process of finding variable addresses, the DSP software establishes a special data structure (*HPI_Block*) so that the variables will be in a known location.

The host software first loads the program onto the DSK, then establishes the location of the *HPI_Block* structure by reading the DSK memory location at 0x00000200. The address of the HPI_Block structure is stored at this location by the code in the `HPI_Block.asm` file. This address can then be obtained in the host application using the function *GetH-piBaseAddress()*. The host application can determine a variable's address by adding the desired variable's offset within the HPI_Block structure to the value returned by GetH-piBaseAddress. This address is then used in the `HPI_Read` and `HPI_Write` functions.

E.2.1   The Basic Windows Control Application

The basic Windows Control application (in the Appendix E `DSK6_CONTROL` directory) implements a simple audio talk-through with a gain control. Highlights of the software are shown below. At least a cursory reading of all the source code is advised for a better understanding of the interface structure.

The HPI_Block Variable

[Refer to `DSK6_CONTROL\DSP\TalkThru\HPI_Block.c` and `HPI_Block.h`] The first six entries in the HPI_Block structure are dedicated to initialization and control variables used by the host software. These must not be changed or deleted. The `HPI_Block.Codec` variable is used to control which codec support software is used by the DSK program. This allows the same program to be used for all of the supported TMS320C6x DSKs and codecs. The host program must set the value of this variable once the program is running.

Since all HPI transfers must be 32-bits, always be sure that any added HPI_Block member variables are at least 32-bits in width. Note that all member variables must be declared *volatile*. For the basic windows control application, the member variable HPI_Block.Gain has been added to control the audio gain.

The Host Software

[Refer to `DSK6_CONTROL\C6X_CONTROLDlg.cpp`] The host software is written to include the proper initialization software for all supported DSKs and codecs. Detailed documentation of the host software function calls is available in the document `DSK6_CONTROL.pdf` in the Appendix E directory.

The `CC6X_CONTROLDlg::OnInitDialog()` function is executed when the application dialog window is created. The function `C6XCONTROL_StartUp()` is called to initialize the host interface software. Then, the interface is set for the desired mode by ensuring that all but the correct pair of lines are commented out. After this, the application waits for the user to click the Reset or Run button. When the Reset button is clicked, the `CC6X_CONTROLDlg::OnResetButton()` function is executed and causes the DSK to be reset. When the Run button is clicked, the `CC6X_CONTROLDlg::OnRunButton()` function is executed, loading a program onto the DSK and starting it running. The host then sets the `HPI_Block.Codec` variable to select the correct codec support software. (The DSK application waits for the host to write the `HPI_Block.Codec` variable before configuring the interface to the codec.)

As the gain slider is moved, the `CC6X_CONTROLDlg::OnReleasedcaptureSlider1()` function is executed when the left button is released. This function gets the current position of the gain slider and maps its position to a floating-point value on a logarithmic scale. This value is then written to the `HPI_Block.Gain` variable on the DSK, controlling the audio gain. Note that the variable address is computed as the sum of `C6XCONTROL_getHPI\` `_BaseAddress()+HPI_Block_Application_Base+0`. The first term gives the address of the HPI_Block structure, the second term accounts for the 6 required member variables, and the third term is the offset of this variable past the required terms. If another float variable was added to HPI_Block, its address would be `C6XCONTROL_getHPI_BaseAddress()` `+ HPI_Block_Application_Base + 4`.

The DSK Software

[Refer to `DSK6_CONTROL\DSP\TalkThru\main.c`] In the `main()` function, the DSP waits for the host to set the `HPI_Block.Codec` variable to a non-negative value. Then, it initializes the codec support, enables interrupts, and goes into an infinite loop. The actual program "work" is then accomplished by the interrupt service routines.

[Refer to `DSK6_CONTROL\DSP\TalkThru\codec.c`] The interrupt service routines in `codec.c` handle the codec input and output data. (The 6211/6711 onboard codec uses McBSP0, the other codecs all use McBSP1.) The incoming data is stored in the variable `PcmDataIn`, with stereo codec data stored as two 16-bit samples in the 32-bit variable. The outgoing data is obtained by multiplying the individual channels of `PcmDataIn` by the `HPI_Block.Gain` variable and storing the result in `PcmDataOut`. `PcmDataOut` is then sent to the codec.

[Refer to `DSK6_CONTROL\DSP\support.c`] If the 6713 DSK is being used, the codec audio source can be selected from one of three possibilities:

- line input,

- microphone input,

- or, microphone input with 20db amplification.

The codec input source can be selected by ensuring that only the desired line for codec register 4 is not commented out in the `Init_AIC23()` function.

Building and Running the Application

To prepare to run the application, first open the DSP project in Code Composer Studio. Make any needed changes for the DSK being used, and then build the project in a *Release* configuration. Then, open the Visual C++ project and ensure the code support is correct for the DSK and codec being used. Build and run the program — you will now be able to control the audio gain of the DSK from the host computer.

E.2.2 Creating an Oscilloscope Application

The basic Windows Control application (`DSK6_CONTROL`) is intended to be a starting point for modification and extension. In this section, the application will be modified to serve as an oscilloscope display of the DSK audio input data. The modifications listed below are to be applied to a clean copy of the `DSK6_CONTROL` project. All pathnames shown in this section are relative to the top-level directory of the project. A complete version with all modifications incorporated is available in `DSK6_CONTROL_OSCOPE`.

To function as a basic oscilloscope, the incoming audio data must first be stored in a buffer on the DSK. The current index value (which is equal to the number of samples currently in the buffer) is stored in an HPI_Block member variable. When the index is equal to the buffer length, the buffer is full and the host computer can read the data. After the host reads the data, it clears the index to zero so that the DSP can store new data into the buffer. The data must then be drawn in the host application dialog. Note: this implementation is intended to be easy to understand, not a high performance application. Therefore, no ping-pong buffering or high speed graphics mechanisms are used.

DSP Code Modifications

Open the DSP project (`.\DSP\TalkThru\talkthru.pjt`) in /ccs/. The first order of business is to define the required variables in the DSP application. First, a data buffer must be declared and allocated — it can be done in several locations but here we choose to use `hpi.h`. The length of the buffer is somewhat arbitrary — a length of 512 sample pairs is chosen here. Add a declaration for the buffer above the HPI_Block instantiation.

```
1  #define BUFFER_LENGTH 512
   extern volatile unsigned int Buffer[BUFFER_LENGTH];
```

Then, the HPI_Block declaration in `hpi.h` needs to be modified to incorporate two new member variables in the application specific data area. One will store the buffer address, and the other will be used to store the current buffer index. (Although the *Gain* member variable will not be used, it will be left in place.)

```
   /* application specific data */
2  volatile float Gain;
   volatile unsigned int Index;
4  volatile unsigned int *BufferAddress;
```

Now, in `hpi.c`, allocate the buffer above the HPI_Block allocation.

```
volatile unsigned int Buffer[BUFFER_LENGTH];
```

Add the initializers for the new member variables. The *Index* variable is set to 0 to indicate that no data has been stored in the buffer. The *BufferAddress* variable is set to the address of the buffer — the host will read this value to get the buffer address.

```
1  // add additional variables below
   1,                    // Gain
3  0,                    // Index
   Buffer                // BufferAddress
```

Now, the interrupt service routines must be modified to store the incoming data in the buffer. The code below is to be inserted into both of the receive ISRs (`McBSP1_Rx_ISR()` and `McBSP1_Rx_ISR()`). The software checks the index value to determine if there is space remaining in the buffer. If space is available, the incoming data is stored in the buffer and the index incremented. Otherwise, the data is not stored and the index is not changed.

```
   // store data until buffer is full
2  if(HPI_Block.Index < BUFFER_LENGTH) {
       Buffer[HPI_Block.Index++] = PcmDataIn.uint;
4  }
```

Build a *Release* version of the project.

Host Code Modifications

Open the host project (`C6X_CONTROL.dsw`) in Visual C++. The modifications to the host project are more involved. The host application needs to load the program and start it running as before, but then it needs to periodically check the DSP to see if the buffer is full. The obvious solution to this would be to stay in a tight loop, reading the HPI_Block.Index variable until it reached the BUFFER_LENGTH value. However, in the Windows environment, remaining in a tight loop will prevent any other applications from receiving messages from the graphical environment, effectively "locking up" the display if something goes wrong. One solution to this is to use a Windows timer to periodically check the DSP state. There are much more sophisticated solutions (i.e., multithreading), but for understandability we will use the timer solution.

In the `C6X_ControlDlg.h` file, we need to add a buffer to store the data read from the DSK. Also, we will define the buffer length as in the DSP project, and define a number for use in activating and identifying our timer messages. Place the code below directly above the `CC6X_CONTROLDlg` class declaration.

```
#define BUFFER_LENGTH 512
#define SCOPE_TIMER_ID 0x12345678
```

Then, add a buffer variable to the `CC6X_CONTROLDlg` class declaration immediately below the *DskIsRunning* variable declaration.

```
unsigned long Buffer[BUFFER_LENGTH];
```

In the `C6X_ControlDlg.cpp` file, add the declarations below for the offsets of the HPI_Block variables. It is good programming practice to put the HPI_Block member variable offset calculations all together in one place to avoid errors. These should be placed below the last *#include* line at the top of the file.

```
// HPI_Block offsets
const unsigned long GainOffset            =
                    HPI_Block_Application_Base +  0*4;
const unsigned long BufferIndexOffset     =
                    HPI_Block_Application_Base +  1*4;
const unsigned long DataBufferAddressOffset =
                    HPI_Block_Application_Base +  2*4;
```

For consistency, we should update the gain slider message handler function. Change the line in `CC6X_CONTROLDlg::OnReleasedcaptureSlider1()` to match the code shown below.

```
if(!C6XCONTROL_WriteFloat(C6XCONTROL_getHPI_BaseAddress() +
                GainOffset, 1, &data)) {
```

To use a Windows timer, several things need to be done. First, we will create a message handler function, the add code to start and stop the timer when necessary. To add the message handler,

- Right-click in the `C6X_ControlDlg.cpp` file, and select *ClassWizard...*,

- Select the *WM_TIMER* message as shown in Figure E.1, and click the `AddFunction` button. This will add the message handler function to the `C6X_ControlDlg.cpp` file.

- Close the ClassWizard dialog.

In this application, the timer should be started when the *DSK Run* button is clicked, and stopped when the *DSK Reset* button is clicked. The `SetTimer()` function is used to start a Windows timer. Place the below at the end of the `CC6X_CONTROLDlg::OnRunButton()` function. This code will cause a timer message to be issued every 200ms, with the message identified by the `SCOPE_TIMER_ID` we defined previously.

```
SetTimer(SCOPE_TIMER_ID, 200, NULL); // timer at 5Hz
```

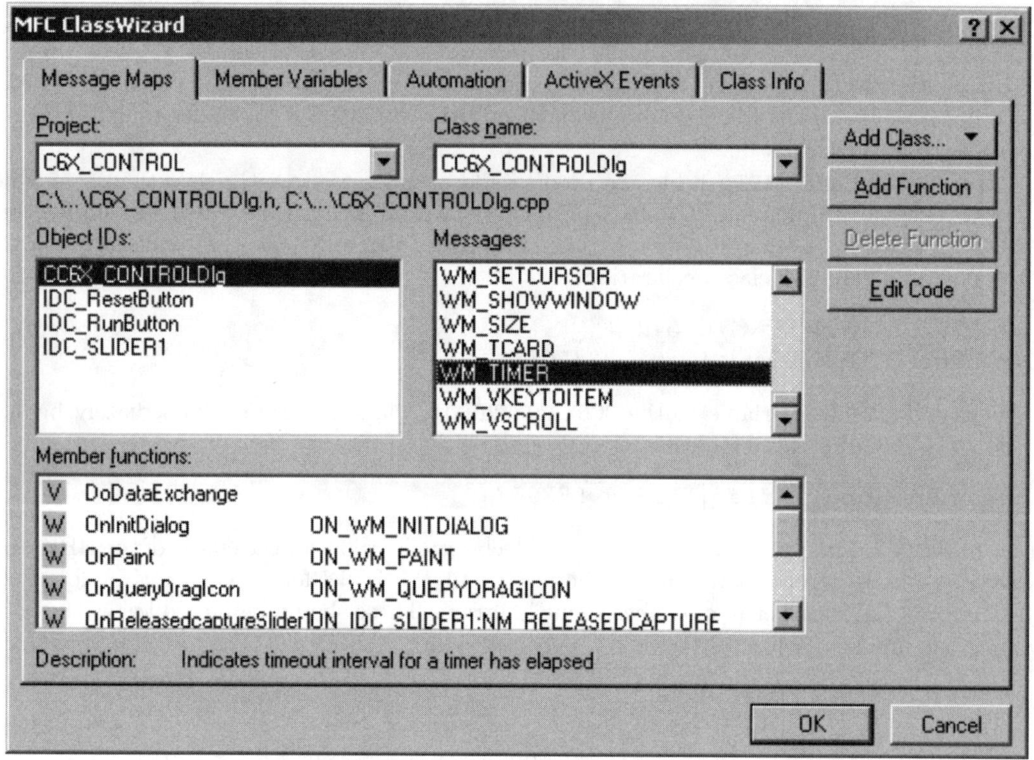

Figure E.1: MFC Class Wizard.

The `KillTimer()` function is used to stop the timer. Place the code below at the beginning of the `CC6X_CONTROLDlg::OnResetButton()` function. We do not need to worry about stopping the timer if the application is closed, this is handled automatically by the Windows framework.

```
1  KillTimer(SCOPE_TIMER_ID);
```

For simplicity, this application will plot the audio data directly on the application's dialog window. To support this, the dialog needs to be made larger and the control buttons moved. This can be accomplished easily in the Visual Studio Resource Editor window. Open the application dialog, drag the dialog to a larger size, and move the buttons to one side. A sample arrangement is shown in Figure E.2.

The remaining task to be accomplished is adding the required code to the timer message handler function to implement the functionality listed below.

- Verify that the timer message is the one we requested.

- Check if data is available on the DSK. If not, exit the message handler function.

- If available, read the data from the DSK and reset the index to 0.

- Erase the display area by drawing a background rectangle.

- Plot the new data. For simplicity, this function will only paint the data for a single audio channel.

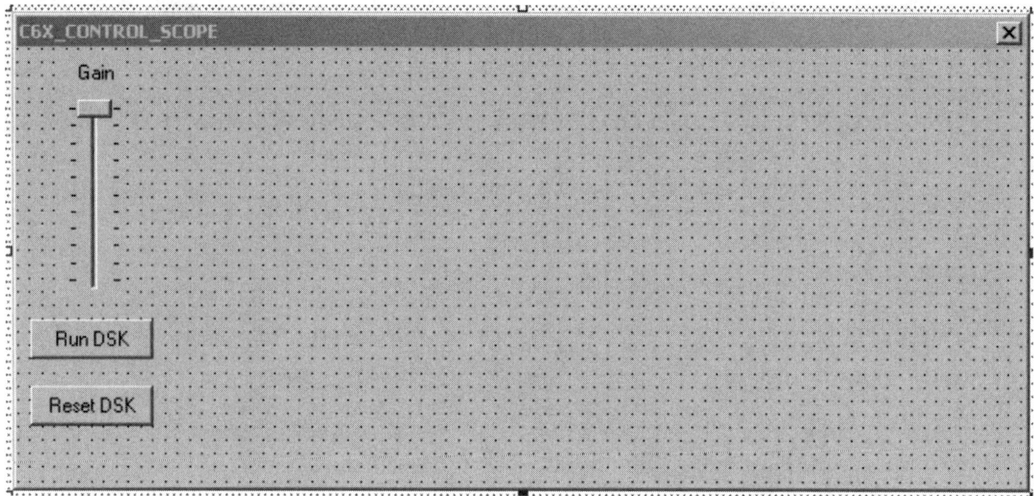

Figure E.2: Sample application dialog.

The code required to implement this functionality is shown below. This should be used to completely replace the body of the `CC6X_CONTROLDlg::OnTimer()` function.

```
if(nIDEvent == SCOPE_TIMER_ID) { // make sure is our timer

    // check if DSK data is ready
    unsigned long index = 0;
    if(!C6XCONTROL_Read(C6XCONTROL_getHPI_BaseAddress() +
            BufferIndexOffset, 1, &index)) {
        AfxMessageBox("DSK Read Failed");
        EndDialog(IDCANCEL);  // close dialog
        return;
    }
    if(index < BUFFER_LENGTH) // is data ready?
        return; // if not, just exit

    // read the buffer address from the DSK
    unsigned long buffer_address = 0;
    if(!C6XCONTROL_Read(C6XCONTROL_getHPI_BaseAddress() +
            DataBufferAddressOffset, 1, &buffer_address)) {
        AfxMessageBox("DSK Read Failed");
        EndDialog(IDCANCEL);  // close dialog
        return;
    }

    // read the buffer data from the DSK
    if(!C6XCONTROL_Read(buffer_address,
            BUFFER_LENGTH, Buffer)) {
        AfxMessageBox("DSK Read Failed");
        EndDialog(IDCANCEL);  // close dialog
        return;
    }
```

```
31      // reset the index on the DSK
        index = 0;
33      if(!C6XCONTROL_Write(C6XCONTROL_getHPI_BaseAddress() +
                BufferIndexOffset, 1, &index)) {
35          AfxMessageBox("DSK Write Failed");
            EndDialog(IDCANCEL);  // close dialog
37          return;
        }
39
        // get a device context for plotting
41      CDC* pDC = GetDC();

43      // clear the plot area
        int xpos = 100, ypos = 132;
45
        CRgn rect;
47      rect.CreateRectRgn( xpos-1, ypos-128,
                            xpos + BUFFER_LENGTH + 1, ypos+129);
49      pDC->PaintRgn(&rect);

51      // plot the data
        // only the upper 8 bits of one channel are displayed
53      pDC->MoveTo(xpos, ypos + (char)(Buffer[0] >> 8));
        for(int i = 1;i < BUFFER_LENGTH;i++)
55          pDC->LineTo(++xpos, ypos + (char)(Buffer[i] >> 8));
        }
57      else
            CDialog::OnTimer(nIDEvent);
```

With these changes made, build and run the program. Click the Run DSK button and
a display similar to Figure E.3 should be seen.

Some further enhancements to consider for the oscilloscope application would be to

- add triggering to synchronize periodic waveforms on the display,

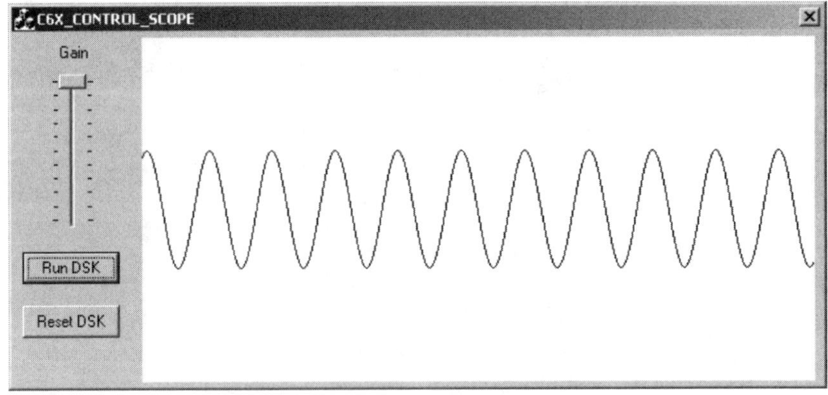

Figure E.3: Oscilloscope application dialog.

- add a gain control to change the display magnitude, or

- plot both channels for stereo codecs.

E.2.3 Creating a Spectrum Analyzer Application

The oscilloscope application provides an excellent base from which to build a spectrum analyzer application. Conceptually, the two are very similar, but instead of reading audio data from the DSK, the host reads spectral data and displays it instead. Most of the modifications are to the DSP application. These include:

- The data buffer must be converted to type float, with the left/right channel data sent to separate buffers. (Or one channel can be discarded and only a single buffer used.)

- The HPI_Block.Index variable is no longer used as the flag to the host indicating that data is ready. Rather, it can be used in the `main()` function's forever loop. When the data buffers are full, they can be processed in the `main()` loop.

- When the `main()` loop processing completes, the host should be notified to read the spectral data from the correct buffer location. When the DSP sees the host has read the software, the index value can be reset to 0 to start gathering a new frame of data.

- A new HPI_Block variable needs to be added to serve as the flag to the host. The flag should be set to 1 when the spectral data is available, and reset to 0 by the host when it has read the data.

Modifications that need to be made to the host computer application include:

- The host computer must read the correct flag in the HPI_Block, not the value of the HPI_Block.Index. When this flag is non-zero, the host should read the data, then reset the flag to 0.

- The data buffer must be converted to type `float`. If two buffers are used on the DSP, there must be two buffers on the host. The data must be read using the `C6XCONTROL_ReadFloat()` function, since it is stored in float format.

- The plotting routine must scale the data appropriately for the plotting. In the oscilloscope application, there are 256 pixels of vertical resolution.

Detailed instructions for making the changes are not given here. However, a completed spectrum analyzer project is available in the `DSK6_CONTROL_SPECTRUM` directory of Appendix E. A sample display from this program is shown in Figure E.4.

E.3 MATLAB Exports

Using the MATLAB program SPTool is a convenient, graphical way to design digital filters. To be able to use those designs in Code Composer Studio (CCS), we need to export them into a C language format. There are four MATLAB m-files discussed in this appendix and included on the CD that can be used to help automate this process. In all cases, two files are created, a C header file (`filename.h`) declaring the variables, and C source file (`filename.c`) defining them. You can use any filename you wish by specifying it in the argument list for the m-file, as shown in the examples below. These files can then be included in your CCS project. In this appendix it is assumed that the reader is familiar with using MATLAB and SPTool. It is recommended that you add a MATLAB path to the directory where you have installed the m-files.

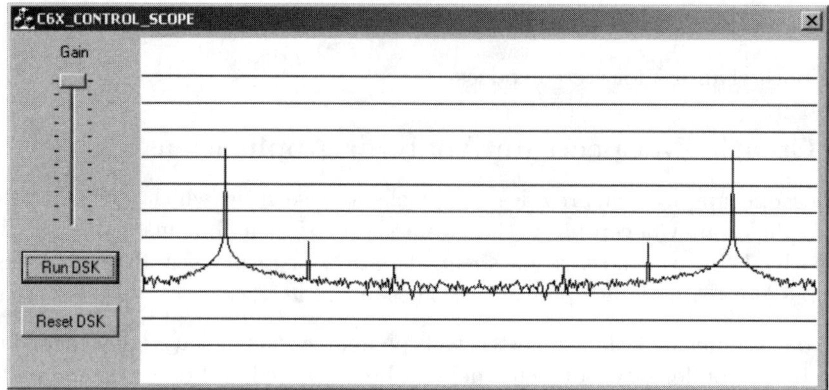

Figure E.4: Spectrum analyzer application dialog.

E.3.1   Exporting Direct-Form II Implementations

The `filt` structure created by MATLAB SPTool contains Direct-Form II numerator coefficients in `filt.tf.num` and denominator coefficients in `filt.tf.den`. If the filter is designed as a finite impulse response (FIR) filter, then only the numerator coefficients need to be exported using `fir_dump2c.m` for floating-point coefficients, or `fir_dump2c_Qxx.m` for fixed-point coefficients.

To use `fir_dump2c.m`, the following steps should be taken:

- Export the filter design from SPTool to the workspace. Ensure that you have specified an FIR filter design. (The remainder of this procedure assumes that filter design was exported with name `filt1`.)

- Execute a MATLAB `cd` command to change to the desired destination directory for the exported files.

- Run the m-file by typing

  `fir_dump2c('coeff','B',filt1.tf.num,length(filt1.tf.num))`

  at the MATLAB command line.

This creates two files, `coeff.c` and `coeff.h`, which declare a `float` array B of length `B_SIZE`. To use `fir_dump2c.m_Qxx`, the following steps should be taken:

- Export the filter design from SPTool to the workspace. Ensure that you have specified an FIR filter design. (The remainder of this procedure assumes that filter design was exported with name `filt1`.)

- Execute a MATLAB `cd` command to change to desired destination directory for the exported files.

- Run the m-file by typing

  `fir_dump2c_Qxx('coeff','B',filt1.tf.num,length(filt1.tf.num),15)`

  at the MATLAB command line.

This creates two files, `coeff.c` and `coeff.h`, which declare a `short` array B of length `B_SIZE`. The last parameter (`Qxx`) determines the location of the binary point. Fixed-point number representations are discussed in Appendix C.

If the filter is designed as an infinite-impulse response (IIR) filter, then the numerator and denominator coefficients can either be exported individually using the methods described for FIR filters above, or both can be exported simultaneously using `df2_dump2c.m`.

To use `df2_dump2c.m`, the following steps should be taken:

- Export the filter design from SPTool to the workspace. (The remainder of this procedure assumes that filter design was exported with name `filt1`.)

- Execute a MATLAB `cd` command to change to desired destination directory for the exported files.

- Run the m-file by typing

  `df2_dump2c('HPF_coeff','HPF',filt1.tf)`

  at the MATLAB command line.

This creates two files, `HPF_coeff.c` and `HPF_coeff.h`, which declare the `float` arrays `HPF_A` of length `HPF_A_SIZE` (denominator coefficients), and `HPF_B` of length `HPF_B_SIZE` (numerator coefficients). The array length is determined by the length of the `filt1` numerator and denominator vectors.

E.3.2 Exporting Second Order Section Implementations

An SPTool filter design can be converted from direct-form II to second order sections by using the MATLAB function `tf2sos` (you may want to type `help tf2sos` in MATLAB for more details). Running `tf2sos` creates an $L \times 6$ matrix, where L is the number of second order sections needed to implement the filter, with each row of the matrix containing the coefficients $(b_0, b_1, b_2, a_0, a_1, a_2)$ for a single second order section. These section order section coefficients can be exported using `sos_dump2c.m`.

To use `sos_dump2c.m`, the following steps should be taken:

- Export the filter design from SPTool to the workspace. (The remainder of this procedure assumes that filter design was exported with name `filt1`.)

- Convert the filter design to second order sections by typing

  `filt1.sos=tf2sos(filt1.tf.num,filt1.tf.den)`

  at the MATLAB command line.

- Execute a MATLAB `cd` command to change to desired destination directory for the exported files.

- Run the m-file by typing

  `sos_dump2c('coeff','bqd_coeff',filt1.sos,size(filt1.sos,1))`

  at the MATLAB command line.

This creates two files, `coeff.c` and `coeff.h`, which declare the 2-dimensional `float` array `bqd_coeff` of size `bqd_coeff_SIZE`-by-5. The a_0 coefficient is assumed to be 1 and is not used in the actual filter implementation, so it is ignored in the export.

E.4 MATLAB Real-Time Interface

The MATLAB real-time interface is a software tool that permits MATLAB to interface directly with a DSK. Data can be imported from the DSK inputs into MATLAB variables, and variables can be written to the DSK outputs. The data transfer capabilities are limited by the bandwidth of the host PC to DSK connection, and the speed of the host computer. At lower sample frequencies, real-time behavior can be maintained. At higher sample frequencies, only a portion of the total codec data stream will be able to be transferred.

An interface that allows the direct importation of real-time DSK data into MATLAB can be used for a number of purposes. The most basic approach is to simply use the DSK as a data acquisition board to obtain live data, and perform all signal processing in MATLAB. This also permits the use of the MATLAB visualization features with real-time DSK data. An interesting example of this approach was the development of real-time, acoustic beamforming systems using multichannel analog input daughtercards on the DSK. Details of these projects are available in a number of references, including [27, 31, 37].

The MATLAB real-time interface driver software and example MATLAB scripts are available in the Appendix E `MatlabInterface` directory. Detailed descriptions of the interface functions are available in the Appendix E `MatlabInterface\Matlab_API.pdf` document.

Appendix F

Programming Perils and Pitfalls

PROGRAMMING in a real-time environment can be challenging even for experienced programmers. This appendix is intended to illustrate some of the common problems that are encountered in this environment, and present practical strategies for avoiding them.

F.1   Debug versus Release Builds

When a project is created in Code Composer Studio, there are two build configurations that are established; *Debug* and *Release*. The debug configuration will embed debugging information in the object file (information that links assembly instructions to the original source code), and also will not optimize the generated code so that there is a direct correspondence between a line of source code and the assembly language that is generated. These permit symbolic debugging, and ensure that the assembly code will execute in the order that the C code was written. The debug configuration is useful when developing software, but the generated code is often significantly slower than the release version. In the release configuration, the compiler attempts to optimize the generated code for best performance, using a number of transformations and algorithms. This means that there may no longer be a 1-to-1 correspondence between the source code and the assembly code; as functionality is moved and reordered, code is reused where possible, and redundancies are eliminated in order to maximize execution speed and/or minimize code size. Debugging the assembly code generated by a release build is a significant challenge for even seasoned programmers. The types and degrees of optimization employed can be controlled on a per-project and per-file basis; further information on this is available in the CCS documentation.

F.2   The Volatile Keyword

Two common situations in real-time DSP programming are variables that directly reference hardware, and variables that are used to communicate between interrupt service routines and the main program. Both of these situations require that the *volatile* keyword be used to control the compiler's optimization of memory references. In the first case, when a pointer variable is dereferenced to access a hardware register, the compiler's optimizer will assume that the transfer is being made to a standard read/write memory location. In the second case, the compiler will assume that the memory location will only change when it is written to in the function being compiled, and that no other access will be made concurrent with that function's execution. For example, suppose there is an integer pointer variable *mcbsp_spcr* that is used to read the McBSP1 receive register status bit as shown below, in order to wait until data is received by the McBSP.

```c
unsigned int *mcbsp_spcr = (unsigned int *)McBSP1_SPCR;

while (!(*mcbsp_spcr & 0x00020000)) // wait for codec ready
    ;
```

In the debug build, the code executes as expected because no optimization is done. However, in the release build, the compiler identifies the expression **!(*mcbsp_spcr & 0x00020000)** to be loop-invariant code, and so pulls it out of the loop and only reads from that location once. While this is generally a very good optimization for true memory locations, in this case we are reading a peripheral register that may change; thus it is NOT loop-invariant. To force the compiler to actually read the location represented by the variable in each loop iteration, we add the qualifier *volatile* to the variable declaration as shown below.

```c
volatile unsigned int *mcbsp_spcr = (unsigned int *)McBSP1_SPCR;
```

The volatile keywords informs the compiler that it may not optimize out any accesses to this variable, so it generates code that actually reads the McBSP register in each iteration. A similar situation occurs when a variable is used to write repeatedly to a hardware register, in this case McBSP1's transmit data register, as illustrated below.

```c
unsigned int *mcbsp_dxr = (unsigned int *)McBSP1_DXR;

*mcbsp_dxr = 1;
*mcbsp_dxr = 2;
*mcbsp_dxr = 3;
```

If this code is compiled under a release build, only a single write of the value 3 will occur. Declaring *mcbsp_dxr* to be volatile will force the compiler to perform the three separate write operations.

When using global variables to communicate between the main program and interrupt service routines (or between interrupt service routines), you should normally declare the global variables to be volatile as well, particularly if they are used in loops. Otherwise, the compiler may optimize out the variable references and you will miss any changes that occur to the variable during an interrupt.

F.3 Function Prototypes and Return Types

If a function is not declared before it is used, the C language requires that the compiler assume that function return type is *int*. This seemingly benign behavior has been observed as the cause of many programs that fail to work properly, because failing to declare a function in C is not an error and so it is not flagged as such. Consider the code below, noting that the sinf function was never declared.

```c
float x;

x = sinf(0);
```

In the C6000 architecture, register A4 is used for the return of 32-bit or smaller values. The sinf function actually returns a single-precision floating-point number in A4. However, since the function was never declared, the compiler assumes that the return type is *int*. So, the compiler assumes A4 contains an integer, and adds code (specifically the INTSP instruction) to convert the value in it to a *float* before storing it in x. As can be imagined, taking a floating-point bit pattern and performing an integer to floating-point conversion on it produces meaningless values. To prevent this situation, it is important (and good programming practice) to declare all functions before use. In the code below, the `math.h` header has been included to ensure the proper declaration of the `sinf` function.

```
#include <math.h>
float x;


x = sinf(0);
```

Now knowing that the `sinf` function returns a float, the compiler will take the return value in register A4 and transfer it into the variable x directly, giving the correct result. Always ensuring that all functions are declared before use will avoid these situations. This can be quite difficult to debug because the code is otherwise correct.

F.4   Arithmetic Issues

A high-level language compiler typically supports a number of arithmetic operations. The compiler will guarantee correct results; however, in real-time software we are also concerned with how long it will take to do the calculation. For an operation that is supported in the processor (i.e., add), the compiler will generate code to use the hardware to perform the calculation. For operations that are not supported in the processor hardware, the compiler will generate software to accomplish the calculation. In general, the software calculations will be much slower than those performed in hardware.

The TMS320C6x DSPs do not have divider hardware, so division should be avoided whenever possible. In the code below, the calculations are numerically equivalent.

```
float x = 100.0F;


x = x / 10.0F;   // calculation A
x = x * 0.1F;    // calculation B
```

Calculation A specifies a division, so the compiler will insert a call to a subroutine to perform the calculation in software. Since the processor has a hardware multiplier, the calculation B can be accomplished much faster. Note that the "F" suffix on the numbers indicates that they are constants of type *float* — otherwise they would be interpreted as type double, requiring the promotion of x to type double before performing the calculation as a double-precision operation.

When using an array variable to implement circular buffering, the index needs to be "wrapped-around" when the end of the buffer is reached. In the code sample below, a buffer and index variable are allocated.

```
#define BUFFER_SIZE 100
float x[BUFFER_SIZE] = {0.0F};
int index = 0;
```

In the following examples, we will assume the index value is being incremented. A decrementing index would be handled in a similar fashion. Perhaps the most immediately intuitive way to accomplish index wrap-around is to simply check the index value and set it back to 0 when it reaches the end of the buffer.

```
index++;
if(index >= BUFFER_SIZE)
    index = 0;
```

Note that this requires a comparison, and is limited to index increments of 1. If arbitrary increments are needed, we need another approach. The *modulus* operation (%) provides a seemingly simple fix.

```
1  index++;
   index = index \% BUFFER\_SIZE;
```

However, the modulus operator computes the remainder of the index value divided by BUFFER_SIZE, so we are implicitly invoking a division operation. As an alternative, note that if the increment is less than BUFFER_SIZE, we can obtain the remainder by subtracting BUFFER_SIZE whenever the index is greater than or equal to BUFFER_SIZE.

```
   index++;
2  if(index >= BUFFER\_SIZE)
       index = index - BUFFER\_SIZE;
```

This reduces the modulus calculation to a simple subtraction operation, which is supported in hardware. However, it still requires a comparison to see if the index needs to be wrapped around. In real-time code, we may find even this operation to be prohibitively expensive. To eliminate the comparison completely, the buffer size is set to 2^n. Then, the wrap-around is accomplished with only a logical AND operation of the index with $2^n - 1$. If the index is less than BUFFER_SIZE, the AND operation will leave it unchanged. If the index is greater than or equal to BUFFER_SIZE, the AND operation will result in same result as the modulus operation.

```
1  #define BUFFER\_SIZE 512 // must be a power of 2
   float x[BUFFER\_SIZE] = {0.0F};
3  int index = 0;

5  index++;
   index = index & (BUFFER\_SIZE - 1);
```

Note that in this implementation the buffer size *must* be a power of 2. This is a classic software trade-off between size and speed, and is often seen in production code.

F.5 Controlling the Location of Variables in Memory

When declaring a variable in software, we normally do not concern ourselves with the actual variable location in memory. Rather, we simply refer to the variable by name. The compiler and linker are responsible for making sure that the correct memory location(s) are accessed. However, there are times when we will want to control where variables are placed in memory. To do that, we need to do two things:

1. instruct the linker where the physical memory is in our system, and

2. tell the compiler which variables we want placed in other than the default locations.

When our code is compiled, the compiler places the output into a number of predefined sections. Global variables are typically placed in the *.data* or *.bss* sections. The linker command file (i.e. `lnk7.cmd`) lists the physical memory available in the system, and indicates which sections are placed into which memory areas. In the linker command files used throughout the text, all compiler output is placed into the DSP's on-chip memory (the *IRAM* area). This is a relatively small memory area, so if we want to have large data buffers we need to place them in the much larger off-chip memory. In the linker command file, this area is designated *SDRAM*, all compiler output in section *"CE0"* will be placed there.

To instruct the compiler to place a given variable into the *"CE0"* section, we use a compiler *pragma*. In general, pragmas are compiler-specific directives that permit detailed control over various aspects of the compiler's operation. To control the section into which a variable is placed, we can use the DATA_SECTION pragma. This instructs the compiler to place the variable named as the first parameter into the section named as the second parameter.

```
#pragma DATA_SECTION (buffer , "CE0"); // allocate buffer in SDRAM
volatile float buffer[BUFFER_LENGTH];
```

Further information on the various pragmas available in Code Composer Studio can be found in the online help and the C compiler user's manual.

F.6 Real-Time Schedule Failures

One of the most difficult challenges in writing real-time software is determining if the software will in fact be able to meet the real-time schedule. In particular, for an interrupt driven system, each interrupt service routine (ISR) must complete its processing before the next interrupt occurs, and the programmer must allow sufficient "slack time" to account for interrupt service overhead. One simple and effective way to measure the time that an ISR takes is to change the state of a logic signal on entering and leaving the ISR, and then monitor that signal with an oscilloscope. On the DSKs, the signals CNTL0 and CNTL1 can be monitored for this purpose on the Peripheral Expansion Interface connector J3.

DSK type	port address	port width	CNTL0 bit	CNTL1 bit
C6711	0x9000800	32-bit	D27	D28
C6713	0x9000801	8-bit	D0	D1

Example code for both 6711 and 6713 DSKs is shown below.

Listing F.1: Example code for 6711 DSK.

```
void MyISR_6711()
{
    *(volatile unsigned int *)IO_PORT = 0x08000000; // CNTL0 high
    // ISR code here
    *(volatile unsigned int *)IO_PORT = 0x00000000; // CNTL0 low
}
```

Listing F.2: Example code for 6713 DSK.

```
void MyISR_6713()
{
    *(volatile unsigned char *)(IO_PORT+1) = 0x01; // CNTL0 high
    // ISR code here
    *(volatile unsigned char *)(IO_PORT+1) = 0x00; // CNTL0 low
}
```

Appendix G

Abbreviations

THIS is a partial list of abbreviations used in the text, provided in the hope that it will be helpful to some readers.

Symbols

( )	used for a continuous function.
[ ]	used for a discrete function.

Greek Letters

λ	wavelength.
π	ratio of a circle circumference to diameter, $3.1415926535897932\ldots$
τ	time constant.
ω	radian frequency.

A

a filter coefficient associated with an output term, y. When used in a transfer function, the a coefficients are associated with the denominator of the transfer function.

A vector or array containing all of the a terms.

AWGN   additive white Gaussian noise.

ADC   analog to digital converter.

AIC   analog interface circuit (see codec).

B

b filter coefficient associated with an input term, x. When used in a transfer function, the b coefficients are associated with the numerator of the transfer function.

B vector or array containing all of the b terms.

BW bandwidth of a bandpass signal.

BP   bandpass.

BPF   bandpass filter.

BPSK   binary phase shift keying.

C

C	capacitor.
CCS	Texas Instruments' Code Composer Studio™.
CISC	complex instruction set computer.
codec	coder-decoder. An integrated circuit that contains both an ADC and a DAC.
CPU	central processing unit.

D

DAC	digital to analog converter.
D.C.	direct current (0 Hz).
DDS	direct digital synthesizer or direct digital synthesis.
DF-I	direct form I.
DF-II	direct form II.
DFT	discrete Fourier transform.
DMA	direct memory access.
DSK	DSP starter kit.
DSP	digital signal processing or digital signal processor.
DTFT	discrete-time Fourier transform.

E

EDMA	enhanced direct memory access.

F

FCC	Federal Communications Commission.
FIR	finite impulse response.
FFT	fast Fourier transform.
FT	Fourier transform.
$\mathcal{F}$	Fourier transform.
$\mathcal{F}^{-1}$	inverse Fourier transform.
f_h	highest or maximum frequency that is present in a signal.
F_s	sample frequency (samples/second) $= 1/T_s$.

H

$H(e^{j\omega})$	discrete-time frequency response.
$H(j\omega)$	continuous-time frequency response.
$h[n]$	discrete-time impulse response or unit sample response.
$h[t]$	continuous-time impulse response.
$H(s)$	continuous-time transfer or system function.
$H(z)$	discrete-time transfer or system function.
HP	highpass.

HPF   highpass filter.

HPI   host port interface.

Hz   hertz (cycles per second).

$\boxed{\text{I}}$

IF   intermediate frequency.

IFFT   inverse fast Fourier transform.

IIR   infinite impulse response.

ISR   interrupt service routine.

$\boxed{\text{J}}$

j $\sqrt{-1}$; identifies the imaginary part of a complex number.

$\boxed{\text{L}}$

$\mathcal{L}$ Laplace transform.

$\mathcal{L}^{-1}$ inverse Laplace transform.

LP   lowpass.

LPF   lowpass filter.

LSB   lower sideband.

$\boxed{\text{M}}$

M   the number of bands in a graphic equalizer.

MA   moving average.

McBSP   multi-channel buffer serial port.

ML   maximum likelihood.

$\boxed{\text{N}}$

n index or sample number.

N often used as filter order; in other contexts it is used for the length of a sequence, or for the length of an FFT.

NCO   numerically controlled oscillator.

$\boxed{\text{P}}$

PC   personal computer.

PCM   pulse code modulation.

PLL   phase-locked loop.

$\boxed{\text{Q}}$

Q quality factor. Q = bandwidth of a BP filter divided by its center frequency. The higher the value of Q, the more selective the BP filter is.

R

r	magnitude of a pole. This is a measure of how far the pole is from the origin.
R	resistor.
RC	resistor-capacitor.
RISC	reduced instruction set computer.
RF	radio frequency.

S

s	the Laplace transform independent variable, $s = \sigma + j\omega$.

T

τ	a dummy variable often used in convolution.
t	time.
T	period of a signal or function.
TED	timing error detector.
T_s	sample period $= 1/F_s$.
TI	Texas Instruments.

U

$u[n]$	discrete-time unit step function.
$u(t)$	unit step function.
U.S.	United States (of America).
USB	upper sideband; also used for Universal Serial Bus.

V

V	voltage in Volts.
V_{in}	input voltage.
V_{out}	output voltage.
VLIW	very long instruction word; this is a type of architecture for DSPs.

W

winDSK	original Windows-based program for the C31 DSK, created by Mike Morrow.
winDSK6	Windows-based program, the follow-on to winDSK, for the C6x DSK series. It was created by Mike Morrow.

X

$X(j\omega)$	result of the Fourier transform $\mathcal{F}\{x(t)\}$; it shows the frequency content of $x(t)$.
$x[n]$	a discrete-time input signal.
$x(t)$	a continuous-time input signal.

$\boxed{\mathbf{Y}}$

$Y(j\omega)$	result of the Fourier transform $\mathcal{F}\{y(t)\}$; it shows the frequency content of $y(t)$.
$y[n]$	a discrete-time output signal.
$y(t)$	a continuous-time output signal.

$\boxed{\mathbf{Z}}$

z	the independant transform variable for discrete-time signals and systems.
z^{-1}	a delay of 1 sample.
Z_c	impedance of a capacitor.
$\mathcal{Z}^{-1}$	inverse z-transform.

References

[1] B. Porat, *A Course in Digital Signal Processing*. John Wiley & Sons, 1997.

[2] A. V. Oppenheim, R. W. Schafer, and J. R. Buck, *Discrete-Time Signal Processing*. Prentice Hall, 2nd ed., 1999.

[3] J. G. Proakis and D. G. Manolakis, *Digital Signal Processing*. Prentice Hall, 3rd ed., 1996.

[4] S. K. Mitra, *Digital Signal Processing: A Computer-Based Approach*. McGraw-Hill, 2nd ed., 2001.

[5] C. Marven and G. Ewers, *A Simple Approach to Digital Signal Processing*. John Wiley & Sons, 1996.

[6] J. H. McClellan, R. W. Schafer, and M. A. Yoder, *DSP First: A Multimedia Approach*. Prentice Hall, 1998.

[7] C. S. Burrus, "Teaching filter design using MATLAB," in *Proceedings of the IEEE International Conference on Acoustics, Speech, and Signal Processing*, pp. 20–30, Apr. 1993.

[8] C. J. McCormack, A. S. Ali, R. L. Haupt, and C. H. G. Wright, "Computer supplements to engineering labs," *ASEE Comput. Educ. J.*, vol. III, pp. 58–62, Apr. 1993.

[9] R. F. Kubichek, "Using MATLAB in a speech and signal processing class," in *Proceedings of the 1994 ASEE Annual Conference*, pp. 1207–1210, June 1994.

[10] R. G. Jacquot, J. C. Hamann, J. W. Pierre, and R. F. Kubichek, "Teaching digital filter design using symbolic and numeric features of MATLAB," *ASEE Comput. Educ. J.*, vol. VII, pp. 8–11, January–March 1997.

[11] M. A. Yoder, J. H. McClellan, and R. W. Schafer, "Experiences in teaching DSP first in the ECE curriculum," in *Proceedings of the 1997 ASEE Annual Conference*, June 1997. Paper 1220-06.

[12] C. H. G. Wright and T. B. Welch, "Teaching real-world DSP using MATLAB," *ASEE Comput. Educ. J.*, vol. IX, pp. 1–5, Jan–Mar 1999.

[13] T. B. Welch, B. Jenkins, and C. H. G. Wright, "Computer interfaces for teaching the Nintendo generation," in *Proceedings of the 1999 ASEE Annual Conference*, June 1999. Paper 3532-02.

[14] T. B. Welch, C. H. G. Wright, and M. G. Morrow, "Poles and zeroes and MATLAB, oh my!," *ASEE Comput. Educ. J.*, vol. X, pp. 70–72, Apr. 2000.

[15] S. D. Stearns, *Digital Signal Processing with Examples in* MATLAB. CRC Press, 2003.

[16] V. K. Ingle and J. G. Proakis, *Digital Signal Processing Using* MATLAB *V.4*. Bookware Companion Series, PWS Publishing, 1997.

[17] J. H. McClellan, C. S. Burrus, A. V. Oppenheim, T. W. Parks, R. W. Schafer, and S. W. Schuessler, *Computer-Based Exercises for Signal Processing Using* MATLAB 5. MATLAB Curriculum Series, Prentice Hall, 1998.

[18] A. Ambardar and C. Borghesani, *Mastering DSP Concepts Using* MATLAB. Prentice Hall, 1998.

[19] D. C. Hanselman and B. L. Littlefield, *Mastering* MATLAB 7. Prentice Hall, 2005.

[20] D. M. Etter, *Engineering Problem Solving with* MATLAB. Prentice Hall, 2nd ed., 1997.

[21] H. V. Sorensen and J. Chen, *A Digital Signal Processing Laboratory Using the TMS320C30*. Prentice Hall, 1997.

[22] R. Chassaing, *DSP Applications Using C and the TMS320C6x DSK*. John Wiley & Sons, 2002.

[23] N. Kehtarnavaz, *Real-Time Digital Signal Processing Based on the TMS320C6000*. Elsevier, 2005.

[24] C. H. G. Wright and T. B. Welch, "Teaching DSP concepts using MATLAB and the TMS320C5X," in *Proceedings of the 1998 Texas Instruments DSP Educators and Third-Party Conference*, August 6–8, 1998.

[25] C. H. G. Wright and T. B. Welch, "Teaching DSP concepts using MATLAB and the TMS320C31 DSK," in *Proceedings of the IEEE International Conference on Acoustics, Speech, and Signal Processing*, Mar. 1999. Paper 1778.

[26] M. G. Morrow, T. B. Welch, and C. H. G. Wright, "An inexpensive software tool for teaching real-time DSP," in *Proceedings of the 1st IEEE DSP in Education Workshop*, IEEE Signal Processing Society, Oct. 2000.

[27] M. G. Morrow, T. B. Welch, C. H. G. Wright, and G. York, "Teaching real-time beamforming with the C6211 DSK and MATLAB," in *Proceedings of the 2000 Texas Instruments DSP Educators and Third-Party Conference*, August 2–4, 2000.

[28] C. H. G. Wright, T. B. Welch, and M. G. Morrow, "Teaching transfer functions with MATLAB and real-time DSP," in *Proceedings of the 2001 ASEE Annual Conference*, June 2001. Session 1320.

[29] M. G. Morrow, T. B. Welch, and C. H. Wright, "An introduction to hardware-based DSP using winDSK6," in *Proceedings of the 2001 ASEE Annual Conference*, June 2001. Session 1320.

[30] T. B. Welch, C. T. Field, and C. H. G. Wright, "A signal analyzer for teaching signals and systems," in *Proceedings of the 2001 ASEE Annual Conference*, June 2001. Session 2793.

[31] G. W. P. York, M. G. Morrow, T. B. Welch, and C. H. G. Wright, "Teaching real-time sonar with the C6711 DSK and MATLAB," in *Proceedings of the 2001 ASEE Annual Conference*, June 2001. Session 1320.

[32] M. G. Morrow, T. B. Welch, and C. H. G. Wright, "A tool for real-time DSP demonstration and experimentation," in *Proceedings of the 10th IEEE Digital Signal Processing Workshop*, Oct. 2002. Paper 4.8.

[33] T. B. Welch, D. M. Etter, C. H. G. Wright, M. G. Morrow, and G. J. Twohig, "Experiencing DSP hardware prior to a DSP course," in *Proceedings of the 10th IEEE Digital Signal Processing Workshop*, Oct. 2002. Paper 8.5.

[34] C. H. G. Wright, T. B. Welch, D. M. Etter, and M. G. Morrow, "A systematic model for teaching DSP," in *Proceedings of the IEEE International Conference on Acoustics, Speech, and Signal Processing*, vol. IV, pp. 4140–4143, May 2002. Paper 3243.

[35] C. H. G. Wright, T. B. Welch, D. M. Etter, and M. G. Morrow, "Teaching hardware-based DSP: Theory to practice," in *Proceedings of the IEEE International Conference on Acoustics, Speech, and Signal Processing*, vol. IV, pp. 4148–4151, May 2002. Paper 4024.

[36] C. H. G. Wright, T. B. Welch, D. M. Etter, and M. G. Morrow, "Teaching DSP: Bridging the gap from theory to real-time hardware," in *Proceedings of the 2002 ASEE Annual Conference*, June 2002.

[37] G. W. P. York, C. H. G. Wright, M. G. Morrow, and T. B. Welch, "Teaching real-time sonar with the C6711 DSK and MATLAB," *ASEE Comput. Educ. J.*, vol. XII, pp. 79–87, July 2002.

[38] T. B. Welch, R. W. Ives, M. G. Morrow, and C. H. G. Wright, "Using DSP hardware to teach modem design and analysis techniques," in *Proceedings of the IEEE International Conference on Acoustics, Speech, and Signal Processing*, vol. III, pp. 769–772, Apr. 2003.

[39] T. B. Welch, M. G. Morrow, C. H. G. Wright, and R. W. Ives, "commDSK: a tool for teaching modem design and analysis," in *Proceedings of the 2003 ASEE Annual Conference*, June 2003. Session 2420.

[40] C. H. G. Wright, T. B. Welch, and M. G. Morrow, "An inexpensive method to teach hands-on digital communications," in *Proceedings of the IEEE/ASEE Frontiers in Education Annual Conference*, Nov. 2003.

[41] C. H. G. Wright, T. B. Welch, D. M. Etter, and M. G. Morrow, "Teaching DSP: Bridging the gap from theory to real-time hardware," *ASEE Comput. Educ. J.*, vol. XIII, pp. 14–26, July 2003.

[42] T. B. Welch, M. G. Morrow, and C. H. G. Wright, "Using DSP hardware to control your world," in *Proceedings of the IEEE International Conference on Acoustics, Speech, and Signal Processing*, vol. V, pp. 1041–1044, May 2004. Paper 1146.

[43] T. B. Welch, M. G. Morrow, C. H. G. Wright, and R. W. Ives, "commDSK: a tool for teaching modem design and analysis," *ASEE Comput. Educ. J.*, vol. XIV, pp. 82–89, Apr. 2004.

[44] T. B. Welch, C. H. G. Wright, and M. G. Morrow, "Caller ID: An opportunity to teach DSP-based demodulation," in *Proceedings of the IEEE International Conference on Acoustics, Speech, and Signal Processing*, vol. V, pp. 569–572, Mar. 2005. Paper 2887.

[45] M. G. Morrow, T. B. Welch, and C. H. G. Wright, "Enhancing the TMS320C6713 DSK for DSP education," in *Proceedings of the 2005 ASEE Annual Conference*, June 2005.

[46] Educational DSP (eDSP), L.L.C., "DSP resources for TI DSKs," 2005. `http://www.educationaldsp.com/`.

[47] M. G. Morrow and T. B. Welch, "winDSK: A windows-based DSP demonstration and debugging program," in *Proceedings of the IEEE International Conference on Acoustics, Speech, and Signal Processing*, vol. 6, pp. 3510–3513, June 2000.

[48] M. G. Morrow, "Software for DSP," 2002. `http://eceserv0.ece.wisc.edu/~morrow/software/`.

[49] R. G. Lyons, *Understanding Digital Signal Processing*. Addison Wesley, 1997.

[50] The MathWorks, Inc., *MATLAB: The Language of Technical Computing*, 2005.

[51] S. M. Kuo and B. H. Lee, *Real-Time Digital Signal Processing: Implementations, Applications and Experiments for the TMS320C55x*. John Wiley & Sons, 2001.

[52] K. C. Pohlmann, *Principles of Digital Audio*. Howard W. Sams & Co., 2nd ed., 1989.

[53] Texas Instruments, Inc., *TMS320C6000 Peripherals Reference Guide*, 2001. Literature Number: SPRU190D.

[54] J. W. Cooley and J. W. Tukey, "An algorithm for the machine computation of complex fourier series," *Mathematics of Computation*, vol. 19, pp. 297–301, Apr. 1965.

[55] S. M. Kay, *Modern Spectral Analysis: Theory and Application*. Prentice Hall, 1988.

[56] S. M. Kay, *Fundmentals of Statistical Signal Processing: Estimation Theory*, vol. 1. Prentice Hall, 1993.

[57] D. G. Manolakis, V. K. Ingle, and S. M. Kogon, *Statistical and Adaptive Signal Processing: Spectral Estimation, Signal Modeling, Adaptive Filtering, and Array Processing*. McGraw-Hill, 2000.

[58] M. R. Schroeder, "Natural sounding artificial reverberation," *Journal of the Audio Engineering Society*, vol. 10, pp. 219–223, 1962.

[59] S. J. Orfanidis, *Introduction to Signal Processing*. Prentice Hall, 1996.

[60] H. Chamberlin, *Musical Applications of Microprocessors*. Hayden Book Company, 2nd ed., 1985.

[61] A. B. Carlson, P. B. Crilly, and J. C. Rutledge, *Communication Systems*. McGraw-Hill, 4th ed., 2002.

[62] L. W. Couch, II, *Digital and Analog Communication Systems*. Prentice Hall, 6th ed., 2001.

[63] M. E. Frerking, *Digital Signal Processing in Communication Systems*. Van Nostrand Reinhold, 1994. 7th printing by Kluwer Academic Publishers, 2000.

[64] S. A. Tretter, *Communications System Desgin Using DSP Algorithms: With Laboratory Experiments for the TMS320C30*. Plenum Press, 1995.

[65] S. A. Tretter, *Communications System Desgin Using DSP Algorithms: With Laboratory Experiments for the TMS320C6701 and TMS320C6711.* Kluwer Academic Publishers (Plenum Press), 2003.

[66] U. Mengali and A. N. D'Andrea, *Synchronization Techniques for Digital Receivers.* Plenum Press, 1997.

[67] B. Sklar, *Digital Communications: Fundamentals and Applications.* Prentice Hall, 2nd ed., 2001.

[68] Texas Instruments, Inc., *TMS320 DSP/BIOS User's Guide*, 2001. Literature Number: SPRU423A.

[69] Texas Instruments, Inc., *TMS320C6000 DSP/BIOS Application Programming Interface (API) Reference Guide*, 2001. Literature Number: SPRU403D.

[70] Institute of Electrical and Electronics Engineers (IEEE), "IEEE standard for binary floating-point arithmetic: ANSI/IEEE STD 754-1985," 1985.

[71] R. W. Hamming, *Numerical Methods for Scientists and Engineers.* McGraw-Hill, 2nd ed., 1973.

[72] S. S. Rao, *Applied Numerical Methods for Engineers and Scientists.* Prentice Hall, 2002.

[73] Texas Instruments, Inc., *TMS320C6000 CPU and Instruction Set Reference Guide*, 2000. Literature Number: SPRU189F.

[74] J. L. Hennessy and D. A. Patterson, *Computer Architecture: A Quantitative Approach.* Morgan Kaufmann Publishers, 3rd ed., 2003.

[75] J. A. Fisher, P. Faraboschi, and C. Young, *Embedded Computing: A VLIW Approach to Architecture, Compilers and Tools.* Morgan Kaufmann Publishers, 2004.

Index